Receptor Purification

Receptor Purification

Volume 2

Receptors for Steroid Hormones, Thyroid Hormones, Water-Balancing Hormones, and Others

Edited by

Gerald Litwack

*Fels Institute for Cancer Research
and Molecular Biology, Philadelphia, PA*

Humana Press
Clifton, New Jersey

Library of Congress Cataloging in Publication Data
(Revised for volume 2)

Receptor purification.

      Includes index.
      Contents: v. 1. Receptors for CNS agents, pituitary
growth factors, hormones, and related substances — v. 2.
Receptors for steroid hormones, thyroid hormones, water-
balancing hormones, and others.
      1. Neurotransmitter receptors—Purification.
2. Hormone receptors—Purification. 3. Cell receptors—
Purification. 4. Receptors, Endogenous Substances—
isolation & purification. I. Litwack, Gerald.
QP364.7.R4278 1990   612.4   90-4689
ISBN 0-89603-167-5 (v. 1)
ISBN 0-89603-183-7 (v. 2)

© 1990 The Humana Press Inc.
Crescent Manor
PO Box 2148
Clifton, NJ 07015

Printed in the United States of America.

Preface

The purpose of these volumes is to provide a reference work for the methods of purifying many of the receptors we know about. This becomes increasingly important as full-length receptors are overexpressed in bacteria or in insect cell systems. A major problem for abundantly expressed proteins will be their purification. In addition to purification protocols, many other details can be found concerning an individual receptor that may not be available in standard texts or monographs. No book of this type is available as a compendium of purification procedures.

Receptor Purification provides protocols for the purification of a wide variety of receptors. These include receptors that bind: neurotransmitters, polypeptide hormones, steroid hormones, and ligands for related members of the steroid supergene family and others including receptors involved in bacterial motion. The text of this information is substantial so as to require its publication in two volumes. Consequently, a division was made by grouping receptors depending upon the nature of their ligands. Thus, in volume 1 there are contributions on serotonin receptors, adrenergic receptors, the purification of GTP-binding proteins, opioid receptors, neurotensin receptor, luteinizing hormone receptor, human chorionic gonadotropin receptor, follicle stimulating hormone receptor, thyrotropin receptor, prolactin receptor, epidermal growth factor receptor, platelet derived growth factor receptor, colony stimulating factor receptor, insulinlike growth factor receptors, insulin receptor, fibronectin receptor, interferon receptor, and the cholecystokinin receptor.

In vol 2, there are contributions on: dexamethasone mesylate, dexamethasone-biotin affinity probes, the purification of different forms of the glucocorticoid receptor, the mineralocorticoid receptor, the androgen receptor, the estrogen receptor, the progesterone receptor, the vitamin D receptor, the retinoic acid binding protein, thyroid hormone receptor, vasopressin VI receptor, protaglandin E1/prostacyclin receptors, angiotensin II receptor, asialoglycoprotein receptor and receptors of bacterial chemotaxis. Information on receptors in the *erb* A super-gene family, were included together in this volume, although the arylhydrocarbon receptor was not included because at the time of solicitation of manuscripts, no purified functional arylhydrocarbon receptor had been reported. Receptors involved in fluid homeostasis were segregated in this volume because of the activities of glucocorticoids and mineralocorticoids in governing sodium ion uptake from lumina into surrounding epithelial cells. Some additional receptors are included in this volume. In several cases, for the better known receptors, more than one manuscript has been sought and included in this collection because of the manifold approaches to the purification of a given receptor. Representation of several approaches could be beneficial and outweigh the costs of increased numbers of pages. Also, for some receptors, purification is tailored for different forms of the receptor.

Many of the contributions to these volumes are from laboratories that lead the field not only in purification but in the phenomenology of receptors, both at the protein level and at the molecular biological level. Any gaps in coverage of the specific receptors the reader may discover usually result from difficulties in obtaining manuscripts from representative laboratories or generate from contracts with laboratories that failed ultimately to provide a manuscript. Nevertheless, the coverage of various receptor types is broad and should be of considerable benefit to researchers in this overall area.

We now begin to witness the overexpression of full-length receptors in bacterial or baculovirus systems. Overexpression solves an important problem in receptor research in that it overcomes the difficulties of concentration typified by many receptors that occur in the nanomolar range. With the advent of overexpression systems, it is now possible to generate receptors in the micromolar range advancing their status from trace proteins to abundant cellular proteins. These proteins need to be purified from the cells overexpressing them. Such purification processes must rely heavily on the information garnered for purification of the same receptor from various cells and tissues. This information is surveyed in this two volume work and appears at a time coinciding with the purification of receptors from overexpression systems.

The presentation of information about each receptor is complete enough so that all of the protocols required for the purification process are available without further pursuit of the literature for precise details. Moreover, in some cases, especially in volume two, there are contributions of specific agents useful for covalent binding of ligands to their receptors. In many cases information on cloning is presented that may be helpful to investigators considering overexpression systems. The material available in this two-volume work should be of great use to biologists, biochemists, pharmacologists, molecular biologists, and endocrinologists, including graduate students as well as sophisticated workers.

I thank the Publisher for responsiveness and rapid generation of these volumes in a diligent manner. I am indebted to the contributors for their conscientious preparation of manuscripts. I, and the contributors, are further indebted to a wide variety of journals and other publications that so freely allowed the reproduction of published data appearing in these volumes. Denise Valentine helped in the preparation of the index and other typing responsibilities.

Gerald Litwack

Contents

Contributors

Ciro Abbondanza • General Institute of Pathology and Oncology, University of Naples, Naples, Italy

E. E. Baulieu • INSERM, Paris, France

J. Edwin Blalock • Departments of Medicine and Physiology and Biophysics, University of Alabama at Birmingham, Birmingham, AL

Francesco Bresciani • General Institute of Pathology and Oncology, University of Naples, Naples, Italy

Thomas A. Brown • Department of Biochemistry, University of Wisconsin-Madison, Madison, WI

Jan Carlstedt-Duke • Department of Medical Nutrition and Center for Biotechnology, Karolinska Institute, Huddinge, Sweden

C. H. Chang • Department of Cell Biology, Baylor College of Medicine, Houston, TX

Janna C. Collins • Departments of Medicine, Biochemistry, and Pediatrics and Liver Research Center, Albert Einstein College of Medicine, Bronx, NY

Karin Dahlman • Department of Medical Nutrition and Center for Biotechnology, Karolinska Institute, Huddinge, Sweden

Leslie J. DeGroot • Department of Medicine, University of Chicago, Chicago, IL

Hector F. DeLuca • Department of Biochemistry, University of Wisconsin-Madison, Madison, WI

Burton F. Dickey • Department of Pulmonary Medicine, University of Massachusetts Medical Center, Worcester, MA

Terry S. Elton • Department of Medicine, University of Alabama, Birmingham, AL

Asim K. Dutta-Roy • Department of Medicine, Wright State University, VA Administration Medical Center, Dayton, OH

Jordan B. Fishman • Department of Pharmacology, University of Massachusetts Medical Center, Worcester, MA

Pierre Formstecher • Université de Nantes, Laboratoire de Biochimie, Nantes, France

John W. Funder • Medical Research Centre, Prince Henry's Hospital, Melbourne, Victoria, Australia

Manjapra V. Govindan • Department of Endocrinology, Le Centre Hospitalier de l'Université Laval, Québec, Canada

Jan-Åke Gustafsson • Department of Medical Nutrition and Center for Biotechnology, Karolinska Institute, Huddinge, Sweden

Tsuneyoshi Horigome • National Institutes of Health, Bethesda, MD

Kazuo Ichikawa • Department of Gerontology, Endocrinology, and Metabolism, Shinshu University School of Medicine, Matsumoto, Japan

M. P. Johnson • Department of Cell Biology, Baylor College of Medicine, Houston, TX

Michimasa Kato • First Department of Internal Medicine, Gifu University School of Medicine, Gifu, Japan

Kenneth S. Korach • National Institutes of Health, Bethesda, MD

Zygmunt S. Krozowski • Medical Research Centre, Prince Henry's Hospital, Melbourne, Victoria, Australia

Gerald Litwack • Fels Institute for Cancer Research and Molecular Biology, Temple University, Philadelphia, PA

Patrick Lustenberger • Université de Nantes, Laboratoire de Biochimie, Nantes, France

Bernhard Manz • Johannes Gutenberg-Universtät, Mainz, FRG

Nicola Medici • General Institute of Pathology and Oncology, University of Naples, Naples, Italy

Bruno Moncharmont • General Institute of Pathology and Oncology, University of Naples, Naples, Italy

Anatol G. Morell • Departments of Medicine, Biochemistry, and Pediatrics and Liver Research Center, Albert Einstein College of Medicine, Bronx, NY

Sherry L. Mowbray • Departments of Biochemistry and Pharmacology, and The Howard Hughes Medical Institute, University of Texas, Southwestern Medical Center, Dallas, TX

Yasutoshi Muto • First Department of Internal Medicine, Gifu University School of Medicine, Gifu, Japan

Masataka Okuno • Department of Laboratory Medicine, Gifu University School of Medicine, Gifu, Japan

Suzanne Oparil • Departments of Medicine and Physiology and Biophysics, University of Alabama, Birmingham, AL

Kunhard Pollow • Johannes Gutenberg-Universtät, Mainz, FRG

Giovanni Alfredo Puca • General Institute of Pathology and Oncology, University of Naples, Naples, Italy

J. M. Renoir • INSERM, Paris, France

D. R. Rowley • Department of Cell Biology, Baylor College of Medicine, Houston, TX

S. Stoney Simons • National Institutes of Health, Bethesda, MD

Asru K. Sinha • Department of Medicine, Montefiore Medical Center, Albert Einstein College of Medicine, Bronx, NY

David F. Smith • Department of Biochemistry and Molecular Biology, Mayo Clinic, Rochester, MN

Richard J. Stockert • Departments of Medicine, Biochemistry, and Pediatrics and Liver Research Center, Albert Einstein College of Medicine, Bronx, NY

Per-Erik Strömstedt • Department of Medical Nutrition and Center for Biotechnology, Karolinska Institute, Huddinge, Sweden

D. J. Tindall • Department of Cell Biology, Baylor College of Medicine, Houston, TX

David O. Toft • Department of Biochemistry and Molecular Biology, Mayo Clinic, Rochester, MN

Wayne V. Vedeckis • Department of Biochemistry and Molecular Biology, Loiusiana State University Medical Center, New Orleans, LA

Nancy L. Weigel • Department of Cell Biology, Baylor College of Medicine, Houston, TX

C. Y. F. Young • Department of Cell Biology, Baylor College of Medicine, Houston, TX

Dexamethasone 21-Mesylate

Applications for Purification and Examination of Glucocorticoid Receptors

S. Stoney Simons, Jr.

1. Introduction

The purification of glucocorticoid receptors has been hampered by the three properties of very low abundance, instability, and absence of activity except when bound with the cognate steroid. Since early purification schemes were required to maintain the receptor in a biologically active form in order to subsequently identify the receptor, the types of procedures that could be employed were severely limited. With the advent of affinity labeling steroids, however, the biologically inactive receptor protein could now be identified, purified, and studied. Because the covalent receptor-steroid complex cannot dissociate, many drastic purification steps (e.g., acid precipitation and SDS polyacrylamide gel electrophoresis [SDS-PAGE]) can also be utilized.

Currently there are three general methods of affinity labeling: Photoaffinity, electrophilic affinity, and chemoaffinity (for review, *see* Simons and Thompson, 1982). So far, the most widely used affinity label for glucocorticoid receptors has been the electrophilic affinity label dexamethasone 21-mesylate (Dex-Mes) (Simons and Thompson, 1981). Dex-Mes labels only thiols (Simons et al., 1980; Eisen et al., 1981; Simons, 1987), no other steroid receptors (Simons et al., 1987), relatively few other proteins (Simons, 1988a), and usually acts as an irreversible antiglucocorticoid (for reviews, *see* Simons and Yen, 1987; Simons et al., 1988). Furthermore, Dex-

Receptor Purification, vol. 2   ©1990 The Humana Press

Mes labeled glucocorticoid receptors retain some biological activity in that their activation* (Simons et al., 1983; Simons and Miller, 1984) and binding to biologically active DNA sequences (Miller et al., 1984) are virtually indistinguishable from that of the noncovalent dexamethasone (Dex) bound receptors, and covalently labeled complexes display nuclear binding (Simons et al., 1983; Sistare et al., 1987; Miller and Simons, 1988b). In this chapter, I will discuss the different experimental techniques that have been employed with Dex-Mes to afford covalently labeled glucocorticoid receptors, which have then been studied at various levels of receptor purification.

2. Methods for Affinity Labeling Glucocorticoid Receptors with Dex-Mes

2.1. Affinity Labeling of Cell-Free Receptors with Dex-Mes

Crude receptor solutions are most easily prepared by freeze-thaw lysis of washed cells followed by dilution with an equal vol of high pH buffer to raise the pH to a level where Dex-Mes rapidly reacts with thiols (Simons et al., 1980; Eisen et al., 1981). After ultracentrifugation (200,000 xg), the cytosolic supernatant can usually be kept in liquid N_2 for many months. As an example of the labeling procedures, HTC cell cytosol, prepared by freeze-thaw lysis of whole cells (Simons and Miller, 1984) was treated for 2.5 h at 0°C in 0.67 vol of pH 9.5 TAPS buffer (25 mM TAPS, 1 mM EDTA, 10% [v/v] glycerol, pH adjusted at 0°C. *Note:* Small amounts of thiols can sometimes be tolerated [Simons et al., 1990], but *no added thiols* should be used, since they selectively and rapidly react with Dex-Mes [Simons et al., 1980; Simons, 1987]) with [^{3}H]Dex-Mes in the absence or presence of 80-fold excess of [^{1}H]Dex (final pH = 8.8 at 0°C). In those experiments, employing $2 \times 10^{-7}M$ [^{3}H]Dex-Mes, the amount of covalently labeled 98kDa** receptor equaled $98 \pm 11\%$ (SD, $n = 8$) of the total available

---

*The terms "activate" and "activation" are used in this paper to describe the currently unknown mechanism which initially formed receptor-steroid complexes, with little or no affinity for DNA or nuclei, are converted by manipulations such as heat, dilution, increased salt, or increased pH to complexes with relatively high affinity for DNA or nuclei.

**The reported mol wts for affinity labeled glucocorticoid receptors vary from ~85 to ~105kDa daltons. Part of this variation is owing to the problems associated with obtaining accurate M_r values from slice-and-counted gels; fluorographic analysis of the intact gel is inherently more accurate. Another major cause for these differences arises from using values other than 97,400 (Titani et al., 1977) for the M_r of phosphorylase b. Since most of the affinity labeled receptors migrate with, or slightly slower than, phosphorylase b, we have used 98kDa (instead of various published values) as the M_r of the receptor throughout this article in order to avoid confusion.

cell-free receptors, as measured by the specific charcoal-resistant binding of $5 \times 10^{-8} M$ [^{3}H]Dex ± 600-fold excess [^{1}H]Dex in a parallel assay solution (Simons and Miller, 1984). The yield of covalent receptors is decreased by incubating for less time, with less Dex-Mes, with more dilute receptor solutions, or at lower pH. Higher pHs and temperatures can be counterproductive because of receptor instability. The optimal pH for cytosols of other cells will vary, depending on receptor stability (Simons et al., 1983).

It has been stated that using frozen livers or cells to prepare steroid receptors increases the proteolytic degradation of the receptors (Sherman et al., 1983). We do see considerable proteolysis of receptors from frozen cells that are also mechanically homogenized in the presence of 2 mM Ca^{2+} (Fig. 1A). However, receptors prepared by freeze-thaw lysis in buffer without Ca^{2+} exhibited no proteolytic fragments (Fig. 1B) unless Ca^{2+} is added (Simons, 1988b).

With some cytosols, such as rat liver cytosol, it is necessary to remove components (probably low molecular weight thiols such as glutathione) that otherwise prevent the covalent labeling of receptors by Dex-Mes. Sephadex G-100 chromatography can be used instead of $(NH_4)_2SO_4$ precipitation to remove these interfering components. In both cases, Na_2MoO_4 is required to preserve steroid binding activity (Eisen et al., 1981). Thus, $(NH_4)_2SO_4$ precipitated liver cytosol (5 mL) was resuspended in 5 mL of pH 8.2 (at 0°C) TAPS buffer (containing 20 mM Na_2MoO_4) in the presence of 1.2 $\times 10^{-7}$ M [^{3}H]Dex-Mes ± 100-fold excess of [^{1}H]Dex for 2 h at 0°C. The yield of covalently labeled receptor is ~80% of the total available specific [^{3}H]Dex binding to cell free receptors (Eisen et al., 1981).

2.2. Affinity Labeling of Whole Cell Receptors with Dex-Mes

For labeling at 37°C, a 1 mL suspension of HTC (or Fu5-5) cells at ~5 $\times 10^6$ (or 10×10^6) cells/mL in Swim's S77 medium (pH 7.75 at r.t.) containing 5% fetal calf serum, 5% newborn calf serum, 0.1% NaHCO$_3$ and 1.4% EtOH plus [^{3}H]Dex-Mes ± 100-fold excess of [^{1}H]Dex (or [^{3}H]Dex ± 600-fold excess of [^{1}H]Dex) was incubated with occasional agitation for 30 min. The binding reaction was terminated by the addition of 3 mL of phosphate buffered saline (PBS, pH 7.8 at r.t.) and the cells were washed (3 × 3mL) with PBS at r.t. The drained cell pellets were ruptured by freezing for 10 s in liquid N$_2$. The frozen, lysed cells were brought to 0°C and then treated with 0.3 mL of pH 8.8 TAPS buffer (25 mM TAPS, 1 mM EDTA, 10% (v/v) glycerol, pH adjusted at 0°C), resuspended by vortexing, and centrifuged (500xg/0°C/10 min) to give cytosol (= supernatant) and nuclear (= pellet)

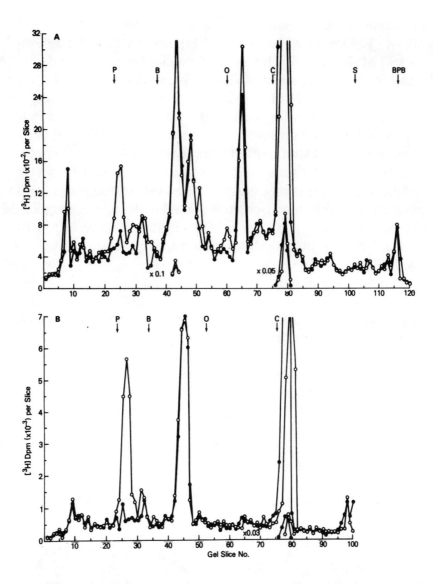

Fig. 1. [³H]Dex-Mes labeling of HTC cell cytosol receptors prepared by different methods. (A) Cytosol was prepared by mechanical homogenization of frozen cells in buffer A (20 mM Tricine, 2 mM CaCl₂, pH 8.0 at 0°C) as previously described (Simons et
al., 1979). (B) Cytosol was prepared by freeze-thaw in pH 9.5 TAPS buffer as described
by Reichman et al. (1984). For both cytosol preparations, covalent labeling of the
receptors (i.e., 2 × 10⁻⁸M [³H]Dex-Mes without (O) and with (●) 1.6 × 10⁻⁶M [¹H] Dex
at 0°C, pH ~8.8, for 2-3 h) and analysis of solutions by SDS-PAGE were conducted under
similar conditions (*see* Simons et al., 1983; Reichman et al., 1984). P, phosphorylase b;
B, bovine serum albumin: O, ovalbumin; C, carbonic anhydrase; S, soybean trypsin inhibitor; BPB, bromophenol blue (*from* Reichman et al., 1984).

fractions. The nuclear fraction was resuspended in 0.5 mL of pH 8.8 TAPS buffer and centrifuged to give a supernatant fraction containing some [³H]Dex bound receptor (~10% of those in the above cytosol fraction) that sometimes was combined with the cytosol fraction for further analysis (*see below*). The washed nuclear pellet was extracted for 30 min at 0°C with 0.5 mL of pH 8.8 TAPS buffer containing 0.3M NaSCN and centrifuged (2100xg/0°C/10 min) to give the nuclear extract fraction. The salt-extracted pellet was washed at 0°C with 0.5 mL of TCM buffer (20 mM Tricine, 2 mM CaCl$_2$, 3 mM MgCl$_2$, pH 7.7 at r.t.) and then incubated at 10°C for 60 min with 0.5 mL of TCM buffer containing 100 μg/mL of DNase I. The pellets of the DNase I-treated nuclei were extracted twice with 0.1% sodium dodecylsulfate (SDS; 100 μL of 0.1% SDS for 30 min followed by 200 μL of 0.1% SDS for 30-45 min, all at 0°C). For cells treated with [³H]Dex or [³H]Dex-Mes, the combined SDS extracts contained about 34 or 84%, whereas the residual pellets contained ≤ 1 or ≤ 5%, respectively, of the nuclear non-NaSCN extractable dpm (Simons and Miller, 1986).

Labeling at 0°C (Foster et al., 1983; Miller and Simons, 1988b) is conducted as described above for 37°C, except that the incubation (for 1 h) and washing of the cells were conducted at 0°C.

2.3. Quantitation of Dex-Mes Labeled Receptors

The quantitation of purified, [³H]Dex-Mes labeled receptors can be performed directly by scintillation counting. More involved procedures are required for crude [³H]Dex-Mes labeled receptors, because the nonspecific binding, and covalent labeling, of crude receptor preparations increases dramatically with higher concentrations of Dex-Mes. Also, some of the nonspecifically labeled proteins are more extensively labeled in the presence of competing [¹H]Dex (Fig. 1 and Simons et al., 1983). To date, SDS-PAGE has proved to be the only method that can reliably be used to quantitate the covalently labeled receptors. However, even this method does not work with very high levels of [³H]Dex-Mes (e.g., ≥ 10^{-6}M in whole HTC cell labeling experiments) because the nonspecific labeling in the region of the 98kDa receptor is so extensive (Simons et al., unpublished results).

For samples containing large amounts of [³H]Dex-Mes labeled receptors (i.e., ≥5000 dpm/20 μL), the sample is diluted with an equal vol of 2 × SDS sample buffer (0.6M Tris [pH 8.85], 2% SDS, 0.2M dithiothreitol, 20% [v/v] glycerol, and bromphenol blue). Dilute solutions of labeled receptor are concentrated by precipitation in 10% trichloroacetic acid (TCA) at 0°C for at least 2 h (samples can be left overnight at 0°C at this stage). If the total protein concentration after addition of TCA is ≤100 μg/mL, soybean trypsin inhibitor (at a final concentration of 100 μg/mL) is added as carrier protein. The precipitate is collected by centrifugation (2000xg/10 min), resuspended in 1 mL of 10% TCA for at least 1 h, and recentrifuged as before, all at 0°C. The supernatant is removed by aspiration to give pellets that can be stored at 0–20°C for a few days. To prepare the acid-precipitated material for polyacrylamide gels, the pellet is dissolved in a minimum of $1M$ NaOH (usually 10–15 μL) and then diluted with up to 100 μL of $2 \times$ SDS sample buffer (Simons et al., 1983).

Acetone has sometimes been used to wash acid-precipitated pellets (Wrange et al., 1979). However, we have found that TCA precipitated, covalently labeled receptors are quite soluble in acetone, even at 0°C (Simons et al., 1983). Cytosol solutions can be stored at –20 or –78°C for several mo before gel analysis. However, once $2 \times$ SDS sample buffer is added to the samples, the abundant proteins and labeled receptor undergo considerable degradation over time, even when stored at –20°C (Simons, unpublished results).

Samples were diluted with an equal vol of $2 \times$ SDS sample buffer, heated at 100°C for 5 min, and applied to constant percentage acrylamide gels (10.5–11% or 15%, overlaid with a 3.6% stacking gel, both with a 1:38.9 ratio of bis-acrylamide to acrylamide), which were run in a water cooled (15°C) Protean or Protean II slab gel apparatus (Bio-Rad). The gels were run at constant current (20 mA/gel while in the stacking gel, then 30 mA/gel for the 10.5–11% gels and 25 mA/gel for the 15% gels). Gels were fixed and stained in 50% methanol/7.5% acetic acid containing 0.02% Coomassie Blue (R250), destained in 10% methanol/7.5% acetic acid, incubated for 1 h in Enhance (New England Nuclear) and 30–60 min in 10% Carbowax PEG 8000 (formerly PEG 6000; Fisher; recommended by New England Nuclear) with constant shaking at room temperature, dried on a Bio-Rad model 443 slab gel drier at 60°C with a sheet of dialysis membrane backing (Bio-Rad) directly over the gel to prevent cracking, and fluorographed for 7–12 d with Kodak X-OMAT XAR-5 film. The positions of the nonradioactive molecular weight markers were visualized on the fluorographs by overlaying the dried gel with a phosphorescent marker (Ult-Emit, New England Nuclear). For the slice and counting of gels, gels were sliced (Bio-Rad gel slicer), digested in 0.5 mL of 30% hydrogen peroxide at 60°C for about 4 h, and counted in 5 mL of Hydrofluor (National Diagnostics) (Reichman et al., 1984; Simons, 1987).

3. Applications of Dex-Mes Labeling of Glucocorticoid Receptors

3.1. Purification of Glucocorticoid Receptor

A common application of affinity labels is to allow the use of conditions that would otherwise cause the dissociation of ligand from the binding macromolecule. Such conditions are particularly helpful during attempts to purify trace proteins such as the receptor. Thus we have used reverse phase HPLC in an effort to isolate protease digestion fragments of Dex-Mes labeled receptors (Simons et al., 1987). The intact, 98kDa [^{3}H]Dex-Mes labeled receptor from WEHI-7 cells has been isolated from modified SDS-PAGE gels with considerable purification and excellent recovery (80–95%) (Smith et al., 1988).

Affinity labeling has greatly facilitated the purification of receptors by providing a means of following the receptor at various stages (Westphal et al., 1981; Payvar et al., 1983; Govindan and Gronemeyer, 1984; Singh and Moudgil, 1984; Wrange et al., 1984; Housley et al., 1985). Affinity labeling has also been used to determine whether the purified receptor is intact or has been degraded (Westphal et al., 1981; Kovacic-Milivojevic et al., 1985) (*see also* Section 3.3.1.).

3.1.1. (NH$_4$)$_2$SO$_4$ Precipitation

[^{3}H]Dex-Mes labeled HTC cell receptors are easily precipitated at 0°C upon adjusting to 40% of saturated (NH$_4$)$_2$SO$_4$ (Reichman et al., 1984). The recovery of labeled receptor is >90% (Simons et al., unpublished results) and the three major, contaminating [^{3}H]Dex-Mes labeled species have been removed or greatly reduced in concentration (Reichman et al., 1984; Simons, 1988a). A 20% (NH$_4$)$_2$SO$_4$ precut can also be used to decrease the recovery of nonreceptor proteins (Miller and Simons, 1988b). Chromatography of the labeled receptors on Sephadex G-100 or G150 prior to (NH$_4$)$_2$SO$_4$ precipitation reduces the final recovery of complexes to ~65% with no appreciable further purification (Simons, 1988a).

3.1.2. Immunoprecipitation

Eisen et al. (1981) were the first to use immunoprecipitation to purify [^{3}H]Dex-Mes labeled receptors. This procedure also confirmed the identification of the [^{3}H]Dex-Mes labeled, 98kDa species as the glucocorticoid receptor. Mendel et al. (1987) have used immunoprecipitation to obtain receptors that were sufficiently purified so that the ratio of [^{32}P]phosphate to [³H]Dex-Mes binding sites could be calculated. Other investigators have used this approach to help obtain receptors pure enough to determine the amino acid labeled by Dex-Mes (Smith et al.,1988), to determine the efficiency of immunoadsorption (Howard and Distelhorst, 1988), and to identify which of several immunoprecipitated species is receptor (Denis et al., 1987; Gustafsson et al., 1987; Rao and Fox, 1987; Rehmus et al., 1987).

3.1.3. DNA-Cellulose Chromatography

Single-stage DNA-cellulose chromatography affords a good, large scale purification of [³H]Dex-Mes labeled receptors in yields of 50–55% (Simons et al., 1983; Reichman et al., 1984; Simons, 1987). A more elaborate, two-stage protocol (DNA-cellulose chromatography of [³H]Dex-Mes labeled receptors both before and after activation) gives receptors that are about 25% pure (Govindan and Gronenmeyer, 1984). It should be noted, however, that the addition of any NaCl to the activated complexes before DNA-cellulose chromatography with buffers of pH around 8.5 can result in a decrease in the amount of complexes retained by the DNA-cellulose column (Simons, Yen, and Eisen, unpublished results). Furthermore, not all buffers can be efficiently used in a two-stage protocol. Thus, about half of those complexes in a pH 8.8 TAPS buffer (25 mM TAPS, 1 mM EDTA, 10% [v/v] glycerol, pH at 0°C) that can be activated were found to bind to DNA-cellulose columns without formally being activated (Simons and Eisen, unpublished results), presumably owing to the relatively rapid activation of complexes in pH 8.8 TAPS buffers at 0°C (Simons and Miller, 1984).

3.2. Identification of Amino Acids in Steroid Binding Cavity

3.2.1. Specificity of Dex-Mes Reactions

Another common use of affinity labels is to identify amino acids in the ligand binding cavity. In preparation for such studies of glucocorticoid receptors labeled with Dex-Mes, we showed that Dex-Mes selectively reacts with thiols under many conditions (Simons et al., 1980, 1983; Eisen et al., 1981; Simons, 1987).

3.2.2. Identification of the Amino Acid of Glucocorticoid Receptors that Reacts with Dex-Mes

Dex-Mes labeling of cell-free, HTC cell receptors is nearly quantitative (98 ± 11%) (Simons and Miller, 1984). The thiol-specific reagent methyl-methanethiolsulfonate (MMTS) is equally effective in preventing [³H] Dex-Mes labeling, and [³H]Dex binding, of cell-free glucocorticoid receptors

(Simons, 1987). These results argue that Dex-Mes labeling of the gluco-corticoid receptor occurs at the same free -SH group(s) that is involved in Dex binding.

Limit proteolysis of [^{3}H]Dex-Mes labeled receptors with trypsin, chymotrypsin, or *S. aureus* V8 protease each produced a unique labeled fragment at an apparent mol wt of 1700–1500 dalton. The V8 protease and chymotrypsin limit digest fragments each contained all of the radioactivity that was initially present in the undigested 98 kDa receptor. Thus, all of the [^{3}H]Dex-Mes labeling of receptor -SH groups appears to occur in one small segment of the receptor (Simons, 1987).

A comparison of the sequential Edman degradation data of each limit protease digest (*see* Simons et al., 1987) with the known sequence of the rat glucocorticoid receptor (Miesfeld et al., 1986) gave Cys-656 as the only possible [^{3}H]Dex-Mes labeled amino acid (Fig. 2) (Simons et al.,1987). These results have recently been confirmed by others (Carlstedt-Duke et al., 1988). Furthermore, photoaffinity labeling of rat liver receptors with tri-amcinolone acetonide has additionally identified Met-622 and Cys-754 as being in the steroid binding cavity (Carlstedt-Duke et al., 1988).

Considering the greater than 95% homology between the steroid bind-ing domains of rat (Miesfeld et al., 1986), mouse (Danielsen et al., 1986), and human (Hollenberg et al., 1985) receptors, it was predicted that Dex-Mes labels Cys-644 of the mouse receptor and Cys-638 of the human receptor (Fig. 2). Dex-Mes labeling of Cys-644 of the mouse receptor has recently been confirmed (Smith et al., 1988). There is also considerable homology between the steroid binding domains of glucocorticoid, progesterone (Loosfelt et al., 1986), estrogen (Krust et al.,1986), and mineralocorticoid (Arriza, 1987) receptors. In each receptor, the amino acid corresponding to Cys-656 of the rat glucocorticoid receptor has been changed to some other amino acid. Consistent with this is the observation that Dex-Mes has not yet been found to affinity label any other class of receptors (W.T. Schrader, personal communication; J.M. Harmon, personal communication).

3.2.3. Involvement of a Second Thiol in Steroid Binding to Receptors

The dose-response curves for [^{3}H]Dex binding to steroid-free recep-tors preincubated with different concentrations of thiol reagents are nor-mally sigmoidal, but unusual biphasic and bimodal curves are obtained after varying lengths of preincubation with MMTS (Fig. 3). The bimodal dose-response curve seen after 2.5h preincubation virtually demands the involvement of two (or more) thiol groups. Those receptors pretreated with intermediate concentrations of MMTS (~1 m*M*) retain ~70% of the

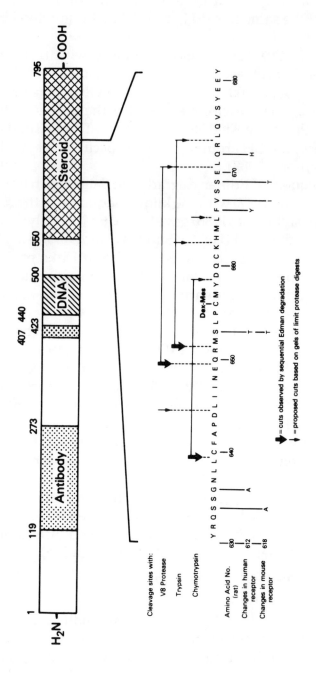

Fig. 2. Protease cleavage sites near [³H]Dex-Mes labeled Cys-656 of the HTC cell glucocorticoid receptor. The various domains of the receptor (antibody binding sites, DNA binding domain, and steroid binding domain) in the intact receptor (Miesfeld et al., 1986) are shown on top. Antibodies 250 (Okret et al., 1984) and BUGR2 (Gametchu and Harrison, 1984) recognize amino acids contained between 119-273 and 407-423, respectively (Rusconi and Yamamoto, 1987). Below is given the single letter amino acid sequence, and numbering, of the region surrounding Cys-656 of the HTC cell receptor (Miesfeld et al., 1986). The heavy arrows (➡) indicate the cleavage sites of V8 protease, trypsin, and chymotrypsin that were identified by sequential Edman degradation. The light arrows (↓) show the probable additional cleavage sites that would account for the observed sizes of the single protease limit digest fragments (indicated by ——) and the other small species seen after single and double limit digests (Simons et al., 1987). Those amino acids which are different in the human (Hollenberg et al., 1985) and mouse (Danielson et al., 1986) glucocorticoid receptors are indicated (*from* Simons et al., 1987).

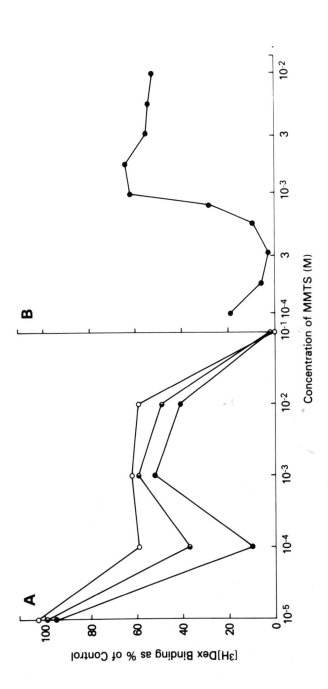

Fig. 3. Inhibition of [³H]Dex binding to HTC cell receptors by preincubation time on inhibition of [³H]Dex binding. **A)** Effect of MMTS preincubation time on inhibition of [³H]Dex binding. Duplicate samples of HTC cell cytosol solution were pretreated with various concentrations of MMTS in absolute EtOH for 0 h (○), 0.5 h (○), or 2.5 h (●) before the addition of [³H]Dex ± [¹H]Dex. Following a further 2.5 h incubation, the specific binding to receptors was determined after the addition of dextran coated charcoal to remove free steroids, expressed as percent of EtOH control binding, and plotted versus the final concentration of MMTS. The data points shown are the average values derived from 1–5 experiments. **B)** Detailed dose-response curve for inhibition of [³H]Dex binding after 2.5 h preincubation with MMTS. Duplicate samples of HTC cell cytosol solution were treated with various concentrations of MMTS for 2.5 h before determining the remaining steroid binding activity of the receptors as in A (*from* Miller and Simons, 1988a).

initial binding capacity and 1/5 the affinity for Dex. Solutions of this low affinity form of receptor contain essentially no accessible -SH groups; and, all of the usual covalent labeling by Dex-Mes of various proteins, including the receptor, is blocked. The facts that this low affinity form of the receptor is not affected by added iodoacetamide, cannot be produced from the non-steroid binding form of receptor simply by adding more MMTS, and displays different kinetics of formation than does the nonsteroid binding form of receptor, all argue that reaction of the receptor with intermediate and low MMTS concentrations occurs *via* different pathways such that the two thiols involved in steroid binding are differently modified at low and intermediate MMTS concentrations. Nevertheless, the effects of both concentrations of MMTS on the receptor are fully reversible with added DTT. The kinetics of inhibition of [³H]Dex binding at low MMTS concentrations are independent of receptor concentration, indicating an intramolecular reaction. Collectively, these data suggest a model of steroid binding involving two thiols, one of which appears to be Cys-656 (Fig. 4). Low concentrations of MMTS induce the formation of an intramolecular disulfide, which prevents steroid binding and Dex-Mes labeling, while the intermediate MMTS concentrations convert both thiols directly to mixed disulfides and steroid binding (but not Dex-Mes labeling) persists. Furthermore, in contrast to accepted tenants, these data indicate that free -SH groups are *not* absolutely required for steroid binding to glucocorticoid receptors (Miller and Simons, 1988a).

3.3. Identification of Receptors

3.3.1. Intact and Proteolyzed Receptors

Covalent labeling with [³H]Dex-Mes has been found not to alter the apparent M_r of the receptor on SDS-PAGE gels (Mendel et al., 1986a). In the absence of proteolysis (*see below*) and mutated receptors (*see* Gehring, 1988; Northrop et al., 1985), only one high affinity, low capacity binding macromolecule with an apparent mol wt of about 98kDa is affinity labeled by Dex-Mes (c.f., Simons and Thompson, 1981; Eisen et al., 1981; Simons et al., 1983; Foster et al., 1983; Thompson et al., 1983; Housley and Pratt, 1983). Dex-Mes labeled receptors have also been invaluable in distinguishing between the 98kDa receptor and the associated nonreceptor molecule hsp90 of similar M_r (Mendel et al., 1986; Housley et al., 1985; Denis et al., 1987).

SDS-PAGE gel analysis of affinity labeled receptors appears to be the safest method to assess proteolysis. Mendel et al. (1985) have found that [³H]Dex-Mes labeled receptors prepared from fresh rat thymus are fully degraded to a 50kDa complex during a 2 h incubation at 3°C, even though

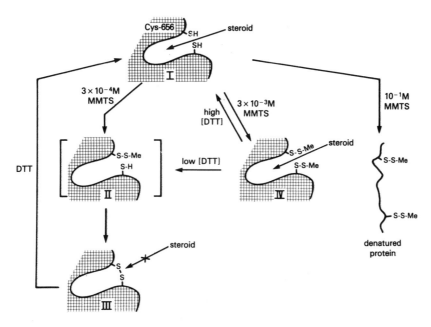

Fig. 4. Model of MMTS/DTT modifications of sulfhydryl groups in the steroid binding cavity of glucocorticoid receptors. A hypothetical steroid binding cavity of the receptor containing part of the receptor protein and Cys-656 is shown. Steroid can bind to the fully reduced, unmodified receptor (I) and to the mixed disulfide form (IV) but not to the intramolecular disulfide form (III) (*from* Miller and Simons, 1988a).

this 50kDa complex appears as an intact 8 nm complex when analyzed by gel filtration. Others have used SDS-PAGE to examine the proteolysis of labeled receptors in the presence of low levels of Ca^{2+} (Simons, 1988b), to examine the consequences of proteases in specific tissues (Hendry et al., 1987; Distelhorst and Miesfeld, 1987), to show that proteolysis is not occurring in given samples (Howard and Distelhorst, 1988), and to validate Western blotting results when proteolysis is known to have occurred (Bresnick et al., 1988). The accuracy of a new technique called protein blotting was ascertained by the demonstration that one of the major proteins separated on SDS-PAGE, transferred to nitrocellulose, and then bound by ^{32}P-MMTV DNA was probably the glucocorticoid receptor since it comigrated with Dex-Mes labeled receptors (Silva et al., 1987).

Affinity labeled receptors are uniquely suited to follow proteolysis because many proteases appear to preferentially attack the amino terminal portion of the glucocorticoid receptor (Reichman et al., 1984; Hoschutzky and Pongs, 1985; Simons et al., 1987) that contains all of the antigenic domains (Miesfeld et al., 1986) of those antibodies (Okret et al., 1984; Gametchu and Harrison, 1984) that have been characterized (Fig. 2). In contrast, the steroid binding domain and the amino acids around the Dex-Mes labeled Cys-656 of the rat receptor are one of the last regions to be digested (Simons, 1987; Simons et al., 1987). Thus, [³H]Dex-Mes labeled receptors were used to help localize the antigenic region for the BUGR2 monoclonal antibody to a 16kDa fragment that contains the DNA binding, but not the steroid binding, region of the receptor (Eisen et al., 1985). This same approach was recently used to identify the 16kDa sequence of Thr_{537}-Arg_{673} in the rat glucocorticoid receptor as being a core unit for steroid binding activity (Simons et al., 1989). Dex-Mes labeled human (Smith and Harmon, 1987) and mouse (Distelhorst et al., 1987b) receptors have been used to confirm the functional domain results first seen with rat receptors (Reichman et al., 1984; Eisen et al., 1985).

3.3.2. Receptors with Unusual Properties

Affinity labeling with Dex-Mes has been used to establish the presence of receptor in tissues with unusual properties (Foster et al., 1983; Gadson et al., 1984; Schlechte et al., 1985; Distelhorst et al., 1985) and to argue that a smaller 72kDa protein that copurifies with the receptor during affinity chromatography is not a proteolytic fragment of the 98kDa receptor (Wrange et al., 1984). Dex-Mes labeling has been used to show that different forms of the receptor on sucrose gradient gels all contained an identically sized receptor monomer (Kasayama et al., 1987) and that the smaller species observed from human leukemia cells after SDS-PAGE were most likely formed by neutrophil elastase digestion of the intact receptor (Distelhorst and Miesfeld, 1987; Distelhorst et al., 1987a). Epstein-Barr virus transformation of human lymphocytes caused a fivefold increase in glucocorticoid receptor number but the M_r of the receptor was unchanged (Tomita et al., 1985). Conversely, the receptors present at reduced levels in Dex-resistant, MMTV infected, WEHI-7 cells were also found to be of the normal size (Rabindran et al., 1987). Similarly, two other Dex-resistant lymphoid cell lines were found to contain Dex-Mes labeled receptors of the same size as the wild type receptors (Schlechte and Schmidt, 1987; Lucas et al., 1988). SDS-PAGE of mesylate labeled, transfected receptors has been used to confirm the size of both wild type (Vanderbilt et al., 1987) and mutant receptors (Miesfeld et al., 1987; Rusconi and Yamamoto, 1987). Comigration of affinity labeled receptors and a ³²P-labeled protein on SDS-PAGE has been taken as evidence that the receptor can be phosphorylated in whole and broken cell systems (Housley and Pratt, 1983; Singh and Moudgil, 1985; Sanchez and Pratt, 1986; Tienrungroj et al., 1987). The covalent nature of Dex-Mes labeled receptors was used to show that the steroid was bound to receptor, not RNA, in a 6S receptor-steroid-RNA complex (Ali and Vedeckis, 1987). Finally, the fact that Dex-Mes specifically labeled three of the five plasma membrane species detected by an antiglucocorticoid antibody strengthened the argument that these membrane-located molecules were a form of glucocorticoid receptors (Gametchu, 1987).

3.3.3. Cloning of the Receptor

Dex-Mes receptors were found to be immunoprecipitated by those antibodies that were then used to select c-DNA clones producing immunoreactive material, thus confirming the specificity of the antibodies for the glucocorticoid receptor and strengthening the conclusion that the immunoselected clones coded for antigenic regions of the receptor (Govindan et al., 1985). Even more conclusive was the elegant demonstration that [^{3}H]Dex-Mes labeling of cytosol extracts of receptor negative HTC cells that had been stably transfected with a suspected receptor cDNA clone afforded a specifically labeled, 98kDa species that could be immunoprecipitated by BUGR2 monoclonal antireceptor antibody (Miesfeld et al., 1986).

3.4. Examination of Receptor Heterogeneity

Receptor heterogeneity has always been a formal possibility to explain the different actions of glucocorticoid steroids in different tissues (Webb et al., 1985). Unfortunately, the ability of SDS-PAGE to discern size differences among various affinity labeled receptors is limited by the resolution of the gels. Such analysis has revealed an apparent mol wt for glucocorticoid receptors (and also for the rabbit progesterone receptor [Logeat et al., 1985; Loosfelt et al., 1986]) that is significantly larger than the calculated value (e.g. [^{3}H]Dex-Mes labeled rat receptor = 98,600 $\pm$ 1,500 dalton [SD, n = 28; Reichman et al., 1984]; calculated M_r = 87,512 [Miesfeld et al., 1986]). This difference is not caused by the covalently bound Dex-Mes (Mendel et al., 1986); it could be an artifact of the SDS-PAGE gel system (Reichman et al., 1984; Simons, 1985; Mendel et al., 1986). Dex-Mes labeled receptors extracted from HeLa cell nuclei with pyridoxal phosphate display a slightly higher M_r on SDS-PAGE (Scheible et al., 1987), presumably owing to the covalent attachment by pyridoxal phosphate. In contrast, attempts to remove phosphate from mesylate labeled receptors (*see below*) from AtT-20 mouse pituitary cells by alkaline phosphate treatment caused no apparent change in M_r, although the sucrose gradient sedimentation value was decreased from 9.1S to 5.2S (Reker et al., 1987).

Protease digestion of Dex-Mes labeled receptors has been helpful in further examining receptor homology. No differences were found between labeled rat liver and HTC cell receptors (Reichman et al., 1984) or HTC and Fu5-5 cell receptors (Miller and Simons, 1988b) but the several differences observed in the immunogenic and steroid binding domains of labeled rat and human receptors (Hoschutzky and Pong, 1985) were subsequently confirmed when the respective receptors were cloned (Miesfeld et al., 1986; Danielsen et al., 1986). A combination of affinity labeling with Dex-Mes and immunoreaction with the antireceptor antibody BUGR2 (Gametchu and Harrison, 1984) were used to conclude that nt^i receptors are both smaller and have an altered sequence in the antigenic region (Rehmus et al., 1987).

When affinity labeled receptors are analyzed on SDS-PAGE gels, there is no difference in the M_r of unactivated and activated complexes (Simons et al., 1983; Reichman et al., 1984; Distelhorst and Benutto, 1985). Thus, within the above noted limitations of gel analysis, one can conclude that activation does not cause any major changes in the size of the denatured, unactivated 98kDa receptor protein. Detailed analysis of the proteolytic digestion patterns of native or denatured, activated and unactivated, [^{3}H]Dex-Mes labeled HTC cell receptors on SDS-PAGE gels revealed no differences down to $M_r = 14{,}000$ (Reichman et al., 1984). These results argue against activation involving oligomer dissociation or a conformational change in the receptor, but it is possible that the first few cuts cause both the activated and unactivated forms of the native receptor to assume a common structure to give identical digestion patterns. Further digestion of the denatured [^{3}H]Dex-Mes labeled, activated and unactivated complexes gave identical patterns in the region of $M_r = 14{,}000–1{,}400$ (Simons, 1987). These results suggest that activation does not involve any covalent modifications in the area of the receptor close to the affinity labeled Cys-656 (Simons, 1987). Similarly, Smith and Harmon (1987) found no difference in the pI of the trypsin 29 and 27kDa meroreceptor fragments of covalently labeled, activated and unactivated complexes and concluded that activation did not produce any charge modifications in this larger steroid binding region of the receptor.

Affinity labeled receptors from a given source may be homogeneous in size but they are heterogeneous in charge. Unactivated [^{3}H]Dex-Mes labeled HeLa cell (human) receptors migrated on 2D-gels as 4–8 species with identical mol wts but with pIs for the native receptors that ranged from 6.5–7.2 (Cidlowski and Richon, 1984). These receptors were obtained from whole cells that were treated with [^{3}H]Dex-Mes but most of the covalent labeling undoubtedly occurred after cell lysis, since the yield of labeled receptors was increased simply by raising the pH of the lysis buffer. In contrast, unactivated IM-9 cell (human) receptors that were labeled by [^{3}H]Dex-Mes under cell-free conditions at 0°C were resolved into three forms (pI = 5.2, 5.7, and 6.0–6.5) of identical M_r on 2D-gels but only when using nonequilibrium pH gel electrophoresis (NEPHGE) of denatured receptors (in 9.5*M* urea) in the first dimension (Smith and Harmon, 1985; Smith et al., 1986). No change in the pI of any of the species was observed after being exposed to activating conditions but only the most basic form was found to bind to DNA (Smith et al., 1986). Protease digestion studies have localized these pI differences to the 27kDa meroreceptor fragment (Smith and Harmon, 1987). A similar charge heterogeneity was noted in native mouse L cell receptors in the range of pI = 5.3–6.2, which is thought to be due to phosphorylation of one or more serines (Housley and Pratt, 1983). Cell-free phosphorylation was seen for steroid-free or steroid-bound receptors in the unactivated or activated state (Sanchez and Pratt, 1986). Unfortunately, the functional significance of receptor phosphorylation and charge heterogeneity is not yet understood.

The pI values for native, Dex-Mes labeled receptors have recently been found to be much more acidic (i.e., 4.4-5.7) on agarose isofocusing gels. The unactivated complexes appear to have a pI of 4.4 (Danze et al., 1987). Agarose gels are thought to circumvent the problems of hydrophobic interactions and protein aggregation that may artifactually yield higher pI values on polyacrylamide gels (Danze et al., 1987). It will be interesting to see whether these more acidic pI values also obtained with denatured complexes or are due to other proteins associated with the native receptor.

4. Conclusions

Dex-Mes has been very helpful in the purification and study of glucocorticoid receptors. Its most frequent future uses will probably be in those situations where one desires to eliminate steroid dissociation from the receptor protein, e.g., ascertaining the presence of the receptor (by its size and labeling specificity) in new target cells, identifying possible new steroid binding forms or protease digestion fragments of the receptor, assisting in development of new receptor purification procedures, and assessing the effects of modifications of the steroid binding domain.

Recent reports indicate that nonreceptor molecules in addition to hsp90 may be associated with receptors in ways that modify the biological activity of the receptor. These nonreceptor molecules can be of cytoplasmic origin, but almost certainly include some chromatin associated molecules that are intimately involved in steroid hormone regulation of gene transcription (c.f., Cordingley et al., 1987). Cross-linking studies with Dex-Mes labeled receptors would be invaluable in investigating these associated nonreceptor molecules. Just such an approach has been used to show that cytoplasmic glucocorticoid receptors are associated with RNA (Economidis and Rousseau, 1985). DNA-protein cross-linking offers a means of directly identifying those nucleotides involved in receptor-steroid complex binding to DNA.

In summary, Dex-Mes has proven to be invaluable in solving many problems concerning glucocorticoid receptor structure, function, and purification, some of which could not be addressed by any other currently available technique. For this reason, Dex-Mes has now become another standard tool for studying glucocorticoid receptors.

Acknowledgments

I thank Brenda Briscoe and Anthony Forman for assistance in the typing of this paper.

References

Ali, M., and Vedeckis, W. V. (1987) *J. Biol. Chem.* **262**, 6771–6777.
Arriza, J. L., Weinberger, C., Cerelli, G., Glaser, T. M., Handelin, B. L., Housman, D. E., and Evans, R. M. (1987) *Science* **237**, 268–275.
Bresnick, E. H., Sanchez, E. R., Harrison, R. W., and Pratt, W. B. (1988) *Biochemistry* **27**, 2866–2872.
Carlstedt-Duke, J., Stomstedt, P. E., Persson, B., Cederlund, E., Gustafsson, J. A., and Jornvall, H. (1988) *J. Biol. Chem.* **263**, 6842–6846.
Cidlowski, J. A. and Richon, V. (1984) *Endocrinology* **115**, 1588–1597.
Cordingley, M. G., Riegel, A. T., and Hager, G. L. (1987) *Cell* **48**, 261–270.
Danielsen, M., Northrop, J. P., and Ringold, G. M. (1986) *EMBO J.* **5**, 2513–2522.
Danze, P.-M., Formstecher, P., Richard, C., and Dautrevaux, M. (1987) *Biochim. Biophys. Acta* **927**, 129–138.
Denis, M., Wikstrom, A.-C., and Gustafsson, J. A. (1987) *J. Biol. Chem.* **262**, 11803–11806.
Distelhorst, C. W. and Benutto, B. M. (1985) *J. Biol. Chem.* **260**, 2153–2159.
Distelhorst, C. W. and Miesfeld, R. (1987) *Blood* **69**, 750–756.
Distelhorst, C. W., Benutto, B. M., and Griffith, R. C. (1985) *Blood* **66**, 679–685.
Distelhorst, C. W., Janiga, K. E., Howard, K. J., Strandjord, S. E., and Campbell, E. J. (1987a) *Blood* **70**, 860–868.
Distelhorst, C. W., Kullman, L., and Wasson, J. (1987b) *J. Steroid Biochem.* **26**, 59–65.
Economidis, I. V. and Rousseau, G. G. (1985) *FEBS Lett.* **181**, 47–52.
Eisen, H. J., Schleenbaker, R. E., and Simons, Jr., S. S. (1981) *J. Biol. Chem.* **256**, 12920–12925.
Eisen, L. P., Reichman, M. E., Thompson, E. B., Gametchu, B., Harrison, R. W., and Eisen, H. J. (1985) *J. Biol. Chem.* **260**, 11805–11810.
Foster, C. M., Eisen, H. J., and Bloomfield, C. D. (1983) *Cancer Res.* **43**, 5273–5277.
Gadson, P. F., Russell, J. D., and Russell, S. B. (1984) *J. Biol. Chem.* **259**, 11236–11241.
Gametchu, B. (1987) *Science* **236**, 456–461.
Gametchu, B. and Harrison, R. W. (1984) *Endocrinology* **114**, 274–279.

Gehring, U. (1988) in *Affinity Labeling and Cloning of Steroid and Thyroid Hormone Receptors* (Gronemeyer, H., ed.), Ellis Horwood Ltd., Chichester, U. K., pp. 144–166.

Govindan, M. V. and Gronemeyer, H. (1984) *J. Biol. Chem.* **259**, 12915–12924.

Govindan, M. V., Devic, M., Green, S., Gronemeyer, H., and Chambon, P. (1985) *Nucleic Acids Research* **13**, 8293–8304.

Gustafsson, J. A., Carlstedt-Duke, J., Poellinger, L., Okret, S., Wikstrom, A.-C., Bronnegard, M., Gillner, M., Dong, Y., Fuxe, K., Cintra, A., Harfstrand, A., and Agnati, L. (1987) *Endocrine Reviews* **8**, 185–234.

Hendry III, W. J., Danzo, B. J., and Harrison III, R. W. (1987) *Endocrinology* **120**, 629–639.

Hollenberg, S. M., Weinberger, C., Ong, E. S., Cerelli, G., Oro, A., Lebo, R., Thompson, E. B., Rosenfeld, M. G., and Evans, R. M. (1985) *Nature* **318**, 635–641.

Hoschutzky, H. and Pongs, O. (1985) *Biochemistry* **24**, 7348–7356.

Housley, P. R. and Pratt, W. B. (1983) *J. Biol. Chem.* **258**, 4630–4635.

Housley, P. R., Sanchez, E. R., Westphal, H. M., Beato, M., and Pratt, W. B. (1985) *J. Biol. Chem.* **260**, 13810–13817.

Howard, K. J. and Distelhorst, C. W. (1988) *J. Biol. Chem.* **263**, 3474–3481.

Kasayama, S., Noma, K., Sato, B., Nakao, M., Nishizawa, Y., Matsumoto, K., and Kishimoto, S. (1987) *J. Steroid Biochem.* **28**, 1–8.

Kovacic-Milivojevic, B., LaPointe, M. C., Reker, C. E., and Vedeckis, W. V. (1985) *Biochemistry* **24**, 7357–7366.

Krust, A., Green, S., Argos, P., Kumar, V., Walter, P., Bornert, J.-M., and Chambon, P. (1986) *EMBO J.* **5**, 891–897.

Logeat, F., Pamphile, R., Loosfelt, H., Jolivet, A., Fournier, A., and Milgrom, E. (1985) *Biochemistry* **24**, 1029–1035.

Loosfelt, H., Atger, M., Misrahi, M., Guiochon-Mantel, A., Meriel, C., Logeat, F., Benarous, R., and Milgrom, E. (1986) *Proc. Natl. Acad. Sci.* **83**, 9045–9049.

Lucas, K. L., Barbour, K. W., Housley, P. R., and Thompson, E. A., Jr. (1988) *Mol. Endo.* **2**, 291–299.

Mendel, D. B., Holbrook, N. J., and Bodwell, J. E. (1985) *J. Biol. Chem.* **260**, 8736–8740.

Mendel, D. B., Bodwell, J. E., Gametchu, B., Harrison, R. W., and Munck, A. (1986) *J. Biol. Chem.* **261**, 3758–3763.

Mendel, D. B., Bodwell, J. E., and Munck, A. (1987) *J. Biol. Chem.* **262**, 5644–5648.

Miesfeld, R., Rusconi, S., Godowski, P. J., Maler, B. A., Okret, S., Wikstrom, A.-C., Gustafsson, J. A., and Yamamoto, K. R. (1986) *Cell* **46**, 389–399.

Miesfeld, R., Godowski, P. J., Maler, B. A., and Yamamoto, K. R. (1987) *Science* **236**, 423–427.

Miller, N. R. and Simons, Jr., S. S. (1988a) *J. Biol. Chem.* **23**, 15217–15225.

Miller, P. A. and Simons, Jr., S. S. (1988b) *Endocrinology* **122**, 2990–2998.

Miller, P. A., Ostrowski, M. C., Hager, G. L., and Simons, Jr., S. S. (1984) *Biochemistry* **23**, 6883–6889.

Northrop, J. P., Gametchu, B., Harrison, R. W., and Ringold, G. M. (1985) *J. Biol. Chem.* **260**, 6398–6403.

Okret, S., Wikstrom, A.-C., Wrange, O., Andersson, B., and Gustafsson, J. A. (1984) *Proc. Natl. Acad. Sci. USA* **81**, 1609–1613.

Payvar, F., DeFranco, D., Firestone, G. L., Edgar, B., Wrange, O., Okret, S., Gustafsson, J. A., and Yamamoto, K. R. (1983) *Cell* **35**, 381–392.

Rabindran, S. K., Danielsen, M., Stallcup, M. R. (1987) *Mol. Cell. Biol.* **7**, 4211–4217.

Rao, K. V. S. and Fox, C. F. (1987) *Biochem. Biophys. Res. Comm.* **144**, 512–519.

Rehmus, E. H., Howard, K. J., Janiga, K. E., and Distelhorst, C. W. (1987) *J. Steroid Biochem.* **28,** 167–177.

Reichman, M. E., Foster, C. M., Eisen, L. P., Eisen, H. J., Torain, B. F., and Simons, Jr., S. S. (1984) *Biochem.* **23,** 5376–5384.

Reker, C. E., LaPointe, M. C., Kovacic-Milivojevic, B., Chiou, W. J. H., and Vedeckis, W. V. (1987) *J. Steroid Biochem.* **26,** 653–665.

Rusconi, S. and Yamamoto, K. R. (1987) *EMBO J.* **6,** 1309–1315.

Sanchez, E. R. and Pratt, W. B. (1986) *Biochemistry* **25,** 1378–1382.

Scheible, P. P., DeLorenzo, T. M., and Cidlowski, J. A. (1987) *J. Steroid Biochem.* **26,** 181–187.

Schlechte, J. A., and Schmidt, T. J. (1987) *J. Clin. Endo. Metab.* **64,** 441–446.

Schlechte, J. A., Simons, Jr., S. S., Lewis, D. A., and Thompson E. B. (1985) *Endocrinology* **117,** 1355–1362.

Sherman, M. R., Morar, M. C., Tuazon, F. B., and Stevens, Y.-W. (1983) *J. Biol. Chem.* **258,** 10366–10377.

Silva, C. M., Tully D. B., Petch, L. A., Jewell, C. M., and Cidlowski, J. A. (1987) *Proc. Natl. Acad. Sci.* **84,** 1744–1748.

Simons, Jr., S. S. (1985) in *Molecular Mechanism of Steroid Hormone Action* (Moudgil, V. K., ed.), de Gruyter, Berlin, pp. 111–140.

Simons, Jr., S. S. (1987) *J. Biol. Chem.* **262,** 9669–9675.

Simons, Jr., S. S. (1988a) in *Affinity Labeling and Cloning of Steroid and Thyroid Hormone Receptors* (Gronemeyer, H., ed.), Ellis Horwood Ltd., Chichester, U. K., pp. 28–54.

Simons, Jr., S. S. (1988b) in *Affinity Labeling and Cloning of Steroid and Thyroid Hormone Receptors* (Gronemeyer, H., ed.), Ellis Horwood Ltd., Chichester, U. K., pp. 109–143.

Simons, Jr., S. S. and Miller, P. A. (1984) *Biochemistry* **23,** 6876–6882.

Simons, Jr., S. S. and Miller, P. A. (1986) *J. Steroid Biochem.* **24,** 25–32.

Simons, Jr., S. S. and Thompson, E. B. (1981) *Proc. Natl. Acad. Sci.* **78,** 3541–3545.

Simons, Jr., S. S. and Thompson, E. B. (1982) in *Biochemical Actions of Hormones*, vol. 9 (Litwack, G., ed.), Academic, New York, pp. 221–254.

Simons, Jr., S. S. and Yen, P. M. (1987) in *Steroid and Sterol Hormone Action* (Spelsberg, T.C. and Kumar, R., eds.), M. Nijhoff, Boston, pp. 251–268.

Simons, Jr., S. S., Thompson, E. B., and Johnson, D. F. (1979) *Biochem. Biophys. Res. Commun.* **86,** 793–800.

Simons, Jr., S. S., Pons, M., and Johnson, D. F. (1980) *J. Org. Chem.* **45,** 3084–3088.

Simons, Jr., S. S., Schleenbaker, R. E., and Eisen, H. J. (1983) *J. Biol. Chem.* **258,** 2229–2238.

Simons, Jr., S. S., Pumphrey, J. G., Rudikoff, S., and Eisen, H. J. (1987) *J. Biol. Chem.* **262,** 9676–9680.

Simons, Jr., S. S., Mercier, L., Miller, N. R., Miller, P. A., Oshima, H., Sistare, F. D., Thompson, E. B., Wasner, G., and Yen, P. M. (1989) *Cancer Res.*, **49,** 2244s–2252s.

Simons, Jr., S. S., Sistare, F. D., and Chakraborti, P. K. (1989) *J. Biol. Chem.* **264,** 14493–14497.

Simons, Jr., S. S., Chakrabarti, P. K., and Cavanaugh, A. H. (1990) *J. Biol. Chem.* **265,** 1938–1945.

Singh, V. B. and Moudgil, V. K. (1984) *Biochem. Biophys. Res. Comm.* **125,** 1067–1073.

Singh, V. B. and Moudgil, V. K. (1985) *J. Biol. Chem.* **260,** 3684–3690.

Sistare, F. D., Hager, G. L., and Simons, Jr., S. S. (1987) *Mol. Endo.* **1,** 648–658.

Smith, A. C. and Harmon, J. M. (1985) *Biochemistry* **24,** 4946–4951.

Smith, A. C. and Harmon, J. M. (1987) *Biochemistry* **26,** 646–652.

Smith, A. C., Elsasser, M. S., and Harmon, J. M. (1986) *J. Biol. Chem.* **261,** 13285–13292.

Smith, L. I., Bodwell, J. E., Mendel, D. B., Ciardelli, T., North, W. G., and Munck, A. (1988) *Biochemistry* **27,** 3747–3753.

Thompson, E. B., Zawydiwski, R., Brower, S. T., Eisen, H. J., Simons, S. S., Jr., Schmidt, T. J., Schlechte, J. A., Moore, D. E., Norman, M. R., and Harmon, J. M. (1983) in *Steroid Hormone Receptors: Structure and Function* (Eriksson, H. and Gustafsson, J. A., eds.), Elsevier, pp. 171–194.

Tienrungroj, W., Sanchez, E. R., Housley, P. R., Harrison, R. W., and Pratt, W. B . (1987) *J. Biol. Chem.* **262,** 17342–17349.

Titani, K., Koide, A., Hermann, J., Ericsson, L. H., Kumar, E. S., Wade, R. D., Walsh, K. A., Neurath, H., and Fischer, E. H. (1977) *Proc. Natl. Acad. Sci.* **74,** 4762–4766.

Tomita, M., Chrousos, G. P., Brandon, D. D., Ben-Or, S., Foster, C. M., de Vough, L., Taylor, S., Loriaux, D. L., and Lipsett, M. B. (1985) *Horm. Metabol. Res.* **17,** 674–678.

Vanderbilt, J. N., Miesfeld, R., Maler, B. A., and Yamamoto, K. R. (1987) *Mol. Endo.* **1,** 68–74.

Webb, M. L,. Miller-Diener, A. S., and Litwack, G. (1985) *Biochemistry* **24,** 1946–1952.

Westphal, H. M., Fleischmann, G., and Beato, M. (1981) *Eur. J. Biochem.* **119,** 101–106.

Wrange, O., Carlstedt-Duke, J., and Gustafsson, J. A. (1979) *J. Biol. Chem.* **254,** 9284–9290.

Wrange, O., Okret, S., Radojcic, M., Carlstedt-Duke, J., and Gustafsson, J. A. (1984) *J. Biol. Chem.* **259,** 4534–4541.

Dexamethasone-Biotin Affinity Probes

Partial Purification
of the Human Spleen Tumor Glucocorticoid Receptor

Bernhard Manz and Kunhard Pollow

1. Introduction

Increasing attention has been devoted to the development of new affinity chromatography methods for hormone receptor purification (Hubert et al., 1978; Manz et al., 1982a). The most promising method for purification of these proteins seems to be the use of hormone-biotin conjugates, which retain their high binding constant toward avidin when bound to the respective hormone receptor. Bifunctional hormone-biotin analogs added during tissue homogenization are not only able to form stable hormone–receptor complexes but also retain their high specificity for immobilized avidin.

We have synthesized a number of steroid derivatives that contain biotin covalently attached to the steroid sidechain. In this report, a methodology for affinity chromatography of glucocorticoid hormone receptors of human white blood cells is described. Nine-fluoro-16α-methyl -11β, 17-dihydroxy-1, 4-androstadiene-3-one-17β-carboxylic acid, a sidechain modified derivative of dexamethasone that is unable to bind to glucocorticoid receptors of different species was coupled to biotin using spacers of

Receptor Purification, vol. 2   ©1990 The Humana Press

various lengths. As shown previously, amidation or methylation of 17β-carboxylic acids of different natural or synthetic glucocorticoids restores their capacity to bind to the receptor. One of these bifunctional derivatives (BioDex 1) binds simultaneously to glucocorticoid receptors and avidin Sepharose and protects the glucocorticoid receptor of human white blood cells against inactivation when previously added during tissue homogenization. Its properties were further tested using anti-dexamethasone antibodies as a model to prove the truly bifunctional nature of this compound. BioDex 1 appears to satisfy the basic criteria required of a good affinity probe for tissues with low receptor content.

2. Experimental Procedures

2.1. Synthesis of Biotin-Dexamethasone Conjugates (BioDex 1-6)

The solvents used to develop TLC plates were ethanol/ethyl acetate/ NH_3 (5 + 5 + 1, by vol) and chloroform/methanol (9 + 1, by vol).

2.1.1. N-*Hydroxysuccinimidyl 9α-Fluoro-16α-Methyl-11β, 17α-Dihydroxy-3-Oxo-1,4-Androstadiene-17β-Carboxyester (Compound II)*

Nine-α-fluoro-16α-methyl-11β,17α-dihydroxy-3-oxo-1,4-androstadiene-17β-carboxylic acid (compound I) was obtained by periodic acid oxidation of dexamethasone, as described previously (Govindan and Manz, 1980). The N-hydroxysuccinamide active ester derivative was prepared from compound I and N-hydroxysuccinamide by coupling with dicyclohexylcarbodiimide. The reactants were dissolved at 8 mM concentration in dimethylformamide and incubated at room temperature for 16 h. The dicyclohexylurea was filtered off and the filtrate was dried on a rotary evaporator. The residue was further purified on preparative TLC (solvent B). Mass spectrum: high-resolution m/z 466.281684 (M ± F), calculated for $C_{25}H_4NO_7$ 466.2805. The same reaction scheme was used to synthesize N-hydroxysuccinimidyl 9α-fluoro-16α-methyl-11β,17α-dihydroxy-3-oxo-1,4-[³H]androstadiene-17β-carboxyester (1 mCi [³H]/g of dexamethasone).

2.1.2. N-*(5-Amino)pentyl 9α-Fluoro-16α-Methyl-11β,17α-Dihydroxy-3-Oxo-1,4-Androstadiene-17β-Carboxamide (III)*

In the second step of synthesis, 40 mg of the active ester derivative II were dissolved in 50 mL of ethyl acetate, 100 μL of pentamethylenediamine were added, and the mixture was allowed to react at room temperature for

16 h. The organic solvent was thoroughly washed with water and removed on a rotary evaporator. The remaining residue was redissolved in benzene and lyophilized. The amine produced one spot on TLC in solvent system A and B, respectively, and was used without further purification.

The synthesis of all other intermediate amino compounds was performed as described above, except that ethylenediamine and cystamine were used.

2.1.3. BioDex 1

[^{14}C]Biotinyl-N-hydroxysuccinamide ester (specific activity 0.2 mCi [^{14}C]/biotin, as well as the unlabeled ester, were prepared from biotin and N-hydroxysuccinamide by coupling with dicyclohexylcarbodiimide. In the next step of synthesis, 20 mg of the amine III were dissolved in 10 mL of ethanol, 29 mg of the biotin reagent (twofold excess) were added and the mixture was allowed to react overnight at room temperature. The crude product was purified by preparatory TLC. The plates were developed in solvent system A, the [^{14}C]-labeled spot was scraped off and extracted with ethyl acetate/ethanol (9 + 1, by vol). The organic solvents were removed under vacuum and the remaining residue was dissolved in benzene and lyophilized. The dexamethasone-biotin conjugate produced one spot on TLC in solvent systems A and B and was used without further purification.

The synthesis of BioDex 4-BioDex 6 was performed as described above except that D-biotinyl-ε-aminocaproic acid N-hydroxysuccinamide ester from Boehringer Mannheim (FRG) was used. To ascertain the molar ratio of BioDex 1, either [^{3}H]-labeled dexamethasone or [^{14}C]-labeled biotin were conjugated with their respective unlabeled partners, quantified using their specific activity (*see above*) and used as competitors.

2.2. Preparation of Antidexamethasone-17β-Carboxamide Antibody

Nine-α-fluoro-16α-methyl-11β,17α-dihydroxy-3-oxo-1,4-androstadiene-17β-carboxylic acid (compound I) was conjugated with BSA according to Geffard et al. (Geffard et al., 1982). Antibody was raised in rabbits by intradermally injecting 1 mg of conjugate, suspended in complete Freund's adjuvant at monthly intervals. After 5 mo, we obtained antibodies applicable to binding studies. The antibody in use was diluted 1000-fold with assay buffer containing 0.3 g rabbit IgG per liter.

2.3. Preparation of Cytosol

Human spleen tumor tissue, a source of large quantities of human white blood cells, was stored at -70°C until use. After thawing at 0°C in homogenization buffer (20 mM Tris/HCl, 20% glycerol, 2 mM CaCl$_2$, 1 mM MgCl$_2$, 5 mM mercaptoethanol, pH 7.4 at 0°C) the tissue was minced in 3 vol buffer and homogenized by 10 strokes of a Teflon™/glass Potter-Elvehjem homogenizer. The homogenate was centrifuged for 45 min at 105,000 × g and the supernatant taken as cytosol (Manz et al., 1982b).

2.4. Quantification of Cytosolic Glucocorticoid Receptor by Saturation Analysis

A five-point titration assay was performed using [³H]dexamethasone as radioactive ligand. To determine total binding, 0.1 mL aliquots of aqueous radioactive solutions in homogenization buffer were pipeted in duplicate into glass tubes (12 × 75 mm) to give final concentrations of 2–32 nM [³H]dexamethasone. For estimation of nonspecific binding, a 200-fold excess of dexamethasone (in 10 μL) was added to a second set of tubes. Aliquots of 0.1 mL cytosol were added to each tube. They were incubated at 4°C for 18 h, followed by 10 min in an ice bath. After addition of 0.5 mL of dextran-coated charcoal suspension (0.05% dextran T 500, 0.5% Norit A in homogenization buffer), the tubes were centrifuged at 1,500 × g for 10 min. The supernatant (0.5 mL) was withdrawn and counted for radioactivity. The data were calculated according to Scatchard (Scatchard, 1949).

2.5. Purification of Dexamethasone Binding Proteins

2.5.1. Glucocorticoid Receptor

Fifty g of human spleen tumors were homogenized in 150 mL ice cold buffer. One mL of an ethanolic solution of BioDex 1 (0.1 mM) was added to 200 mL of the crude homogenate to give the final concentration of 0.5 μM. The homogenate was centrifuged at 105,000 × g for 45 min, the supernatant carefully collected and kept for further 3 h at 4°C. Five mL of carefully washed avidin-Sepharose were added and the incubation mixture was rotated for 1 h. The gel suspension was poured into a siliconized glass funnel (diameter 5 cm), the eluant was collected and the gel was washed in the cold with buffer until A$_{280}$ was negligible. A control incubation was performed by rotating the same cytosol without BioDex 1 in the presence of avidin-Sepharose. Aliquots of supernatants from the control and the BioDex 1 incubations were treated with dextran-coated charcoal to remove excess steroid and were quantified by Scatchard analyses using [³H]dexamethasone. The amount of receptors bound to the affinity resin was then determined by difference.* The receptors were eluted by incubation of the

---

*Between control cytosol and avidin-Sepharosetreated cytosol.

resin for 18 h at 4°C with 5 mL of buffer containing 3.3 mM dexamethasone plus 32 nM [³H]dexamethasone. The amount of receptors eluted was quantified by molecular sieving.

2.5.2. Antidexamethasone Antibodies

Three mL of antiserum were diluted fivefold with buffer. One hundred μL of an ethanolic solution of BioDex 1 (0.1 mM) was added to give the final concentration of 0.5 μM. Five mL of avidin-Sepharose were added and the incubation mixture was rotated for 1 h. The gel suspension was poured into a siliconized glass funnel and the gel was washed in the cold with buffer until A$_{280}$ was negligible. A control incubation was performed by rotating the same diluted antiserum without BioDex 1 in the presence of avidin-Sepharose. The antidexamethasone antibodies were eluted with 15 mL of 1M acetic acid. The acid was removed by gel filtration (PD 10 columns, Pharmacia, Uppsala) and the amount of antibody eluted was quantified by Scatchard analysis using [³H]dexamethasone as ligand (data not shown).

2.6. DEAE-Cellulose Chromatography

DEAE-cellulose (Whatman DE 52) was washed extensively with buffer until the conductivity and pH of the washes were the same as those of the buffer. The ion-exchange resin was used in a column of dimensions 0.5 × 2 cm. Proteins were eluted at 4°C with a linear gradient of 0–0.5M NaCl in buffer.

2.7. Competitive Protein Binding Analysis

The procedure for competitive protein binding analysis is based on those described earlier.

2.7.1. Glucocorticoid Receptor

Aliquots (0.1 mL) of spleen tumor cytosol were incubated in triplicate together with 32 nM [³H]dexamethasone and increasing concentrations (3.2–3200 nM) of competitors. After incubating for 18 h at 4°C, the unbound steroid was removed with dextran-coated charcoal. The measured radioactivity values were expressed as percentage binding of the noncompetitive control.

2.7.2. Human Corticosteroid Binding Globulin

Human pregnancy serum, stripped of endogenous steroids with charcoal, in a 1:5 dilution with buffer and [³H]cortisol were used for incubation with the competitors. Incubation was carried out for 30 min at room temperature then 10 min in an ice bath followed by treatment with dextran-coated charcoal.

2.8. Radioimmunoassay

We prepared five standard solutions of dexamethasone in buffer. Goat anti-rabbit-IgG antiserum was diluted 25-fold with assay buffer and mixed 1 + 1 with a solution of 12% of polyethylene 4000 (w/v) in assay buffer prior to use. The assay procedure is as follows: mix 100 µL of standard or dexamethasone derivative with 50 µL of [^{3}H]dexamethasone (approx 30,000 dpm) and 50 µL of diluted antiserum. Incubate for 1 h at room temperature. Add 1 mL of the goat anti-rabbit-IgG/polyethylene 4000 reagent, incubate for 10 min at room temperature, centrifuge for 10 min at 1,500 × g, then measure the radioactivities of the precipitates with a β-counter.

We checked the specificity of antibody by the method of Abraham (Abraham, 1974).

2.9. [^{3}H]Thymidine Incorporation Assay

Peripheral blood lymphocytes from healthy donors were isolated using a Ficoll Hypaque double-gradient technique (Agnado et al., 1980). The cells were washed three times in Hanks balanced salt solution and resuspended at a concentration of 10^8 cells/mL in medium 199 (Flow Laboratories, Meckenheim, FRG) supplemented with 5% heat-inactivated AB serum, 2 mM L-glutamine, 100 U/mL penicillin and 100 µg/mL streptomycin. More than 95% of the lymphocytes were viable as determined by Trypan Blue exclusion. Two hundred µL of culture medium containing 2 × 10^5 cells were added to each well of round-bottomed microtiter plates, 10 µL of phytohemagglutinin P dilutions (Difco Laboratories, Detroit, MI) (5 µg phytohemagglutinin/mL) and 10 µL of steroid solution of decreasing concentration dissolved in 0.9% NaCl were added to each well and incubated for 72 h at 37°C in a humidified atmosphere of 5% CO_2 and 95% air. Lymphocyte stimulation was assessed by the incorporation of [^{3}H]thymidine into the DNA during the last 4 h of culture (1.0 µCi/mL) (Ling, 1968; Stewart et al., 1975; Berger et al., 1982).

2.10. Determination of Protein Concentration

Protein concentrations were measured directly by the use of the Bio-Rad protein assay reagent with bovine serum albumin as standard. When the protein concentration was extremely low (less than 1 µg/mL), the receptor solutions were dialyzed and then freeze-dried. Dodecylsulfate/ polyacrylamide gel electrophoresis was then performed on the concentrated receptor in parallel with protein standards in the concentration range 0.1–2 µg. The intensity of the silver stained bands (BioRad, Richmond, VA) was compared by densitometry (Gilford spectrophotometer).

3. Results

3.1. Competitive Protein Binding Analysis of BioDex 1–6

In order to assess the interaction of dexamethasone-biotin conjugates with the spleen glucocorticoid receptor, we first examined the ability of the derivatives to compete with [³H]dexamethasone for binding to the receptor in crude cytosol. The results are summarized in Table 1. Periodic acid oxidation of the side chain of dexamethasone to the corresponding 17β-carboxylic acid analogue (compound I) decreases the binding affinity by more than three orders of magnitude. The affinity, however, can be partially restored by amidation (Rousseau et al., 1979). The sequence of binding potency was shown to be the same in all spleen cytosols and was found to be in the following order: dexamethasone > BioDex 1 > BioDex 2 > BioDex 2 and 5 >> BioDex 4, 6, and compound I. Neither dexamethasone nor one of its side chain modified derivatives compete with [³H]hydrocortisone for binding to corticosteroid binding globulin in diluted human pregnancy serum (data not shown).

3.2. Glucocorticoid Receptor Binding to BioDex 1 in the Presence of 2 µM Avidin

The dose-dependent displacement of specifically bound [³H]dexamethasone from spleen tumor glucocorticoid receptors by dexamethasone or BioDex 1 is shown in Fig. 1. Dexamethasone and BioDex displaced practically all the specifically bound [³H]dexamethasone when added at sufficiently high concentrations (procedure I). Pretreatment of the buffered solutions of competitors with avidin (final concentration 2 µM) for 1 h dramatically reduced the affinity of the resulting BioDex 1-avidin complexes vs the receptors. Dexamethasone itself remains unaffected (procedure II). In contrast to these findings, avidin is not able to reduce strongly the affinity of BioDex 1 already bound to the receptor (procedure III). Dexamethasone again remains unaffected.

Table 1
Effect of Substitutions in the Sidechain of Dexamethasone
on Receptor Binding and Glucocorticoid Activity

Derivative	Relative binding affinities, %	Inhibition of [³H]thy-midine incorporation at 5 μM steroid, %
Dexamethasone	100	100
Compound I	0.1	i.s.
BioDex 1	12	i.s.
BioDex 2	5.5	i.s.
BioDex 3	2	i.s.
BioDex 4	0.5	i.s.
BioDex 5	2	i.s.
BioDex 6	0.1	i.s.
Amide A	27	85
Amide B	3.2	50

Human spleen tumors (Hodgkins disease I) were homogenized in Tris/HCl buffer and centrifuged for 45 min at $105,000 \times g$. The supernatant was taken as cytosol. The cytosols were incubated in triplicate with 32 nM [³H]dexamethasone either alone or in the presence of various concentrations of competitors for 18 h at 4°C. The unbound steroid was removed with dextran-coated charcoal. The relative binding affinities listed are means of three determinations and calculated from ratios of steroid concentrations giving a 50% inhibition of [³H]dexamethasone binding. Value for dexamethasone was taken as 100%. No differences in relative binding affinities of BioDex 1, [³H]BioDex 1, or [¹⁴C]BioDex 1 could be detected (*see* Section 2.). To determine inhibition of [³H]thymidine incorporation, lymphocytes were isolated from peripheral blood, as described. Cell count was adjusted to 2×10^5 cells/culture vol (200 μL). Ten μL of phytohemagglutinin (5 μg/mL) and 10 μL of increasing amounts of steroids dissolved in 0.9% NaCl were added to each culture followed by 72-h incubation. Lymphocyte stimulation was assessed by the incorporation of [³H]thymidine into the DNA during the last 4 h of culture. i.s. = Insignificant. Amide A = *N*-benzyl 9-fluoro-16α-methyl-11β,17-dihydroxy-3-oxo-1,4-androstadiene-17β-carboxamide. Amide B = 2'-(9-fluoro-16-α-methy-11β,17-dihydroxy-3-oxo-1,4-androstadiene-17β-carboxamido) glutaric acid *bis*(*tert*-butyl)ester (data taken from literature).

3.3. Binding of [³H]BioDex 1 to Avidin

The bifunctional nature of BioDex 1 was further established by competitive binding experiments to soluble avidin (Fig. 2). Gel filtration of equimolar [³H]BioDex 1/avidin solutions (on the assumption that 1 molecule avidin binds 4 molecules of biotin) on Sephadex G-25 shows a sharp peak of protein-bound radioactivity whereas in the presence of a 100-fold excess of biotin, very little radioactivity was detectable in the protein fraction.

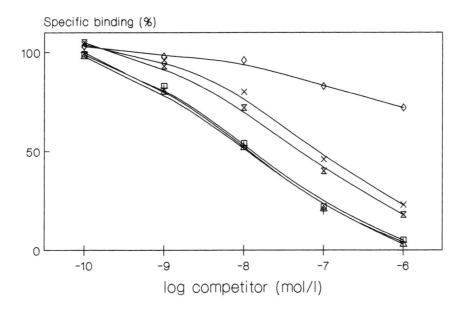

Fig. 1. Competition for dexamethasone binding to human spleen tumor cytosol. (I) The cytosol was incubated with 32 nM [^{3}H]dexamethasone in the absence (control) or presence of the indicated concentration (3.2–3200 nM) of nonradioactive steroid as described in Section 2. After incubating for 5 h at 4°C, the unbound steroids were removed with dextran-coated charcoal. The specific binding is expressed as a percentage of the uncompeted control. (☐) Dexamethasone; (×) BioDex 1. (II) To 25 µL of buffered solutions of competitors (6.4–6400 nM) 25 µL of a buffered avidin solution (16 µM) was added and the mixtures kept for 1 h at 4°C. Thereafter, 50 µL of 128 nM of [^{3}H]dexamethasone (final concentration 32 nM) and 100 µL of cytosol were added. After incubating for 5 h at 4°C the samples were processed as described above. (+) Dexamethasone; (×) BioDex 1–avidin complex. (III) Twenty-five µL of increasing concentrations of competitors (6.4–6400 nM), 50 µL of 128 nM [^{3}H]dexamethasone and 100 µL of cytosol were incubated for 4 h at 4°C. Twenty-five µL of the avidin solution were added and, following an incubation of 1 h, the samples were processed as described above. (Δ) Dexamethasone; (◊) BioDex 1.

3.4. Inhibition of Phytohemagglutinin-Mediated Blastogenesis of Normal Human Peripheral Blood Lymphocytes

The inhibition of [^{3}H]thymidine incorporation in phytohemagglutinin-stimulated peripheral blood lymphocytes by dexamethasone and its derivatives is shown in Table 1. In contrast to benzyl amides A and B,

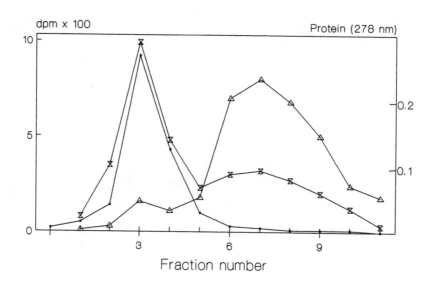

Fig. 2. Sephadex G-25 chromatography of the avidin–BioDex I complex. A 5 μM solution of avidin in buffer was incubated with either 5 μM [³H]BioDex 1 alone or in the presence of a 100-fold excess of biotin. One mL of the incubation mixtures was submitted to gel filtration on a Sephadex G-25 column to remove unbound BioDex 1. [³H] Bio-Dex 1 bound/100 μL, alone (X—X) or in the presence of biotin (Δ—Δ). Protein was determined as A_{278} (●—●).

that effectively inhibit the [³H]thymidine uptake and whose order of binding affinity toward the human glucocorticoid receptor closely resembles that of inhibition potency (Manz et al., 1984), the biotin conjugates are inactive. The original cell number did not change significantly during the first 3 d of culture and there was no hint that the survival of peripheral blood lymphocytes was negatively affected by any of the steroids used in this study.

3.5. Purification of Dexamethasone Binding Proteins

Based on Scatchard analyses of the glucocorticoid receptor content of the cytosol, prior to avidin-Sepharose treatment, and after removal of the affinity gel by filtration, more than 80% of the glucocorticoid receptors were bound to the gel. The receptors were eluted by incubation of the resin with 3.2 μM dexamethasone (containing 32 nM of [³H]dexamethasone), *see* Section 2. Final yield of dexamethasone receptors, determined as protein-bound [³H]dexamethasone, assuming a steroid-receptor ratio of 1:1 and an M_r of 90,000, is usually more than 60% with an approx 2000-fold increase in bound radioactivity over the starting material (Table 2). When submit-

Table 2
Affinity Chromatography Purification of Dexamethasone-Binding Proteins

Fraction	Volume, mL	Total protein, mg	Receptor, μg	Total bound hormone, nM	Recovery, %	Purification, (–fold)
Cytosol	200	10,031	45	0.6	100	1
Affinity eluate	5	0.31	27	0.36	60	>2000

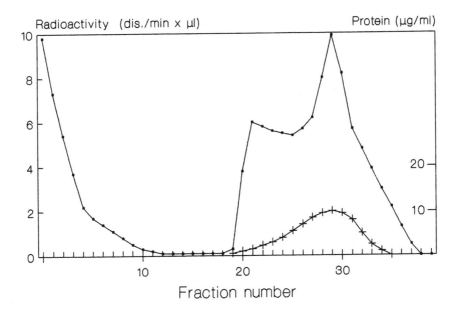

Fig. 3. DEAE-cellulose (DE-52) chromatography of the affinity eluate. Two mL of the affinity eluate, containing approximately 100 μg of protein, were submitted to gel filtration on a Sephadex G-25 column to remove unbound dexamethasone and then submitted to DEAE-cellulose chromatography as described in Section 2. (●—●) [³H]Dexamethasone bound/100 μL; (+ — +) protein.

ted to DEAE-cellulose and developed by a linear NaCl gradient, the affinity purified proteins resolve into two [³H]dexamethasone binding fractions that elute with 0.1 and 0.2*M* salt, respectively (Fig. 3).

4. Discussion

Because of the difficulties anticipated in purification of the glucocorticoid receptor (for review *see* Baxter and Rousseau, 1979), a rapid and effective initial purification step is mandatory. With this objection in mind,

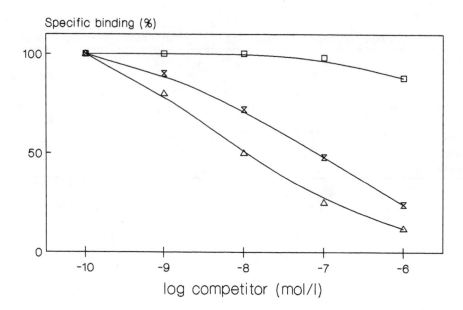

Fig. 4. Competition of BioDex-avidin complexes for [³H] dexamethasone binding to anti-dexamethasone antibodies. (1) The diluted antiserum was incubated with 32 n*M* [³H]dexamethasone in the absence (control) or presence of the indicated concentration (3.2–3200 n*M*) of nonradioactive steroid as described in Section 2. BioDex-avidin complexes. 25 µL of buffered solutions of competitors (6.4–6400 n*M*) 25 µL of a buffered avidin solution (16 µ*M*) was added and the mixtures kept for 1 h at 4°C. Therefore, 50 µL of 128 n*M* of [³H]dexamethasone (final concentration 32 n*M*) and 100 µL of antiserum were added. After incubating for 1 h t at 4°C, the samples were processed as described above. (Δ—Δ) dexamethasone; (X—X) BioDex 1–avidin complex; BioDex 2–avidin complex (□—□).

we have synthesized six dexamethasone-biotin conjugates differing in length and polarity of the respective linkage. Only one of the derivatives, BioDex 1, fullfilled the criteria of binding to both proteins at the same time. In case of BioDex 2, the second derivative with sufficiently high affinity to the receptor, the spacer was obviously too short to bind both proteins. All other derivatives were of no use, most likely because of their low affinities to the receptor. These findings were confirmed by experiments using anti-dexamethasone antibodies instead of glucocorticoid receptor. Only the BioDex 1-biotin complex was able to compete with [³H] dexamethasone for binding to the antibody (Fig. 4).

Interestingly, none of the BioDex derivatives is able to inhibit the phytohemagglutinin-stimulated blastogenesis of normal human peripheral blood lymphocytes. This is surprising as previous studies clearly showed that other steroid 17β-carboxamides behave like glucocorticoid agonists depending on their binding affinity to the receptor. Using avidin-labeled with fluorescein isothiocyanate, we could clearly demonstrate that Bio-Dex 1 is obviously unable to enter the target cell (unpublished results). To demonstrate that biotin-steroid conjugates are indeed excellent candidates for affinity purification of their respective receptors, a purification scheme was established that allows a high enrichment of protein-bound [^{3}H]dexamethasone in one step.

Because of the very low glucocorticoid-receptor content of human spleen tumor cytosol and the high contamination with blood proteins, the affinity purified receptor is far from being pure. Using an improved protein detection procedure (silver stain) we must observe that the staining technique (Coomassie Brilliant Blue) used in our previously published paper (Laemmli, 1970) was unable to detect all copurified or contaminating proteins. Thus the published purification factor has to be corrected for nearly one order of magnitude (Table 2). Nevertheless, we think our results give encouraging indications that steroid-biotin conjugates, together with immobilized avidin, are a promising approach in the field of steroid-hormone receptor purification.

References

Abraham, G. E. (1974) *Acta Endocrinol. Suppl.* **183**, 1.

Agnado, M. T., Pujol, N., Rubiol, E., Tura, M., and Celado, A. (1980) *J. Immunol. Meth.* **32**, 41.

Baxter, J. D. and Rousseau, G. G. (1979) *Glucocorticoid Hormone Action*, Springer-Verlag, Heidelberg, New York, pp. 1–24.

Berger, N. A., Berger, S. J., Sikorski, G. W., and Catino, D. M. (1982) *Exp. Cell Res.* **137**, 71.

Geffard, M. R., Puizillout, J. J., and Delaage, M. A. (1982) *J. Neurochem.* **39**, 1271.

Govindan, M. V. and Manz, B. (1980) *Eur. J. Biochem.* **108**, 47.

Hubert, P., Mester, J., Dellacherie, E., Neel, J., and Baulieu, E.-E. (1978) *Proc. Natl. Acad. Sci. USA*, **75**, 3143.

Laemmli, V. K. (1970) *Nature (Lond.)* **227**, 680.

Ling, N. R. (1968) in *Lymphocyte Stimulation*, North-Holland, Amsterdam, pp. 72–94.

Manz, B., Grill, H.-J., Kreienberg, R., and Pollow, K. (1982a) *Eur. J. Cell Biol.* **27**, 21.

Manz, B., Grill, H.-J., and Pollow, K. (1982b) *J. Steroid Biochem.* **17**, 335.

Manz, B., Grill, H.-J., Kreienberg, R., Rehder, M., and Pollow, K. (1984) *J. Clin. Chem. Clin. Biochem.* **22**, 209.

Rousseau, G. G., Kirchhoff, I., Formstecher, P., and Lustenberger, P. (1979) *Nature (Lond.)* **279**, 158.

Scatchard, G. (1949) *Ann. N.Y. Acad. Sci.* **51**, 660.

Stewart, C. C., Cramer, S. F., and Stewart, P. G. (1975) *Cell Immunol.* **16**, 237.

Purification of the Unactivated/ Nontransformed Glucocorticoid Receptor Complex from Rat Tissues

Gerald Litwack

1. Introduction

For a number of years, the difficulties in purifying a protein complex occurring in nanomolar amounts in liver or other tissues by conventional fractionation methods were painfully obvious. The step required to make such a purification possible proved to be the development of affinity resins that could be used to concentrate the receptor complex at an early stage in purification on the order of 100–200-fold. Only through this strategy was it possible to produce reasonable yields of near homogeneous material.

There were clear advantages to be gained by the purification of the unactivated/nontransformed glucocorticoid receptor complex compared tô the activated/transformed complex. In particular, the availability of the unactivated complex could facilitate experiments on the mechanism of the conversion to the activated form as well as a means to identify various regulators involved in controlling the activation process. The process of activation/transformation in vitro can be written simply by the following expression:

Receptor Purification, vol. 2   ©1990 The Humana Press

Unactivated receptor complex (GR_U)————————→ activated receptor
                                                  complex (GR_A)
              (cannot bind to DNA)   (can bind to DNA)

Later, it became clear that the activation process in vitro was more complicated and could be subdivided into two discrete steps (Schmidt et al., 1985).

$$\text{S-GR}_U \xrightarrow{\text{modulator}} \text{S.GR}^1 \underset{\longleftarrow}{\xrightarrow{\hspace{2cm}}} \text{S-GR}_A$$
                                       intermediate

In this expression, S refers to the steroidal glucocorticoid ligand; GR_U, the unactivated glucocorticoid receptor complex; GR', an intermediate, relatively nonDNA binding complex; and GR_A, the activated glucocorticoid–receptor complex.

The properties of the GR_U are now generally known. The complex has a mol wt between 300,000–320,000kDa (Grandics et al., 1984). The Stokes radius is in the range of 7.1 nm and sedimentation coefficient 8–10 S (Grandics et al., 1984). The physical properties of the GR_U in crude cytosol are equivalent to those of the purified GR_U, although there is a lack of clear understanding as to the composition of this complex. Originally, many believed the GR_U to be a tetramer of the GR_A (Sherman et al, 1983).

A minority opinion held that the GR_U was probably a trimer based on measurements of hydrodynamic parameters (Webb and Litwack, 1986). Subsequently, data were introduced that the 90kDa heat shock protein (hsp 90) was a component of the GR_U (Joab et al., 1984), probably in dimer form (Denis et al., 1987). Since the GR_U and the GR_A showed no difference in the extent to which a monoclonal antibody interacted, it was concluded that the GR_A contained only one molecule of the 94kDa glucocorticoid receptor (Okret et al., 1985). Currently, there is wide support for the content of the hsp 90 in the unactivated complex to serve the role of maintaining this form (Baulieu, 1987). Activation by this postulation would then involve the dissociation of the hsp 90 allowing the DNA-binding form of the receptor to function (Mendel et al., 1986; Sanchez et al., 1987). Modulator is a low mol wt inhibitor that stabilizes the unactivated steroid receptor complex and prevents activation (Bodine and Litwack, 1988a). Its dissociation from the unactivated receptor complex is linked to the activation process (Bodine and Litwack, 1988a). Modulator has now been purified and characterized as a water-soluble ether glycerophosphoserine derivative (Bodine and Litwack, 1988a) that appears to be the endogenous factor whose action is mimicked by molybdate (Bodine and Litwack, 1988b). A speculative mechanism is pictured in Fig. 1.

We have advanced the theory that modulator acts as a crosslinking agent to hold the subunits of the GR_U together by electrostatic forces. Presumably, modulator could make contacts between molecules of the receptor and the other proteins, possibly hsp 90, in the complex. So far, hsp 90 has not been established definitely to be a component of the purified, unactivated receptor complex, leaving open the possibility that the purified complex is a trimer of receptor subunits. Whatever the case, the purification of the unactivated complex is a requirement for the further analysis of its structure.

2. Purification from Rat Liver

2.1. Background to Procedure

The critical step of the purification of the unactivated receptor complex involves affinity chromatography of the unoccupied (absence of ligand),unactivated receptor complex using deoxycorticosterone derivatized agarose and subsequent elution with a ligand of higher affinity [³H]triamcinolone acetonide (TA). This form of affinity resin was first used successfully by Failla et al. (1975) and the method was later modified by our laboratory (Grandics et al., 1984) based on the development of a suitable extender (Grandics et al., 1982). In brief, the procedure requires the preparation of cytosols in the cold (usually in the presence of molybdate to stabilize the unactivated complex) from rats adrenalectomized for about three days in order to minimize endogenous corticosterone. The cytosols are incubated in the cold in the presence of the affinity resin in a slurry. The resin is eluted with [³H]TA, which competes the unactivated receptor complex off from the lower affinity deoxycorticosterone derivative and the [³H]TA-receptor complex is obtained in the unactivated, nonDNA-binding form in about 150-fold purification from the cytosol. A purification step of this magnitude is required early in the purification procedure in order to yield an appreciable amount of the final product. Two subsequent steps will yield the unactivated receptor complex in approximately 4000–5000-fold purification over the cytosol. Subsequent activation of the receptor to the DNA-binding form and repetition of the final purification step yields the activated form at about 10,000-fold or higher purification over the cytosol. This is possible because of the large shift in pI value between the unactivated and activated complexes.

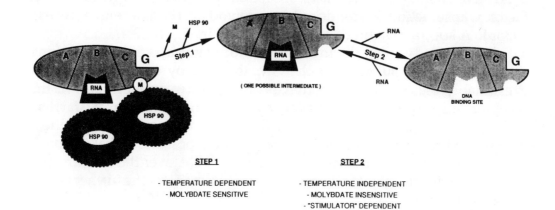

	UNACTIVATED	INTERMEDIATE(S)	ACTIVATED
SEDIMENTATION COEFFICIENT:	8 - 10 S	5 - 7 S	4 - 5 S
STOKES RADIUS:	70 - 80 Å	60 - 80 Å	50 - 60 Å
MOLECULAR WEIGHT:	300,000 - 310,000	130,000 - 180,000	94,000
DNA BINDING:	5 %	10 - 20 %	30 - 40 %
DEAE ELUTION POSITION:	250 mM KP	50 mM KP	50 mM KP

Fig. 1. Speculative in vitro activation mechanism of the glucocorticoid receptor complex. M = modulator; HSP90 = 90kDa heat shock protein; A = N terminal domain of receptor; B = DNA binding domain of receptor; C = steroid binding domain of receptor; G = glucocorticoid agonist. The proposed unactivated receptor complex is shown on the left consisting of one steroid receptor subunit and a dimer of the HSP90 protein crossbinded noncovalently to the receptor by the modulator, (M). An RNA, approx 120 nucleotides, is thought to be complexed with the DNA-binding domain of the receptor through noncovalent interactions. In step 1, the rate-limiting process may be the dissociation of modulator, which would cause the release of the nonhomologous proteins of the complex giving rise to the RNA-receptor that would have little DNA-binding ability. The second step, independent of temperature and molybdate/modulator proceeds by the removal of the RNA to release the fully active receptor able to translocate to the nucleus and bind to DNA. Physical properties of the three forms of the receptor during activation are shown at the bottom and are based on experimental observations. This figure was prepared by Andrew Maksymowych of this laboratory.

After the affinity gel step, gel filtration is carried out on a column of Bio-Gel A1.5 m. This reduces the number of sodium dodecyl sulfate-polyacrylamide gel electrophoresis (SDS-PAGE) stainable bands from about 22, visualized after the affinity step, to about 6 following gel filtration. By this second step, most of the lower mol wt proteins are removed. In the final step, the unactivated complex is chromatographed on DEAE-cellulose where the highly acidic unactivated complex binds tightly to the resin and is separated from most of the residual proteins in the preparation. This fraction shows a major band at 94kDa on SDS-PAGE and a few minor bands, which may be degradation products or other constituents of the complex, or both. This complex can be heat-activated after removal of molybdate to an activated form of lesser negative charge, which, when rechromatographed on DEAE-cellulose, appears as a single band on SDS-PAGE at 94kDa. Conversion of the unactivated form to the DNA binding form proceeds to a lesser extent than visualized in cytosol leading to experiments suggesting that other cytosolic factors are required for full activation. The purified unactivated, nonDNA-binding form may be heat-activated and the resulting activated form subsequently isolated by DEAE-cellulose chromatography to yield a virtually homogeneous receptor.

2.2. Specific Purification Procedure

2.2.1. Preparation of Affinity Resin

The deoxycorticosterone-derivatized agarose affinity resin was synthesized by Grandics et al. (1982) and varies somewhat from the similar affinity resin synthesized earlier by Failla and coworkers (1978). The structure of the affinity resin is shown in Fig. 2. For purification of the spacer arm, 130 mL of $0.5M$ NaOH and 25 mL of 1,4-*bis*(epoxypropoxy)butane are added to 100 mL Sepharose 2B or 4B. This is stirred gently at room temperature for 2 h. The mixture is washed to neutrality with distilled water and then washed with 300 mL acetone to remove traces of epoxide compounds. Washing with distilled water follows and 100 mL $12M$ ammonia solution are added and allowed to stand at room temperature for 5 h or overnight. The material is washed to neutrality with distilled water. These reactions are recapitulated in Fig. 3. The amino group content of the amino gel is determined as outlined below and the gel is stored at 4°C.

2.2.2. Analysis of Amino Derivatized Agarose

To 0.3–0.6 mL of a threefold diluted suspension (0.1 g resin + 0.2 mL water) of the packed amino derivatized agarose in a 15 mL centrifuge tube, add 15 mg picryl sulfonic acid and 10 mL $0.2M$ $Na_2B_4O_7$ at pH 9.0. Let stand at room temperature for 2 h and centrifuge. Wash with distilled water until the wash shows no absorbance at 340 nm. Dilute the packed gel to 5 mL with 50% acetic acid and dissolve by heating at 75°C for 2 h. The amine concentration is calculated using $\varepsilon = 1.4 \times 10^4$ when measured at A_{340nm}.

2.2.3. Preparation of the Steroidal Ligand

To 20 g deoxycorticosterone are added 7 mL methane sulfonyl chlorine in 80 mL pyridine. After incubation for 2 h at 0°C, the reaction mixture is poured slowly onto a tenfold volume excess of ice cold water with con-

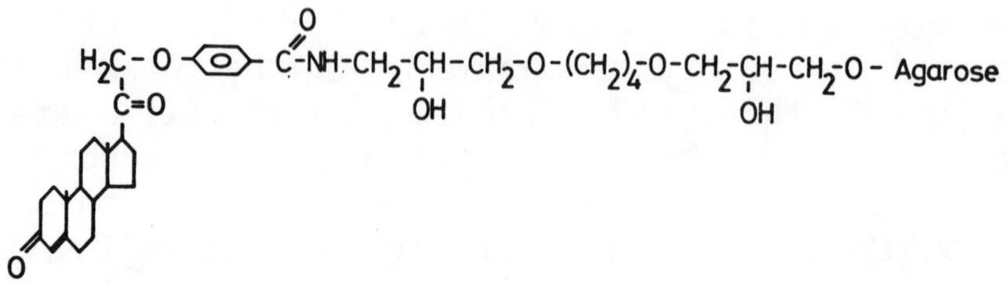

Fig. 2. Structure of the affinity resin used to purify the glucocorticoid receptor.

tinuous stirring. The precipitate is collected by filtration, washed with ice cold water to remove traces of pyridine, and dried under an infrared lamp.

2.2.4. Preparation of p-Hydroxymethyl Benzoate Sodium Salt (HMB/Na)

To 50 mL absolute methanol is added 1.448 g sodium metal to form sodium methylate. 9.416 g *p*-Hydroxymethylbenzoate is added to the solution and methanol is evaporated under a stream of nitrogen gas. The white solid is dissolved in 100 mL *N,N'*-dimethylformamide.

2.2.5. Reaction of Deoxycorticosterone-21-Mesylate with HMB/Na

Deoxycorticosterone-21-mesylate (22.9 g) are added in small portions to *N,N'*-dimethylformamide solution containing HMB/Na. The mixture is stored for 3 h and the reaction may be monitored by thin layer chromatography using benzene:acetone (5:1) as developer. The reaction mixture is poured onto a tenfold volume excess of ice cold water with continuous stirring. The precipitate is filtered, washed with ice cold water to remove traces of *N,N'*-dimethylformamide, and dried under an infrared lamp. The product is recrystallized in 50 mL methanol under reflux conditions and cooled to 0°C; the precipitate is collected. The filtrate is cooled to –20°C to collect additional product.

2.2.6. Hydrolysis with K_2CO_3

K_2CO_3 (5.13 g) is dissolved in 130 mL methanol:water (4:1) by gently heating the solution. This is poured onto 14.4 g 21-(4-carbomethoxy-phenoxy)-progesterone in a round bottom flask and refluxed for 24 h. This is poured onto a tenfold volume excess of ice cold water with continuous stirring and the pH is adjusted to 2.0 by dropwise addition of HCl. The precipitate is collected by filtration, washed to neutrality with cold distilled water, and dried under an infrared lamp.

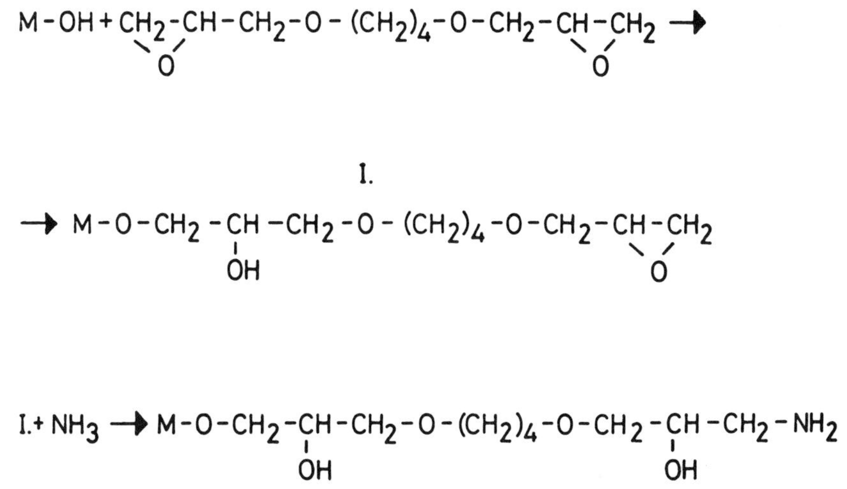

$$M-OH + CH_2-CH-CH_2-O-(CH_2)_4-O-CH_2-CH-CH_2 \longrightarrow$$
$$\backslash O / \qquad\qquad\qquad\qquad \backslash O /$$

I.

$$\longrightarrow M-O-CH_2-CH-CH_2-O-(CH_2)_4-O-CH_2-CH-CH_2$$
$$| \qquad\qquad\qquad\qquad\qquad\qquad \backslash O /$$
$$OH$$

$$I. + NH_3 \longrightarrow M-O-CH_2-CH-CH_2-O-(CH_2)_4-O-CH_2-CH-CH_2-NH_2$$
$$| \qquad\qquad\qquad\qquad\qquad | $$
$$OH \qquad\qquad\qquad\qquad\qquad OH$$

Fig. 3. Chemical reactions involved in the synthesis of the extender of the affinity resin (Grandics, 1981).

2.2.7. Recrystallization

Ten g of the product (above) is dissolved in 120 mL $CHCl_3$; 300 mL *n*-hexane is poured into the solution and cooled to 0°C. With continuous scraping yellow crystals are collected by filtration, washed with 10 mL *n*-hexane, and dried under an infrared lamp.

2.2.8. Preparation of "Active Ester"

Progesterone derivative (1.356 g) is dissolved in 62 mL dioxane. 0.385 g *N*-hydroxysuccinimide and 0.619 g *N,N'*-dicyclohexylcarbodiimide are added and the solution stands at 0–4°C overnight.

2.2.9. Coupling of "Active Ester" and Amino Gel

Sixty-two g wet amino gel are transferred into dioxane:water (1:1) and the solution is adjusted to pH 8.0 with $1M$ $NaHCO_3$. The "active ester" solution is filtered to remove precipitated dicyclohexylurea and the filtrate con-taining "active ester" is added to the amino gel solution with continu-ous stirring. The mixture is stirred at room temperature for 4 h maintaining the pH at 8.0 and filtered. The filtrate is washed with 1 L dioxane:water (1:1), then 500 mL dioxane, then distilled water, and finally with $1M$ NaCl. Allow the gel to stand in $1M$ NaCl overnight.

2.2.10. Acetylation of Resin

The gel is washed with distilled water and transferred into $0.1M$ potassium phosphate buffer, pH 7.5. Acetic anhydride (1.25 mL) is added and the suspension is stirred for 30 min. The resin is washed extensively with water and stored in 50 mM monobasic potassium phosphate, 10 mM Na$_2$MoO$_4$ and 0.01% sodium azide at 4°C.

2.2.11. Regeneration of Affinity Resin After Use

Used resin is washed with 2–3M NaCl, distilled water, CHCl$_3$: CH$_3$OH:Triton X-100:H$_2$O (30:45:10:15), 90% methanol, distilled water, and 50 mM potassium phosphate-10 mM Na$_2$MoO$_4$ at pH 7.0 with 0.02% azide in sequence, and stored in the last buffer.

2.2.12. Sources of Reagents

Deoxycorticosterone, N-hydroxysuccinimide and picrylsulfonic acid are obtained from Sigma. Methane sulfonyl chloride and 1,1,1,3,3,3-hexamethyl-disilaxane are obtained from Kodak. Pyridine, p-hydroxymethyl-benzoate and N,N'-dicyclohexylcarbodiimide are obtained from Fisher. One, four-Butanediol diglycidyl ether is obtained from Aldrich. Sepharose 4-B or 2-B is obtained from Pharmacia.

2.2.13. Adsorption of Unactivated Unoccupied Glucocorticoid Receptor from Adrenalectomized Rat Liver Cytosol to Affinity Resin

Male rats (150–200 g) are adrenalectomized for 3 d prior to sacrifice. Livers are perfused *in situ* via the portal vein with cold 0.9 NaCl and then with buffer A (50 mM potassium phosphate, 10 mM sodium molybdate, 10 mM thioglycerol, pH 7.0 at 22°C). Five mM dithiothreitol can replace 10 mM thioglycerol. Livers are excised and minced with scissors, homogenized in one vol of buffer A with a Teflon™-glass Potter-Elvehjem homogenizer and centrifuged at 12,000 × g for 30 min. The upper lipid layer is aspirated and the supernatant fraction centrifuged at 105,000 × g for 1 h to obtain cytosol. All procedures are done at 0–4°C. Cytosol can be stored in a liquid nitrogen freezer in the gas phase. Twenty mL of cytosol are mixed with 8 mL of affinity resin and carefully stirred at 0°C for 2 h. After centrifugation, the supernatant fraction is removed and assayed for specific binding (*see below*) and protein content. The gel is washed batchwise 10 times with 20 mL of buffer A over a 2-h period. Each washing entails gentle stirring for 5 min and centrifugation. Then, 8 mL of buffer B (10 mM potassium phosphate, 10 mM sodium molybdate, 10 mM thioglycerol, and 10% glycerol, pH 7.0 at 22°C) containing 100 nM to 2 µM [^{3}H]TA is added and the mixture is mildly agitated on a rotator for 16 h at 0°C. The slurry is centrifuged or filtered on a Teflon™ filter. The absorbent is washed with 4 mL of buffer B and the combined eluate is assayed for specific binding and protein content.

2.2.14. Determination of Specific Binding

Cytosol or a given fraction is incubated with 50 nM [^{3}H]TA (unless the fraction is already radioactively labeled) in the presence or absence of a 500-fold excess of nonradioactive TA for 2 h at 0–4°C. Specific binding in cytosol and in purified samples is determined using a hydroxylapatite technique (Erdos et al., 1970). Nonspecific binding to hydroxylapatite is estimated in purified samples by heat denaturation of the steroid–receptor complexes at 37°C for 3 h before testing.

2.2.15. Bio-Gel A1.5 m Chromatography

The affinity gel eluate is filtered on a Bio-Gel A1.5 m column of 220 mL bed volume with an inner diameter of 2.5 cm. The column is equilibrated with buffer A at 0–4°C. Eight mL fractions are collected and aliquots are taken for measurement of radioactivity. The receptor peak fractions are combined and used in the next purification step. A typical gel filtration pattern is shown in Fig. 4. If the affinity resin is washed extensively, this step can be omitted.

2.2.16. DEAE-Cellulose Chromatography

Whatman DE-52 DEAE-cellulose is swollen at neutral pH for 5 h at room temperature. The swollen material is washed with 0.5M HCE, with water, with 0.5M NaOH on a sintered glass Buchner filter until the filtrate is free of chlorine; then it is washed to neutrality with water and finally with buffer A to pH 7.0 of the filtrate. A 3 mL bed volume column is packed in a 5 mL disposable plastic syringe barrel and the combined receptor peak fractions from the previous purification step are loaded. The column is washed with 10 mL buffer A. Receptor is eluted with a 50–500 potassium phosphate gradient containing 10 mM sodium molybdate and 10 mM thioglycerol, pH 7.0, at a flowrate of 30 mL/h. A typical chromatogram of the purified unactivated receptor is shown in Fig. 5. A gel filtration experiment to determine the Stokes radius of the purified unactivated receptor complex is shown in Fig. 6. A one-dimensional denaturing gel, stained with silver, illustrating the maximal number of peptide components at each fractionation step is shown in Fig 7. The overall 3-step purification procedure is summarized in Table 1.

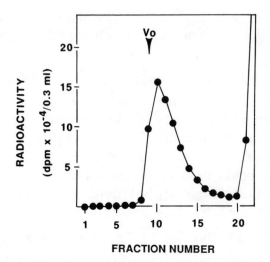

Fig. 4. Bio-Gel A-1.5 m column chromatography of affinity gel eluted [³H] triamcin-olone acetonide-glucocorticoid receptor. (Affinity eluate was applied to a Bio-Gel A-1.5 m column (220 mL bed vol) equilibrated with buffer A and chromatographed. Eight-mL fractions were collected and 0.3 mL aliquots were taken for counting radioactivity. Fractions 9–15 were subsequently combined. Reproduced from Grandics et al. (1984).

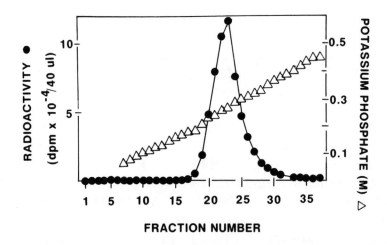

Fig. 5. DEAE-cellulose chromatography of purified [³H]triamcinolone acetonide-glucocorticoid receptor. Pooled receptor peak fractions from the Bio-Gel chromatography step were passed through a DEAE-cellulose column (3 mL bed volume) and eluted with a 50–500 mM phosphate gradient containing 10 mM molybdate and 10 mM thioglycerol pH 7.0. One-mL fractions were collected and 40 μL aliquots were taken for counting radioactivity. Conductivity in each fraction was measured using a conductivity meter. Reproduced from Grandics et al. (1984).

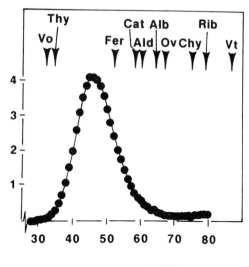

Fig. 6. Gel filtration analysis of DEAE-cellulose purified [³H]triamcinolone acetonide-glucocorticoid receptor on Bio-Gel A-1.5 m agarose. The final receptor was run on a Bio-Gel A-1.5 m agarose. The column was previously standardized with proteins of known Stokes radii. Two-mL fractions were collected and solubilized in Liquiscint scintillation fluid for counting radioactivity. Thy, thyroglobulin; Fer, ferritin, Cat, catalase; Ald, aldolase; alb, albumin; Ov, ovalbumin; Chy chymotrypsinogen A; Rib, ribonuclease. Reproduced from Grandics et al. (1984).

3. Activation of the Purified
Unactivated Receptor Complex

Confirmation that the receptor complex purified is the unactivated, nonDNA-binding form comes from various data (9.4 S; $R_s = 7.3$ nm; calculated mol wt, 303,000; inability to bind to DNA but activation by heat to a DNA-binding form) (Grandics et al., 1984).

To remove salt and retain the unactivated state of the purified receptor complex, the final receptor preparation from DEAE-chromatography is filtered through a Sephadex G-25 (PD-10) column equilibrated with buffer C (10 mM Mes, 0.5 mM EDTA, 0.5 mM dithiothreitol, and 2 mg/mL of BSA, pH 6.5 at 4°C) containing 10 mM Na_2MoO_4 at 0–4°C. There was low binding to DNA-cellulose of this preparation (Table 2). Filtration of the preparation through a Sephadex G-25 column without molybdate enhanced ability to bind to DNA- cellulose (Table 2) aligning with the known effects of gel filtration in less purified systems. The DNA-cellulose binding of the gel filtered preparation was, however, lower than that previously observed for receptors in crude homogenates (Grandics et al., 1984).

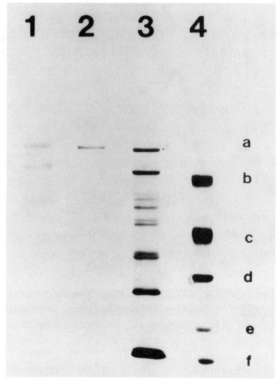

Fig. 7. Silver-stained denaturing electrophoretic gel of three purification steps. Molecular weight standards in lane 4 are indicated by positions a–f: a = phosphorylase b(97.4kDa); b = bovine serum albumin (68 kDa); c = ovalbumin (43kDa); d = chymotrypsinogen (25.7 kDa); e = soybean trypsin inhibitor (2.1 kDa) ; f = ribonuclease (14.4 kDa). Lane 3 is proteins eluted from affinity resin by [³H]triamcinolone acetonide. Lane 1 is peak from Bio-Gel A-1.5 m eluate; although not readily visible, there are 6 bands in this lane. Lane 2 is unactivated receptor complex from final DEAE-cellulose fractionation step (refer to Table 1). Unpublished Experiment of Grandics and Litwack (1983).

Table 1
Three-Step Purification of Unactivated
Rat Hepatic Glucocorticoid Receptors

Fraction	Yield, %	Fold purification[a]
Crude cytosol	100	
Affinity gel eluate	60	144
Bio-Gel filtrate	50	566
DEAE-cellulose eluate	34	4295

[a]Average data for three separate experiments from Grandics et al., 1984.

Table 2

In Vitro Activation of Purified Glucocorticoid-Receptor Complexes*

Treatment	Number of experiments	[³H] Triamcinolone acetonide-receptor[a] dpm added	DNA-cellulose bound dpm	Percent of total bound to DNA-cellulose
Gel filtration in the presence of 10 mM Na$_2$MoO$_4$	1	111,600	1,000	0.9
	2	29,700	270	0.9
Gel filtration	1	86,900	6,200	7.2
	2	28,600	1,320	4.6
Gel filtration + heat (25°C, 30 min)	1	84,000	15,200	18.1
	2	24,100	4,470	18.5
Gel filtration + 10 mM Na$_2$MoO$_4$ + heat (25°C, 30 min)	1	86,500	7,100	8.2
	2	25,000	2,300	9.2
Gel filtration + cytosol + heat (25°C, 30 min)	1	85,400	39,000	45.6
	2	24,900	10,700	42.8
Gel filtration + cytosol + 10 mM Na$_2$MoO$_4$ + heat (25°C, 30 min)	1	69,500	9,800	14.1
	2	25,500	2,960	11.6

*Receptor was from the pooled radioactivity peak fractions of DEAE-cellulose chromatograpy.
[a]After each treatment [³H]triamcinolone acetonide-receptor was assayed by the hydroxylapatite technique (Erdos et al., 1970). Reproduced from Gandics et al. (1984).

4. Purification of Unactivated Kidney
Cortex Receptor Complex

Essentially the same procedure for purification of the unactivated receptor complex from rat liver cytosol was applied to rat kidney cortex cytosol (Webb et al., 1985). After the affinity gel step a 30–40% recovery was obtained with a 100–200-fold purification as measured by specific steroid binding per mg protein. As before, with the liver preparation, the affinity resin eluate was filtered on a preparative Bio-Gel A1.5 m column (Fig. 8). Receptor-containing fractions were recovered in the first peak of radioactivity; the second peak contained free [^{3}H]TA. Recovery of receptor complex at this point was 20% and the purification increased another three– to fourfold. Receptor-containing fractions from the Bio-Gel A1.5 m column were combined and chromatographed on a DEAE-cellulose ion exchange column (Fig. 9). Elution of bound glucocorticoid-receptor complexes was performed with a 50–500 mM potassium phosphate gradient. A single peak of radioactivity was observed at a salt concentration of 200–300 mM potassium phosphate. Final recovery of receptor was 5–10% of cytosolic receptor. Purity increased another twofold to a final fold purification of approx 1000. Samples from this step were analyzed further by physicochemical techniques and for functional properties. The difference in final fold purification achieved in the preparations from liver and kidney may be owing, in part, to differences in cellular protein content. Denaturing gel electrophoresis of the kidney receptor preparation is visualized in Fig. 10. For these gels, samples were precipitated at various stages of the fractionation process and after running the SDS-PAGE gels, they were stained with Coomassie blue or double stained with Coomassie Blue and silver (Irie et al., 1982). In our hands, the double-staining gel heightens the sensitivity of detection of protein bands. A heavily stained band at 90–94kDa is apparent in the Bio-Gel A1.5 m and DEAE-cellulose steps of the purification procedure and this band becomes more predominant as the purification progresses. Although other bands can be visualized in the DEAE-cellulose eluate, they do not correspond to the pattern of bound radioactive ligand. Only the 90–94kDa band is clearly present in the [^{3}H]TA-containing fractions from the DEAE-cellulose column and absent in the non-[^{3}H]TA-containing fractions. That the 90–94kDa band was related to the receptor was confirmed by presaturating the cytosol with radioinert TA prior to incubation with the affinity resin. Simultaneous electrophoretic analysis of cytosols presaturated in the presence or absence of $10^{-7}M$ radioinert TA demonstrates the absence of the M_r 90–94kDa band in the presaturated cytosol (Fig. 11).

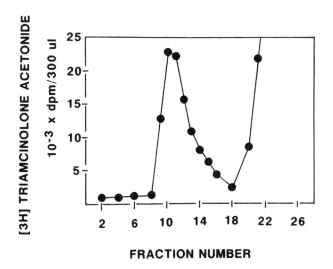

Fig. 8. Gel filtration of affinity resin eluate on a preparative Bio-Gel A.1.5 m column. The column was preequilibrated with 50 mM potassium phosphate, 10 mM sodium molybdate, and 10 mM thioglycerol. Peak glucocorticoid-receptor fractions (9–13) were combined and loaded on the DEAE-cellulose column. Reproduced from Webb et al. (1985).

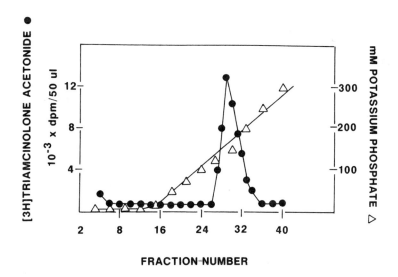

Fig. 9. Ion-exchange chromatography on DEAE-cellulose of the Bio-Gel-eluted glucocorticoid-receptor fractions. Elution was performed with a linear 50–500 mM potassium phosphate gradient containing 10 mM sodium molybdate and 10 mM thioglycerol. Reproduced from Webb et al. (1985).

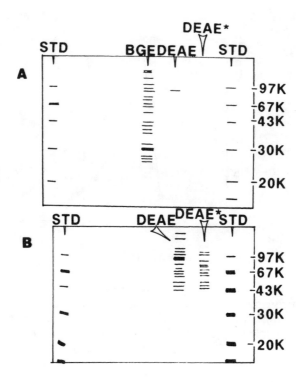

Fig. 10. Electrophoretogram of samples during sequential purification steps and stained with Coomassie Blue (A) or Coomassie Blue and silver (B). Abbreviations: std, standard (97 K, phosphorylase b, M_r 97000; 67 K, albumin, M_r 67000; 43 K, ovalbumin, M_r 43000; 30 K carbonic anhydrase, M_r 30000; 20 K, trypsin inhibitor, M_r 20100; BGE, BioGel eluate; DEAE, DEAE-cellulose eluate from a peak radioactivity region of the column; DEAE*, DEAE-cellulose eluate from a nonpeak radioactivity region of the column. This figure is redrawn from photographs in Webb et al. (1985). The single band in DEAE lane in A represents the glucocorticoid receptor as does the heavy band in the DEAE lane in B.

4.1. Physicochemical Characterization

When the purified receptor was rechromatographed on an analytical Bio-Gel A1.5 m column, a single peak of radioactivity eluted with an R_s of 6.4 ± 0.07 nm ($n = 5$), as shown in Fig. 12. Analysis under hypertonic conditions with 400 mM KCl added to the buffer did not alter the R_s value, and this value was in agreement with the Stokes radius obtained for the unactivated, unfractionated cytosolic receptor complex. Purified kidney receptor from DEAE-cellulose chromatography sedimented at 10.5 S under low salt conditions, as shown in Fig. 13, which was similar to the sedimentation value for the purified liver receptor complex. The apparent mol wt calculated from these data was about 295,000.

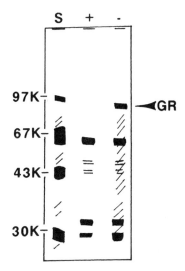

Fig. 11. Electrophoretogram of DEAE-cellulose eluates from purified cytosols pre-saturated in the presence (+) or absence (-) of $10^{-7}M$ radioinert triamcinolone acetonide prior to incubation with affinity resin. S is the standards lane. GR = glucocorticoid receptor. This figure was redrawn from a photograph in Webb et al. (1985). Diagonal lines represent heavy background in the photograph.

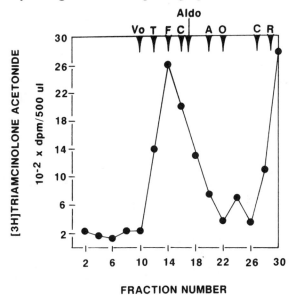

Fig 12. Gel filtration analysis of glucocorticoid-receptor complex. DEAE-cellulose eluted complexes were rechromatographed on an analytical Bio-Gel A-1.5 m column previously calibrated with the following protein standards: T, thyroglobulin (R_s = 8.50 nm); F, ferritin (R_s = 6.10 nm); C, catalase (R_s = 5.22 nm); Aldo, aldolase (R_s = 4.81 nm); A, albumin (R_s = 3.55 nm); 0, ovalbumin (R_s = 3.05 nm); Ch, chymotrypsin (R_s = 2.09 nm); R, ribonuclease (R_s = 1.64 nm). Elution was performed with 50 mM potassium phosphate, 10 mM sodium molybdate, and 10 mM thioglycerol, pH 7.0 at 4°C, buffer. Reproduced from Webb et al. (1985).

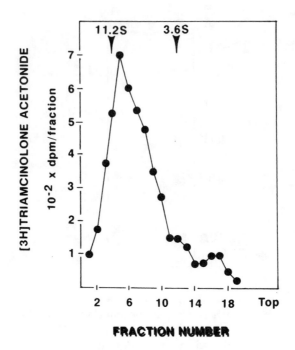

Fig. 13. Density gradient ultracentrifugation of purified glucocorticoid-receptor complex. DEAE-cellulose-eluted complexes were layered onto linear glycerol gradients made in 10 m*M* potassium phosphate, 10 m*M* sodium molybdate, 10 m*M* thioglycerol, and 40 m*M* potassium chloride. External standards were catalase (11.2 S) and ovalbumin (3.6 S). Reproduced from Webb et al. (1985).

4.2. In Vitro Activation of Purified Rat Kidney Cortex Receptor Complex

For studies on activation, molybdate is first removed by filtering the purified receptor preparation through a Sephadex G-25 or G-75 column preequilibrated in 10 mM Mes, 50 mM KF, 0.5 mM EDTA, and 0.5 mM dithiothreitol, pH 6.5 ± 2 mg bovine serum albumin (BSA)/mL. Activation was estimated by the extent to which a preparation, after treatment, bound to DNA-cellulose (Table 3). At 0°C, binding to DNA was low confirming the unactivated form of the purified receptor complex. After heating at 25°C for 30 min, binding to DNA doubled the control value and this activity was inhibited by addition of 10 mM sodium molybdate prior to heating. Chromatography on DEAE-cellulose after heating verified the presence of an activated peak. Following gel filtration to remove molybdate, receptor complexes, maintained in the cold, were confirmed as unactivated and eluted in the appropriate position of the unactivated complex (250 mM) upon DEAE-cellulose chromatography (Fig. 14, left figure). DNA binding was 5.8%. Heat treatment shifted the DEAE-cellulose elu-

Table 3
Activation of Purified Glucocorticoid-Receptor Complexes from Rat Kidney Cortex Cytosol[a]

Treatment	Expt	Hydroxyapatite bound, dpm	Cellulose bound, dpm	Percent binding to DNA-cellulose	Change, x-fold
0°C, 30 min	1	3,803	219	5.8	1.0
	2	7,141	364	5.1	1.0
	3[b]	3,494	470	13.5	1.0
25°C, 30 min	1	2,083	2,074	8.9	+1.6
	2	5,169	418	8.1	+1.5
25°C, 30 min, +10 mM sodium molybdate	3[b]	4,004	757	18.9	-1.4
25°C, 30 min, + RK_cC[c]	1	2,204	498	21.6	+2.4
	3[b]	3,857	1,295	33.6	+1.2
25°C, 30 min, + RLC[d]	1	2,923	627	21.4	+2.4
	3[b]	4,397	1,666	37.9	+1.4
25°C, 30 min, +90°C treated RK_cC	1	3,348	842	25.1	+2.8
	3[b]	3,733	2,115	56.7	+2.1
25°C, 30 min, +90°C treated RLC	1	3,508	1,051	30.0	+3.4
	2	5,203	2,836	54.5	+6.7
	3[b]	3,737	1,939	51.9	+2.0

[a] Glucocorticoid-receptor complexes were passed through a gel filtration column prior to treatment to remove sodium molybdate. Unheated cytosols were presaturated with 10^{-7} M TA for 2 h at 4°C prior to incubation with purified receptor complexes.
[b] Denotes receptor purified through Bio-Gel A-1.5 M chromatography stage.
[c] RK_cC, rat kidney cortex cytosol.
[d] RLC, rat liver cytosol. Reproduced from Webb et al. (1985).

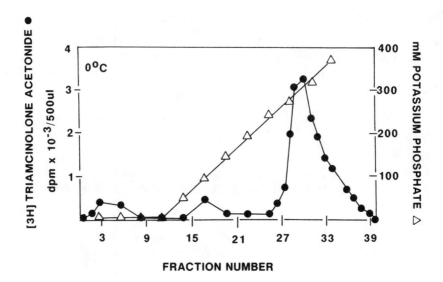

Fig 14. DEAE-cellulose chromatography of thermally activated glucocorticoid-re-ceptor complex. DEAE-cellulose-eluted complexes were filtered through a Sephadex G-25 column to remove sodium molybdate and subsequently were treated at 0°C for 30 min. Samples were then rechromatographed on DEAE-cellulose and eluted with a 5–500 mM potassium phosphate buffer containing 10 mM sodium molybdate and 10 mM thioglycerol. The Sephadex G-25 column buffer was 10 mM Mes, 0.5 mM EDTA, and 0.5 mM dithiothreitol with 5 mg of albumin/mL. Reproduced from Webb et al. (1985).

tion profile to a salt concentration of 5 and 50 mM characteristic of activated forms (Fig. 14, right figure). DNA binding was 8.9%. Although binding to DNA-cellulose nearly doubled compared to unheated controls, it was substantially less than one sees with cytosolic preparations (40–50% bind-ing to DNA after heat activation). Binding of the purified heat-treated receptor to DNA-cellulose could be increased to the level observed with cytosol by the addition of presaturated ($10^{-7}M$ TA) or heated (90°C, 30 min) cytosol from kidney or liver. This observation was similar to activation of purified receptor complexes from liver.

5. Cytosolic Factor Stimulating Activation of the Purified Glucocorticoid Receptor

We have consistently found that purified unactivated receptor com-plexes do not exhibit the capacity to bind to DNA-cellulose exhibited by unactivated cytosolic receptor complexes. Moreover, we have routinely been able to reconstitute the DNA-binding ability of the "activated" puri-fied receptor complexes by addition of receptor-neutralized or receptor-

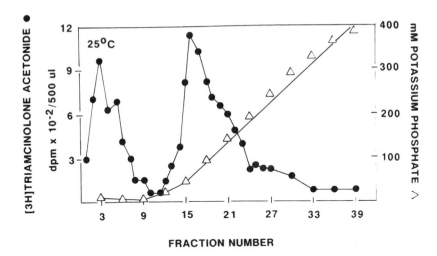

Fig. 14. DEAE-cellulose chromatography of thermally activated glucocorticoid-receptor complex. DEAE-cellulose-eluted complexes were filtered through a Sephadex G-25 column to remove sodium molybdate and subsequently were treated at 25°C for 30 min. Samples were then rechromatographed on DEAE-cellulose and eluted with a 5–500 mM potassium phosphate buffer containing 20 mM sodium molybdate and 20mM thioglycerol. The Sephadex G-25 column buffer was 10 mM Mes, 0.5 mM EDTA, and 0.5 mM dithiothreitol with 5 mg of albumin/mL. Reproduced from Webb et al. (1985).

denatured cytosol. By examination of the unactivated receptor-complex at each purification step, it became clear that the cytoplasmic "stimulator" is resolved from the receptor complex at the first affinity gel step. Indeed, the affinity gel supernatant (wash) was found to contain the stimulator activity. Subsequent purification steps through gel filtration and ion exchange produced complexes with reduced DNA-binding abilities compared to the cytosolic forms. In all cases, reconstitution was possible by addition of the stimulator fraction. The results in Table 4 demonstrate that the affinity gel supernatant contains cytoplasmic stimulator (3 vs 5). Obviously, it should be possible to isolate and characterize stimulator using this assay and starting with the affinity gel supernatant. That the stimulator activity in cytosol survived incubation at 37°C for 30 min, which inactivates unoccupied receptor, is indicative of its thermostability.

Cytoplasmic stimulator enhances activation of partially purified [³H]TA-receptor complexes in a dose-dependent manner (Schmidt et al., 1985) as shown in Table 5. Hepatic cytosol was subjected to gel filtration on Sephadex G-25 to remove the 10 mM sodium molybdate from the homogenization buffer and the resulting eluate was then diluted with buffer A devoid of molybdate and BSA. Diluted cytosol was then tested for its ability to stimulate thermal activation of the partially purified [³H]TA-

Table 4
Thermal Activation
of Purified [³H]TA-ReceptorComplexes in Reconstituted Systems*

Treatment	Total binding of [³H]TA, dpm/100 µL	DNA-cellulose binding, dpm/100 µL	DNA-cellulose binding, percent	Increase in DNA-cellulose binding, -fold
Unactivated control (0°C, 30 min)	28,700	1,630	5.7	1.0
Thermal activation (25°C, 30 min)	14,400	1,980	13.7	2.4
Thermal activation (25°C, 30 min) plus unfractionated cytosol	16,900	5,270	31.2	5.5
Thermal activation (25°C, 30 min) plus unfractionated cytosol pretreated at 37°C	20,500	5,950	29.0	5.1
Thermal activation (25°C, 30 min) plus affinity gel supernatant	18,400	5,020	27.2	4.8

*[³H] TA-receptor complexes (pooled DEAE-cellulose fractions) were purified in the absence of Na₂MoO₄, and spontaneous activation during the purification procedures was minimized by maintaining the pH at 6.5.
ᵃ(DNA-cellulose bound [³H]TA/total receptor bound [³H]TA) × 100.
ᵇFold increase in DNA-cellulose binding when compared to unactivated control. Reproduced from Schmidt et al. (1985).

receptor complexes in the eluate from Bio-Gel A1.5 m. Undiluted cytosol (Table 6) stimulated DNA-binding approx tenfold when compared to the unheated control and about fourfold when compared to the heated control. The stimulation was inhibited by molybdate added prior to heating. The filtered cytosol, diluted 1:10 (v/v) stimulated activation maximally, whereas cytosol that was subsequently diluted 1:1000 (v/v) failed to stimulate activation above that achieved by heat alone. Incubation of undiluted cytosol with purified receptors at 0°C instead of 25°C did not result in increased activation. The hydroxylapatite binding data demonstrate that, in the absence of exogenous cytosolic proteins, heat-induced dissociation of [³H]TA from partially or fully purified receptors takes place. Thus, DNA-cellulose binding must be expressed as a percentage of receptor-bound [³H]TA for that particular treatment. When buffers containing

Table 5
Dose–Response Relationship for the Activation of Partially Purified
[³H]TA-Receptor Complexes by Unfractionated Cytosol*

Treatment	Total binding of [³H]TA, dpm/100 μL	DNA-cellulose binding, dpm/100μL	DNA-cellulose binding, percent	Increase in DNA-cellulose binding, -fold
Unactivated control (0°C, 30 min)	28,600	1,020	3.6	1.0
Thermal activation (25°C, 30 min)	11,800	900	7.6	2.1
Cytosol (undiluted)ᵃ	48,900	16,100	33.0	9.2
Cytosol (undiluted) plus 10 mM Na₂MoO₄	51,200	3,180	6.2	1.7
Diluted cytosol (1:2)	43,700	14,200	32.5	9.0
Diluted cytosol (1:5)	41,100	12,600	30.7	8.5
Diluted cytosol (1:10)	40,900	12,700	31.0	8.6
Diluted cytosol (1:20)	36,300	10,000	27.5	7.6
Diluted cytosol (1:50)	32,000	7,260	22.7	6.3
Diluted cytosol (1:100)	21,800	4,310	19.8	5.5
Diluted cytosol (1:500)	16,200	1,860	11.5	3.2
Diluted cytosol (1:1000)	13,100	1,240	9.4	2.6

*Partially purified [³H] TA-receptor complexes were eluted from affinity resin and subsequently filtered on Bio-Gel A-1.5m.

ᵃUndiluted or diluted cytosol presaturated with nonradioactive TA was filtered on a column of Sephadex G-25 and then was mixed 1:1 (v/v) with partially purified [³H]TA-receptor complexes, and the reconstituted mixtures were then heated at 25°C for 30 min. Reproduced from Schmidt et al. (1985).

5 mg BSA/mL are added to the control reaction, the heat-induced dissociation of [³H]TA can be reduced significantly. When BSA is added at 30mg/mL, a protein concentration similar to unfractionated cytosol, heat-induced activation is not stimulated beyond that detected in the absence of BSA (Table 7). The dose–response data in Table 5 thus reflect the activity of a cytoplasmic stimulator(s).

The stability of stimulator was tested at various temperatures. It appears to be stable at 90°C. Incubation of heat-treated stimulator with trypsin prevented the action of stimulator in increasing activation as measured by ability of the receptor complex to bind to DNA. Clearly, stimulator is a heat-stable cytoplasmic protein. When partially purified glucocorticoid-receptor complexes are heat-activated, there is about a 2.5-fold increase in resulting ability to bind to DNA. When a preparation of stimulator was added, compared to an appropriate control, DNA binding ability was doubled further. Moreover the action of stimulator seems to involve a second

Table 6
Heat Stability of Cytoplasmic Factor(s) that Stimulates Binding
of Purified [³H]TA-Receptor Complexes to DNA-Cellulose

Treatment	Total binding of [³H]TA, dpm/100 μL	DNA-cellulose binding, dpm/100 μL	DNA-cellulose binding, percent	Increase in DNA-cellulose binding, -fold
Unactivated control (0°C, 30 min) plus buffer C (5 mg BSA/mL)	6,070	407	6.7	1.0
Thermal activation (25°C, 30 min) plus buffer C (5 mg BSA/mL)	6,260	1,130	17.9	2.7
Unactivated control (0°C, 30 min) plus buffer C (130 mg BSA/mL)	7,560	642	8.5	1.3
Thermal activation (25°C, 30 min) plus buffer C (30 mg BSA/mL)	6,830	1,330	19.5	2.9
Thermal activation (25°C, 30 min) plus unfractionated cytosol	3,950	1,420	35.8	4.2 (compared to Treatment 3)
Thermal activation (25°C, 30 min) plus cytosol preheated at 90°C for 30 min	5,670	2,220	39.2	5.9 (compared to Treatment 1)

Reproduced from Schmidt et al. (1985).

step of activation in which heat-dependence or inhibition by molybdate is not characteristic, but is needed in the first step of activation. The near future should provide evidence of the purification and characterization of this protein.

6. Conclusion

Since the early 1980s, receptor purification has progressed rapidly. In the case of the glucocorticoid receptor the purified unactivated receptor complex has made possible the study of in vitro activation, converting it to

Table 7
Trypsin Sensitivity of Heat-Stable Activating Factor(s)[*]

Treatment	Total binding of [³H]TA, dpm/100 µL	DNA-cellulose binding, dpm//100µL	DNA-cellulose binding, percent	Increase in DNA-cellulose binding, -fold
Unactivated control (0°C, 30 min) plus buffer C (5 mg BSA/mL)	6,070	407	6.7	1.0
Thermal activation (25°C, 30 min) plus buffer C (5 mg BSA/mL)	6,260	1,130	17.9	2.7
Thermal activation (25°C, 30 min) plus 90°C treated cytosol pretreated (37°C, 30 min) with trypsin (10 µg/mL) followed by a 10-fold excess of soybean trypsin inhibitor (37°C, 30 min)	4,430	723	16.3	2.4
Thermal activation (25°C, 30 min) plus 90°C treated cytosol pretreated (37°C, 30 min) simultaneously with trypsin (10 µg/mL) and 10-fold excess of soybean trypsin inhibitor	6,450	1,950	30.3	4.5

Reproduced from Schmidt et al. (1985).
[*]Aliquots of partially purified [³H]TA-receptor complexes eluted from Bio-Gel A-1.5m were diluted 1:1 (v/v) with Buffer C or with 90°C treated cytosol that had been preincubated with trypsin and soybean trypsin inhibitor as indicated.

the DNA-binding form. It appears that the activation process is highly regulated and complex and the availability of purified unactivated receptor will faciliate the purification and characterization of these regulators, information that is not likely to be generated from studies with the cloned receptor gene.

Acknowledgment

An affinity resin similar to the one described here can be obtained commercially from VWR Scientific Co., and from Sterogene Bioseparations, Inc., 140 E. Santa Clara St., Arcadia, CA 91006.

Research in this laboratory is supported by Research grants DK 13531 from the National Institutes of Health, DCB-8517711 from the National Science Foundation, 87A19 from the American Institute for Cancer Research, DK42353 from the NIH and Core Grant CA 12227 from the National Cancer Institute to the Fels Institute for Cancer Research and Molecular Biology.

References

Baulieu, E.-E. (1987) *J. Cell. Biochem.* **35**, 161–174.

Bodine, P. V. and Litwack, G. (1988a) *J. Biol. Chem.* **263**, 3501–3512.

Bodine, P. V. and Litwack, G. (1988b) *Proc. Natl. Acad. Sci. (U.S.A.)* **85**, 1462–1466.

Denis, M., Wikstrom, A.-C., and Gustafsson, J.-A. (1987) *J. Biol. Chem.* **262**, 11803–11806.

Erdos, T., Best-Belpomme, M., and Bessada, R. (1970) *Anal. Biochem.* **37**, 244–252.

Failla, D., Tomkins, G. M., and Santi, D. V. (1978) *Proc. Natl. Acad. Sci. (U.S.A.)* **72**, 3849–3852.

Grandics, P. (1981) *Biochem. Biophys. Acta* **16**, 51–55.

Grandics, P., Gasser, D. L., and Litwack, G. (1982) *Endocrinology* **111**, 1731–1733.

Grandics, P., Miller, A., Schmidt, T. J., Mittman, D., and Litwack, G. (1984) *J. Biol. Chem.* **259**, 3173–3180.

Irie, S., Sezaki, M., and Kato, Y. (1982) *Anal. Biochem.* **126**, 350–354.

Joab, I., Radanyi, C., Renoir, M., Buchou, T., Catelli, M.-G., Binart, N., Mester, J., and Baulieu, E.-E. (1984) *Nature* **308**, 850–853.

Mendel, D. B., Bodwell, J. E., Gametchu, B., Harrison, R. W., and Munck, A. (1986) *J. Biol. Chem.* **261**, 3758–3763.

Okret, S., Wikström, A.-C., and Gustafsson, J.-A. (1985) *Biochemistry* **24**, 6581–6586.

Sanchez, E. R., Meshinchi, S., Tienrungroj, W., Schlesinger, M. J., Toft, D. O., and Pratt, W. B. (1987) *J. Biol. Chem.* **262**, 6986–6991.

Schmidt, T. J., Miller-Diener, A., Webb, M. L., and Litwack, G. (1985) *J. Biol. Chem.* **260**, 16255–16262.

Sherman, M. R., Moran, M. C., Tuazon, FeB., and Stevens, Y.-W. (1983) *J. Biol. Chem.* **258**, 10366–10377.

Webb, M. L. and Litwack, G. (1986) in *Biochemical Actions of Hormones* vol. 13 (Litwack, G., ed.), Academic, Orlando.

Webb, M. L., Miller-Diener, A. S., and Litwack, G. (1985) *Biochemistry* **24**, 1946–1952.

Purification of the Glucocorticoid Receptor and Its Ligand-Binding Domains

Jan Carlstedt-Duke, Karin Dahlman, Per-Erik Strömstedt, and Jan-Åke Gustafsson

1. Introduction

The receptor proteins for steroid hormones are a family of transcriptional regulatory proteins belonging to a super family encompassing receptors for steroids, thyroid hormone, and fat-soluble vitamins (Evans, 1988). One of the best characterized members of this family is the glucocorticoid receptor protein (GR). One of the reasons for this is the purification of this receptor at an early stage, resulting in preparations of purified material that appeared to retain its biological activity (Wrange et al., 1979). The material obtained was sufficient for further characterization of the protein (Wrange et al., 1984,1986; Carlstedt-Duke et al., 1987,1988) as well as for studying protein–DNA interactions (Payvar et al., 1981,1982,1983, Sakai et al., 1988). The purified GR preparations were also used to raise polyclonal and monoclonal antibodies (Okret et al., 1981,1984). Among other things, these antibodies have been used to clone cDNA for the receptor (Miesfeld et al., 1984,1986).

The knowledge obtained from these initial studies of purified GR has enabled the development of further techniques for the purification of specific fragments of GR in much larger quantities. This chapter describes

Receptor Purification, vol. 2   ©1990 The Humana Press

the original purification technique for rat liver GR (Wrange et al., 1979), with some minor modifications, as well as the purification of various domains of the receptor, either from rat liver GR or by using recombinant DNA-techniques.

One of the main reasons for purifying GR today, apart from characterization of the protein itself, is for use as a model for studying gene regulation in eukaryotic cells. The receptor protein binds to specific enhancer sequences called glucocorticoid response elements (GREs), in the vicinity, usually upstream, of promoters that are regulated by glucocorticoids. This results in the stimulation of transcriptional activity (Yamamoto, 1985), although there are also examples of specific inhibition of transcriptional activity (Sakai et al., 1988; Akerblom et al., 1988). The transcriptional regulatory activity is allosterically regulated by the binding of hormone to the receptor (Yamamoto, 1985). The steroid receptor proteins are just one class of transcriptional factors and there are interactions between different transcriptional factors (including between GR and other factors) resulting, ultimately, in an effect on RNA-polymerase (Yamamoto, 1985; Strähle et al., 1988). Thus, GR is one of the more readily available transcriptional factors today, and is, therefore, of general interest within the field of gene regulation.

2. Purification
of the Rat Glucocorticoid Receptor

2.1. Strategy

Initial studies of steroid receptors showed that these proteins could be recovered in the soluble cytosolic fraction of cellular extracts from target tissues (Baxter et al., 1972; Rousseau et al., 1975). However, these proteins occur at very low concentrations, even in target cells, and make up 0.001–0.01% of the total cytosolic proteins. Thus, it would be very difficult to purify these proteins using conventional techniques based on separation by size or charge. In order to obtain a sufficient degree of purification, some form of affinity chromatography was necessary. The receptor has two different ligand affinities that are of potential use for such chromatography, steroid-binding and DNA-binding.

The kinetics of steroid-binding appears to be more complicated than that of DNA-binding. Prior to activation, binding of the steroid ligand to the receptor appears to be a simple equilibrium reaction. However, following the binding of steroid, the receptor appears to undergo some form of conformational change that results in a tighter binding of the steroid. This process is called activation or transformation and appears to be irreversible. Activation is dependent on the prior binding of steroid ligand (Higgins et al., 1973; Kalimi et al., 1975), and is stimulated by increased temperature or ionic strength. This greatly complicates affinity chromatography based on steroid-binding and we therefore concentrated on affinity chromatography based on the second ligand, DNA. At the time of the development of this purification scheme, affinity chromatography based on specific DNA-sequences was not available, but has been developed since then (Kadonaga and Tjian, 1986). However, in addition to the sequence-specific interaction with GREs, GR also binds in a sequence-independent manner to all forms of DNA. It was, in fact, this nonspecific interaction with DNA that was originally used to describe the activation process (Baxter et al., 1972; Rousseau et al., 1975).

The interaction of GR with DNA (both specific and nonspecific) is allosterically regulated by the primary ligand, the steroid. It is only after the binding of steroid and subsequent activation of the complex that it will interact with DNA (Denis et al., 1988). This is the basis of the purification strategy (Wrange et al., 1979). Rat liver cytosol was incubated with a synthetic glucocorticoid, but at 0°C to prevent activation. The synthetic glucocorticoid is radiolabeled, and this enables the detection of the steroid-GR-complex throughout the purification by simply measuring radioactivity. The cytosol is then passed through phosphocellulose and DNA-cellulose in order to remove all DNA-binding proteins. Phosphocellulose mimics DNA but has a much higher binding capacity than DNA-cellulose. Since the cytosol is maintained at low temperature, the receptor is not activated despite the binding of steroid. Thus, it will not bind to either DNA or phosphocellulose. We have subsequently modified the purification scheme, eliminating this first DNA-cellulose column (Wrange et al., 1984). It is sufficient to pass the cytosol through phosphocellulose only, and no further proteins appear to be removed by the DNA-cellulose. The receptor is then activated by incubating the flowthrough volume from the phosphocellulose column at 25°C. Following heat activation, the flowthrough volume is passed through a DNA-cellulose column to which the receptor will now bind. The column is eluted with a discontinuous salt gradient and the fractions containing GR recovered. Final purification is obtained by ion-exchange chromatography. Thus, the basis for this purification strategy is the differential affinity of the receptor for DNA before and after activation.

2.2. *Preparation of Cytosol*

In order to obtain preparations of purified GR that correspond as closely as possible to the protein in its natural environment, there are certain basic considerations to remember when working with the protein. Like all steroid receptors, GR is highly susceptible to digestion by a great variety of proteases. Care must be taken to minimize proteolytic activity during the preparation of cytosol. This is achieved by perfusing the liver with ice cold homogenization buffer that is hypertonic (Carlstedt-Duke et al., 1977). If the buffer is not hypertonic, the lysosomes will be ruptured during tissue preparation resulting in a high level of proteolytic activity (Carlstedt-Duke et al., 1979). The perfusion rapidly chills the organ as well as removing all blood from the organ. In all subsequent steps during the purification it is important to maintain the temperature at 2°C (except during activation) and to work as rapidly as possible to minimize proteolysis.

In addition to proteolysis, the receptor is also highly susceptible to oxidation. Free thiol groups are required for both steroid and DNA-binding (Grippo et al., 1986). Oxidation of these vital thiol groups is prevented by the addition of dithiothreitol (DTT) to all buffers, starting with 1 mM in the initial buffers and increasing the concentration during the procedure. Another important preventive measure with regard to oxidation is the degassing of all buffers followed by saturation with nitrogen by bubbling the gas through the buffer. The buffers are made from autoclaved stock solutions (0.5–5M) and diluted freshly for each purification. DTT is added from a 1M stock solution immediately prior to use. The DTT stock solution is stored at –20°C. The receptor is further stabilized by the addition of glycerol to the buffers, presumably since hydrophobic interactions are of importance for steroid-binding. The glycerol is also responsible for the hypertonicity of the buffer. The buffers used throughout the purification are shown in Table 1.

After perfusion of the liver, it is removed onto ice, rinsed with ice cold buffer A, and then finely minced in 10 mL buffer A. For each purification, we usually start with 8 livers (8-wk-old male Sprague-Dawley—adrenalectomized two days in advance to remove endogenous ligand). The minced livers are homogenized with three strokes of a tightly-fitting glass-Teflon™ Potter-Elvhjem homogenizer. The speed of rotation of the pestle was chosen to minimize disruption of lysosomes and the homogenizer is packed in ice during the procedure. The homogenate is then centrifuged at 150,000 × g for 70 min, the floating lipids are removed, and the supernatant (cytosol) is recovered.

Table 1
Buffers Used During the Purification Procedures

Buffer A	20 mM phosphate, pH 7.0, 50 mM NaCl, 1 mM EDTA, 10% (w/v) glycerol, 1 mM DTT
Buffer B	20 mM Tris-HCl, pH 7.8, 1 mM EDTA, 10% (w/v) glycerol, 10 mM DTT
Buffer C	20 mM Tris-HCl, pH 8.0, 50 mM NaCl, 1 mM EDTA
Buffer D	50 mM Tris-HCl, pH 8.0, 150 mM NaCl, 1 mM EDTA
Buffer E	50 mM Tris-HCl, pH 8.0
Buffer F	20 mM phosphate, pH 7.4, 50 mM NaCl
Buffer G	20 mM phosphate, pH 7.4, 350 mM NaCl

Following incubation with 100 nM ^{3}H-triamcinolone acetonide for 30 min, the cytosol is passed through a column of phosphocellulose (Table 2) equilibrated in buffer A. The pH of the effluent is checked immediately prior to application of the cytosol to ensure that the column really is equilibrated. It is most important to make sure that the receptor does not become activated at this stage. This is achieved by maintaining the temperature at 0°C and by performing the chromatography as fast as possible (flowrate >20 mL/cm^2/h). At slower flowrates, the ionic strength of the matrix is sufficiently high to induce activation of the receptor that results in its binding to the phosphocellulose (Atger and Milgrom, 1976). Prior to this chromatographic step, the cytosol has a dark red color. Following passage through phosphocellulose, the color of the flowthrough volume is straw-colored.

2.3. DNA-Cellulose Chromatography

The initial chromatographic step removes DNA-binding proteins while maintaining the receptor in a non-DNA-binding state. The next step involves the activation of GR to a DNA-binding state that is achieved by incubation at 25°C for 30 min. The efficiency of this process is about 60%. If the flowthrough volume is diluted with an equal volume of buffer A prior to incubation at 25°C, the efficiency of the activation process is increased to >90% (Wrange et al., 1986). Following activation, the diluted flowthrough volume is applied on a 40 mL DNA-cellulose column (Table 2) equilibrated in buffer B containing 90 mM NaCl. Calf thymus DNA-cellulose can either be prepared as previously described (Alberts and Herrick, 1971; Potuzak and Dean, 1978) or purchased from Sigma (St. Louis, MO). We usually prepare our own DNA-cellulose and mix it with an equal amount of commercially prepared matrix, since we have found this to give optimal flow properties.

Table 2
Column Dimensions and Conditions
for the Chromatographic Procedures for the Purification of Rat GR

	Height × diameter, mm	Volume, mL	Equilizing buffer	Flowrate mL/cm²/h
Phosphocellulose	30 × 90	190	A	max (>20) (>20 mL/min)
DNA-cellulose	20 × 50	40	A	6 (2 mL/min)
DEAE-Sepharose	15 × 16	3	B	12 (0.4 mL/min)

Care should be taken to ensure that the column does not run dry. If this happens, the matrix usually shrinks away from the wall of the column, allowing a large part of the sample volume to pass by the matrix. Occasionally, the rate of flow through the column can diminish or stop all together during sample application. This can be taken care of by very carefully stirring up the DNA-cellulose at the top of the column, making sure that the column does not run dry.

After sample application, the walls of the column are carefully dried and washed repeatedly with buffer B containing 90 mM NaCl. The column is then washed with 2 volumes 90 mM NaCl, 1.5 vol 115 mM NaCl, and 1 vol 0 mM NaCl, all in buffer B. The receptor is then eluted from the column by 27.5 mM MgCl$_2$ in buffer B. The effluent is collected in fractions of about 5 mL and GR detected by measuring the amount of radioactivity in each fraction. The pool of GR at this stage is 30–40% pure, and the recovery is 50–90 μg (starting material, eight livers). The purification is about 6,000 × and the GR thus obtained is adequate for many studies, such as DNA-binding studies. However, for analysis of the protein, further purification is usually required.

2.4. Ion-Exchange Chromatography

Final purification of GR is achieved by anion-exchange chromatography using two alternative methods, conventional chromatography on DEAE-Sepharose or FPLC on Mono Q (Pharmacia, Uppsala, Sweden). The former is a simple and rapid procedure, although the latter has the advantage of increased resolution with more concentrated peaks. Since the GR is eluted from the DNA-cellulose at a relatively low concentration of MgCl$_2$ (27.5 mM), the ionic strength is sufficiently low to allow chromatography directly on the ion-exchange column without sample dilution or change of buffer.

The pool of fractions containing GR eluted from DNA-cellulose is applied on a 3 mL column of DEAE-Sepharose (Table 2) using a closed chromatography system. Once the sample volume has run into the column, the valve is switched and the column eluted with a linear 0–0.5M NaCl gradient in buffer B with 20 mM DTT instead of 10 mM (total volume 50 mL). Fractions of about 1 mL are collected and assayed for radioactivity. The receptor is eluted at about 120 mM NaCl. Fractions containing GR (6–8 mL) are pooled. The recovery is usually 40–60 µg and the purity is >80%, based on SDS-polyacrylamide gel electrophoresis, Coomassie staining, and densitometry.

If FPLC instrumentation is available, this enables the preparation of GR with higher purity and concentration. Using DEAE-Sepharose, the peak fraction contains about 10–12 µg GR/mL. Following chromatography on Mono Q, the peak fraction can contain 40–50 µg GR/mL under optimal conditions. Under these conditions, the purity of the peak fraction is 95–98%. The pooled fractions from DNA-cellulose are applied on a Mono Q 5/50 column and eluted with a 0–1 M NaCl gradient as shown in Fig. 1. The receptor is eluted at about 180 mM NaCl and fractions of 0.3 mL are collected in this region (Carlstedt-Duke and Wrange, unpublished results). As seen in Fig. 2, the major peak of radioactivity contains only the intact GR (M_r 94,000). The two following peaks of radioactivity contain a proteolytic fragment of GR (M_r 79,000) and an associated protein unrelated to GR (M_r 72,000). However, the recovery on Mono Q is usually somewhat lower than on DEAE-Sepharose, and the efficiency of the method is entirely dependent on the condition of the FPLC instrument. As recommended by the manufacturer, Mono Q should not be exposed to higher DTT concentrations than 5 mM.

2.5. Storage of Purified Preparations of Receptor

Following recovery of the appropriate preparation of GR for the studies in question (pooled fractions from DNA-cellulose, DEAE-Sepharose or Mono Q), the receptor can be stored at –80°C for one year or more, without any appreciable loss of DNA-binding activity. The receptor preparations can be frozen directly in the elution buffer, but the following should be considered. The buffer should contain at least 10% (w/v) glycerol to stabilize the protein. The glycerol concentration can be increased up to 50% without any adverse effects. Repeated freezing and thawing of the receptor preparations should be avoided. An alternative is to scrape off an appropriate volume from the frozen stock, using a clean spatula. This can be facilitated by increasing the glycerol concentration, although it is not absolutely necessary. The DTT concentration should be at least 20 mM

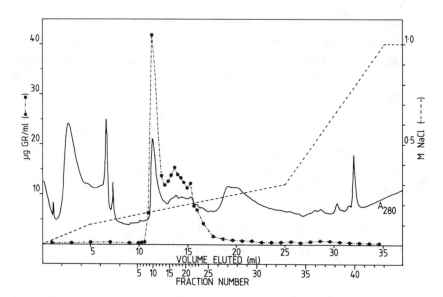

Fig. 1. Final purification of GR by FPLC on Mono Q. The pooled fractions from DNA-cellulose were applied on a Mono Q 5/50 column equilibrated in buffer B containing 5 mM DTT (flowrate 1 mL/min). The column was eluted with a programmed 0–1M NaCl gradient as shown and fractions of 0.3–2 mL collected. The concentration of GR (–•–•–) was calculated from the radioactivity in each fraction. A_{280} (—) was monitored continuously and (NaCl) (---) was determined by the gradient programmer.

prior to freezing as the receptor is inactivated by oxidation. In fact the receptor can be exposed to 100 mM DTT without any adverse effects. For DNA-binding experiments, it is advantageous to add a neutral carrier protein prior to freezing. We routinely add 200 μg insulin/mL as carrier, since this has no apparent effect on DNA-binding (Payvar et al., 1983). Albumin is an equally good alternative (Payvar et al., 1981). Long-term exposure of the receptor to CO_2 by storage on dry ice should be avoided since this results in some loss of DNA-binding activity. If the receptor must be shipped on dry ice, it should first be hermetically sealed.

3. Purification of Individual Receptor Domains

3.1. Domain Structure

The steroid receptors are multifunctional proteins in which the various activities are coordinated for the necessary biological activity. Typical for multifunctional proteins is that the different activities are located within different functional domains within the protein. This is the case for GR. Limited proteolysis of the protein can separate at least three functional

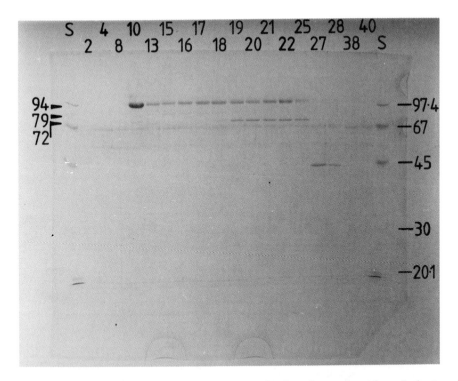

Fig. 2. Analysis of fractions from Mono Q by SDS-polyacrylamide gel electrophoresis. Aliquots from each fraction from the chromatogram shown in Fig. 1 were precipitated with 20% (w/v) trichloroacetic acid and analyzed by electrophoresis. The numbers at the top are the fraction numbers (cf., Fig. 1). The numbers at the right are $M_r \times 10^{-3}$ for the standard proteins (S). The numbers at the left are the corresponding values calculated for the sample proteins.

domains responsible for steroid-binding, DNA-binding, and transcriptional modulation (Carlstedt-Duke et al., 1982; Dellweg et al., 1982; Wrange et al., 1984). *N*-terminal sequence analysis of the proteolytic products has identified the location of the different domains within the primary structure of the protein (Fig. 3). The results are in concordance with the domain definition based on deletion and insertional mutation of GR cDNA (Giguère et al., 1986; Hollenberg et al., 1987; Rusconi and Yamamoto, 1987).

3.2. Purification of Receptor Lacking the N-Terminal Domain

Specific deletion of the *N*-terminal domain can be achieved by incubating purified GR with chymotrypsin (Wrange et al., 1984). The *N*-terminal domain appears to be totally digested by the protease, since no specific fragment corresponding to this domain is recovered. However, when analyzed under nondenaturing conditions, digestion of GR in crude cytosol

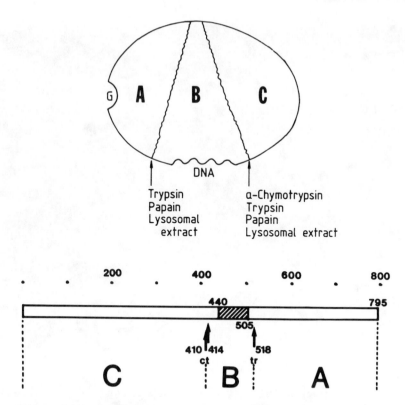

Fig. 3. Domain Structure of Rat GR. The domain structure based on limited proteolysis is shown in the upper part of the figure. The location of the cleavage sites (*N*-ter-
minal residues of proteolytic fragments) is shown in the lower part of the figure
(Carlstedt-Duke et al., 1987). Abbreviations: A, steroid-binding domain; B, DNA binding domain; C, transcriptional modulatory domain; G, glucocorticoid.

gives rise to a specific fragment corresponding to the *N*-terminal domain
(Carlstedt-Duke et al., 1982).

Purified preparations of the *N*-terminal deleted receptor can be obtained by incubation of the pooled fractions from DNA-cellulose (containing intact GR) with 1 µg chymotrypsin/µg GR for 30 min at 10°C. The
inubation is terminated by the addition of chymotrypsin inhibitor or 100
µ*M* ZnCl$_2$ or ZnSO$_4$. The exact concentration of chymotrypsin giving optimal digestion of GR (complete digestion to an M_r 39,000 fragment) varies
from preparation to preparation of chymotrypsin and must be titrated
individually. The concentration of chymotrypsin inhibitor (polypeptide)
depends on the concentration of chymotrypsin and should follow the
manufacturer's recommendations.

Following digestion, the GR fragment is separated from the protease
and any other contaminating proteins, including the chymotrypsin inhibitor and the M_r 72,000 protein, by rechromatography on DNA-cellulose.

However, the sample must first be diluted with an equal vol of buffer B prior to chromatography, otherwise the GR fragment will not bind to the matrix. The sample is applied on a 3–5 mL column of DNA-cellulose equilibrated in buffer B. The column is then washed with two volumes of buffer B and two volumes of buffer B containing 150 mM NaCl. Finally, the column is eluted with 350 mM NaCl in buffer B and 1 mL fractions collected. The pooled fractions containing radioactivity corresponding to the GR fragment appear to contain a homogeneous preparation of the M_r 39,000 protein.

3.3. Purification of the DNA-Binding Domain

Using a similar strategy as that described above, the isolated DNA-binding domain can be purified. However, the efficiency of this procedure is much lower than the purification of the *N*-terminal deleted fragment. The initial steps of the purification are exactly as described above for the *N*-terminal deleted receptor. After digestion with chymotrypsin and application on the DNA-cellulose column, the column is washed with 150 mM NaCl. A solution containing 0.25 µg trypsin/mL buffer B is then applied on the DNA-cellulose column until the entire column contains trypsin. The flow through the column is then stopped and the column incubated at 4°C for 45 min. The column is then washed with 3 volumes of buffer B containing trypsin inhibitor. The isolated DNA-binding domain can then be eluted with 350 mM NaCl in buffer B.

3.4. Purification of the Steroid-Binding Domain

The procedure described above for the purification of the DNA-binding domain results in the simultaneous preparation of the steroid-binding domain. This domain (M_r 27,000) is eluted together with the trypsin after the incubation of the DNA-cellulose column with the protease. In order to avoid further digestion of the domain, trypsin inhibitor should be added to the fractions immediately after elution from the column. Thus, this procedure results in a preparation containing both trypsin and trypsin inhibitor in addition to the GR fragment. An alternative procedure for the preparation of this domain is the incubation of purified GR with Sepharose-bound trypsin. Carrier protein (insulin) must be added first to avoid adsorption of the receptor to the Sepharose. DNA-binding fragments of GR can be removed with DNA-cellulose after the protease digestion. Since the protease is bound to Sepharose, it can easily be removed by centrifugation after digestion is complete. Thus, the preparation obtained is free of protease, which can be advantageous for some experiments.

4. Recombinant Receptor Preparations

Purification of small quantities of GR from natural sources has facilitated cloning and characterization of GR from several species, but these quantities do not allow extensive structural analysis of the molecule. The expression of cloned genes is possible today in a variety of prokaryotic and eukaryotic organisms. Using *E. coli* in particular, both efficient and controlled production of recombinant polypeptides can be achieved (Marston, 1986).

We describe here the purification of milligram amounts of a recombinant form of the DNA-binding domain of the human GR (DBD$_r$). The protein preparation is essentially homogenous as determined by SDS-polyacrylamide gel electrophoresis, *N*-terminal sequence analysis, and total amino acid analysis (Dahlman et al., 1989). Furthermore, the recombinant protein appears to have DNA-binding properties very similar to the native GR purified as described above.

4.1. Recombinant DNA-Binding Domain

For expression of large amounts of GR protein, we have utilized an expression system originally described by Uhlén and coworkers at the Royal Institute of Technology, Stockholm (Nilsson et al., 1985). The general characteristics of the expression system are described below.

The expression system has been designed for intracellular expression of recombinant proteins in *E. coli*. Inserted DNA-sequences will be expressed as fusion proteins with protein A under control of the phage Lambda P_R promoter (temperature inducible in strains containing a temperature sensitive Lambda repressor). Our initial studies had suggested that protein A protected the recombinant protein from degradation by endogenous *E. coli* proteases (Bonifer et al., 1988). Furthermore, the high affinity binding between protein A and IgG can be used for rapid purification of the recombinant protein. We employ an intracellular expression system, since early attempts to get GR-containing peptides exported through the *E. coli* membrane were unsuccessful. There are very few reports of successful export of recombinant proteins in *E. coli* that are normally intracellular.

In the majority of cases, eukaryotic polypeptides expressed in *E. coli* are found to accumulate in an insoluble form (Marston, 1986). Although these polypeptides can usually be solubilized, this involves the use of denaturants to unfold the polypeptide, followed by removal of the denaturant and refolding of the polypeptide. The extent to which a molecule is correctly folded cannot be predicted. DBD$_r$ in fusion with protein A was almost exclusively soluble when expressed in *E. coli,* whereas most other GR containing peptides were largely insoluble (Bonifer et al., 1988).

A novel and nondenaturing purification procedure has been developed to purify DBD_r, essentially to homogeneity, from the soluble extract of *E. coli.* The detailed protocol is given in Table 3. A detailed map of the expression plasmid is given in Fig. 4. Following induction by increased temperature, the fusion protein is expressed and can be recovered in the soluble fraction of the cell lysate (Fig. 5).

The clarified extract is applied on an IgG Sepharose column and the column washed extensively with buffer. Initially, the fusion protein was eluted with 0.5*M* acetic acid followed by renaturation. As described above, GR can be cleaved with chymotrypsin to give a fragment that retains DNA- and steroid-binding properties (Wrange and Gustafsson, 1978). The cleavage site has been identified with cleavage occurring after amino acids 409 and 414 in rat GR (corresponding to amino acids 389 and 393 in human GR) (Fig. 3). We reasoned that it should be possible to use this cleavage site to release DBD_r from the fusion protein (Fig. 6). Although it was possible to perform this cleavage after elution with acid followed by renaturation, we reasoned that it would be advantageous to avoid the denaturation/renaturation step and therefore tested cleavage of the protein while immobilized on IgG Sepharose. Surprisingly, cleavage under these conditions resulted in unique cleavage after amino acid 393. The chymotrypsin concentration used has to be very carefully standardized for each individual IgG Sepharose column, in relation to the concentration of fusion protein that varies from column to column. DBD_r is eluted together with chymotrypsin and the two proteins can be separated by DNA-cellulose chromatoghraphy. One hundred μM Zn^{2+} is added immediately after elution from the IgG Sepharose to inhibit chymotrypsin. The recovery from the DNA-cellulose column could be increased significantly by incubating the effluent from IgG Sepharose with 0.3*M* NaCl followed by dilution. The protein expressed in *E. coli* binds to the bacterial DNA that is digested with DNase prior to clarification. However, the segments of DNA bound to the fusion protein are protected from DNase digestion. These oligonucleotides block the DNA-binding site of DBD_r and must be removed prior to DNA-cellulose chromatography. The protein eluted from DNA-cellulose is apparently homogeneous, as seen in Fig. 7.

The recombinant protein described above consists of the entire DNA-binding domain and two residues from the steroid-binding domain (human GR residues 394–500). In addition, there are eight additional residues at the C-terminal, derived from the polylinker of the expression vector. A more recent expression plasmid with only one additional residue

Table 3
Purification Scheme for DBD$_r$[a]

1. Resuspend cells from 3 L fermentation culture in 200 mL of buffer C, 1 mM PMSF, 1 mM leupeptin, 10mg trypsin inhibitor/mL, 1 mM DTT.
2. Add lysozyme to a final concentration of 0.5 mg/mL and incubate on ice for 30 min with gentle stirring.
3. Add MgCl$_2$, MnCl$_2$, DNase I, and RNase A to final concentrations of 10 mM, 1 mM 10 µg/mL and 10 µg/mL, respectively. Incubate on ice for 30 min.
4. Centrifuge at 150,000 × g for 30 min.
5. Apply the clarified extract on a column (diameter 50 mm) containing 50 mL IgG Sepharose 6 Fast Flow (Pharmacia, Uppsala, Sweden) equilibrated in buffer D (flowrate 21 mL/cm²/h).
6. Wash column with buffer D supplemented with 100 µM PMSF, 10 µM leupeptin, 10 µg trypsin inhibitor/mL, and 1 mM DTT until the A$_{280}$ of the effluent is zero.
7. Reequilibrate the column with 3 volumes of buffer D. Apply a solution consisting of buffer D supplemented with 4 µg chymotrypsin/mL until the whole column is filled. Incubate the column for 1 h and elute with buffer D using reverse flow. Collect 50 mL.
8. Add ZnSO$_4$ to a final concentration of 100 µM.
9. Add NaCl to a final concentration of 300 mM (buffer D contains already 150 mM).
10. After dilution with 100 mL buffer E, apply the total vol on a column of DNA-cellulose (height 28 mm, diameter 26 mm) equilibrated with buffer F (flowrate 10 mL/cm²/h).
11. Wash column with 3 vol of buffer F.
12. DBD$_r$ is eluted with buffer G.

[a]All steps are performed at 4°C unless otherwise stated.

at the C-terminus has been constructed. This recombinant protein behaves identically to DBD$_r$. Both proteins bind to DNA with the same specificity as the intact GR. Thus, the additional amino acid residues at the C-terminus do not appear to play any functional role.

4.2. Other Recombinant Receptor Fragments

As mentioned above, most other GR constructions expressed in bacterial systems result in the expression of insoluble protein, using a variety of expression systems. Some of these proteins can undergo denaturation/renaturation without any apparent problems. For the DNA-binding domain, this results in a functional protein. However, this is not possible with the steroid-binding domain. Renaturation does not result in the restoration of steroid-binding activity. Since we do not have any assay for the other potential functions of the receptor, e.g., transcriptional activation and modulation, the effect of denaturation/renaturation of these parts of the receptor protein cannot be assessed as yet.

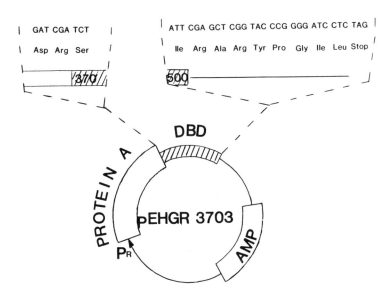

Fig. 4. Plasmid for expression of DBD$_r$ in *E. coli*. The shaded region represents GR derived sequences. The cloning created an in frame fusion between the codon for amino acid 500 of the human GR and the polylinker of the vector resulting in the fusion of eight additional amino acids, derived from the vector, at the C-terminus of the fusion protein. The plasmid pRSV HGR that contained the cDNA sequence for the human GR was obtained from B. L. West, University of California, San Francisco. This plasmid was digested first with Xho II, the sticky ends filled in with the Klenow fragment of DNA polymerase I, followed by a second digestion with Xba I and purification of the approx 1600 bp Xho II/Xba I fragment of the GR gene. The vector, pRIT 33, was obtained from M. Uhlén, Royal Institute of Technology, Stockholm, Sweden. The vector was first cut with Bam HI, the sticky ends filled in with Klenow, and then cut again with Xba I. Ligation with the Xho II/Xba I receptor fragment gave the plasmid pEHGR 3700. This plasmid was cut with Eco RI and the approx 400 bp fragment corresponding to amino acid 370–500 of the human GR was purified and ligated into Eco RI digested pRIT 33 to give the plasmid pEHGR 3703.

The steroid-binding domain of human GR has been expressed as a fusion protein using a similar strategy to that described above (Ohara-Nemoto, Strömstedt, Dahlman, Nemoto, Gustafsson and Carlstedt-Duke; submitted). The entire steroid-binding domain, including the trypsin-cleavage site at residue 498, is expressed as a fusion protein with protein A. Less than 10% of the protein expressed is soluble and only a few percent of the soluble protein retains its steroid-binding activity. The affinity for glucocorticoids is 10–100-fold lower than for the intact GR, but the specificity is similar. Thus, this method appears to be of potential interest for the expression of a functional steroid-binding domain for further structural studies.

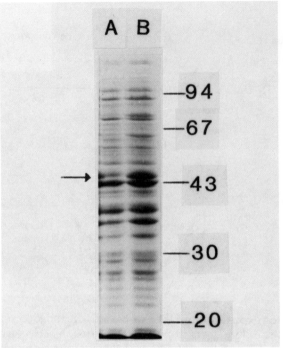

Fig. 5. Growth of bacteria and expression of the fusion protein. Cultures of *E. coli* containing the expression plasmid were grown and induced as described below and analyzed by SDS-polyacrylamide gel electrophoresis followed by Coomassie staining. Lane 1, uninduced and lane 2, induced cultures. The arrow marks the position of the fusion protein. The numbers refer to the size of the reference proteins. The plasmid pEHGR 3703 was maintained in *E. coli* strain JM 109 together with a compatible plasmid expressing the temperature sensitive cI 857 Lambda repressor and kanamycin resistance. Cells were grown overnight in 160 mL of Luria broth supplemented with 1% (w/v) glucose, 1% (w/v) casamino acids, 100 µg ampicillin/mL, and 50 µg kanamycin/mL at 31°C in a shaking flask. The overnight culture was diluted into a 12 L fermentor containing 8 L of the same medium and growth was continued at 31°C until late log phase (OD$_{600}$= 2.5). The pH was maintained at 7.2–7.4 by the addition of 5M NaOH and the airflow was kept at 10 L/min. Production of the recombinant protein was induced by the addition of 4 L medium preheated to 65°C and increasing the temperature to 42°C. Cells were harvested by centrifugation after 90 min incubation at 42°C, washed in 50 mM phosphate, pH 7.4, 150 mM NaCl, 1 mM EDTA, and stored at –70°C.

5. Characterization
of Purified Glucocorticoid Receptor

Rat GR contains a single polypeptide with apparent M_r 94,000 based on SDS-polyacrylamide gel electrophoresis (Wrange et al., 1984), which is larger than the mol wt calculated from the primary structure (Miesfeld et al., 1986). Comparison of the amount of radioactivity (steroid) and protein in the purified preparations shows that there is one single steroid-binding

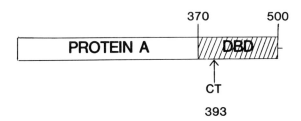

Fig. 6. Chymotrypsin cleavage site in the DBD$_r$ fusion protein. The arrow marks the chymotrypsin cleavage site utilized to release DBD$_r$ from protein A.

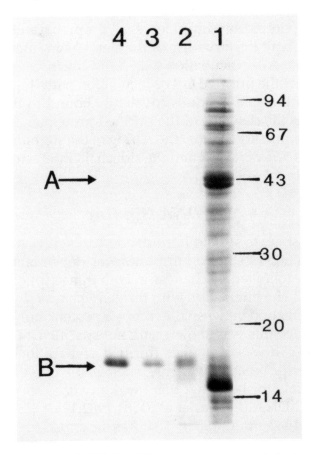

Fig. 7. Analysis of purified DBD$_r$ by SDS-polyacrylamide gel electrophoresis. Aliquots of the protein preparation at the different chromatographic steps were analyzed by electrophoresis followed by Coomassie staining. Lane 1, clarified extract. Lane 2, material eluted from the IgG Sepharose column. Lanes 3 and 4, fractions eluted from the DNA-cellulose column. The arrows mark the position of the fusion protein (A) and DBD$_r$ (B). The numbers refer to the size of the reference proteins.

site on each receptor molecule (Wrange et al., 1979). Following purification, the receptor retains its high affinity for DNA, both sequence-specific and nonspecific.

5.1. Steroid-Binding

The steroid-binding domain occurs at the C-terminal end of the protein (Fig. 3). This domain is relatively highly conserved among the steroid receptors, but not to the same extent as the DNA-binding domain. Rat GR is purified in an activated form, bound to the steroid. All attempts to remove the steroid or to exchange it have resulted in the total loss of activity of the protein, possibly by denaturation. Thus, this preparation is not suitable for studying steroid-binding. However, the preparation has proved to be useful for probing the structure of the steroid-binding site (Carlstedt-Duke et al., 1988). After incubation with ^{3}H-triamcinolone acetonide and photoactivation of the purified GR or incubation with ^{3}H-dexamethasone mesylate, the amino acid residues covalently bound to the steroid were identified by specific cleavage of the purified protein and radiosequence analysis. The results showed that the A-ring of the steroid interacted with Met-622 and Cys-754 of rat GR and the sidechain interacted with Cys-656 (Carlstedt-Duke et al., 1988).

5.2. DNA-Binding

At an early stage in steroid hormone research, it was shown that the mechanism of action of steroid hormones was dependent on interaction with the genome followed by changes in both transcriptional and translational processes. The interaction with the genome was shown to be probably a direct interaction between the receptor protein and DNA, although in the first instance this was shown with nonspecific DNA (Mainwaring and Irving, 1973). It was first when purified preparations of GR that retained their activity became available, such as the one described above, that it was possible to show that there was any base-specific interaction between GR and DNA (Payvar et al., 1981,1982,1983; Geisse et al., 1982; Govindan et al., 1982; Pfahl, 1982). This was described first for a variety of genes that were under positive regulation by glucocorticoids. The mouse mammary tumor virus (MTV) is the most commonly used model in this case. It has subsequently been shown that also negative gene regulation by glucocorticoids is dependent on sequence-specific interaction with DNA, resulting in competition for other positive-acting transcriptional factors that bind to the same segment of DNA (Sakai et al., 1988; Akerblom et al., 1988). Thus, GR preparations such as the rat GR described above can be shown to bind specifically to DNA using a variety of techniques such as nitrocellulose filter binding, gel retardation, electron microscopy, and DNase I, Exo III, and free radical footprinting. Based on the different GR binding sites studied, a consensus sequence GGTACANNNTGTTCT has been established (Jantzen et al., 1987; Beato et al., 1987). In the normal physiological situation, the interaction of GR with the genome appears to be entirely dependent on steroid-binding to the receptor, and this has been confirmed at the DNA level by in vivo footprinting (Becker et al., 1986). However, using a variety of different receptor preparations, various groups have reported steroid-independent binding of GR to DNA in vitro (Willmann and Beato, 1986). The reason for this discrepancy is still unclear.

Deletion mutation of the glucocorticoid receptor based on the cDNA clones available, both rat and human, has shown that in the absence of the steroid-binding domain, the DNA-binding causes constitutive activation of glucocorticoid-regulated genes (Giguère et al., 1986; Godowski et al., 1987; Hollenberg et al., 1987; Miesfeld et al., 1987; Rusconi and Yamamoto, 1987). This constitutive activity is also independent of the N-terminal domain, although it is more active in the presence of the N-terminal domain (Giguère et al., 1986; Hollenberg et al., 1987). Not surprisingly, therefore, the recombinant DNA-binding domain (DBD_r) described above was found to interact with exactly the same DNA-sequences as the intact GR (Dahlman et al., 1983). However, the apparent affinity of DBD_r for DNA seems to be about ten-fold lower, based on the amount of protein on a molar basis that is required to give a footprint of the same intensity (Dahlman et al., 1989). Thus, the DNA-binding domain appears to be entirely responsible for the sequence-specific interaction with DNA, but the other domains of the protein can modulate this interaction.

When the first cDNA clone for a steroid hormone receptor was described, it was noted that the putative DNA-binding domain contained an unusual structure with repetitive pairs of Cys residues (Hollenberg et al., 1985). This structure was reminiscent of the DNA-binding domain of *Xenopus* transcription factor TFIIIA. This protein has been shown to contain repetitive segments containing one pair of Cys and one pair of His residues that bind Zn^{2+} (Miller et al., 1985). The DNA-binding properties of TFIIIA are entirely dependent on the Cys/His coordinated zinc. The similarity between the structure of the steroid receptors and transcription factors such as TFIIIA led to the proposition that the steroid receptors were also metalloproteins (Berg, 1986). However, we found that incubation of both rat GR and DBD_r with 100 mM DTT, 50 mM o-phenanthroline or 250 mM EDTA had no effect at all on the DNA-binding properties, which was in contrast to the other transcription factors described above (Carlstedt-Duke et al., 1989). It was when DBD$_r$ was first denatured with urea in the presence of EDTA that any effect was seen on DNA-binding. Renatura-tion in the absence of metal resulted in the loss of DNA-binding. Renaturation in the presence of Zn^{2+}, Cd^{2+}, Mn^{2+}, or Ni^{2+}, however, resulted in the restoration of DNA-binding (Carlstedt-Duke et al., 1989). Similar results have been described with a slightly larger form of GR, including the DNA-binding domain (Freedman et al., 1988). In that study, it was also shown that the recombinant GR preparation contained two zinc ions/molecule and that the zinc was tetravalently coordinated with Cys. Thus, the steroid receptors are metalloproteins with two "zinc fingers" within the DNA-binding domain.

5.3. Dimer Formation

As the consensus sequence for GR binding became apparent the two-fold symmetry of the partial palindromic sequence was noticed and it was proposed that the receptor bound as a dimer (Scheidereit and Beato, 1984). Using the recombinant GR preparation DBD$_r$, it could be seen that the protein bound first to one half of the GRE and then to the other half in a cooperative manner (Tsai et al., 1988). This was shown by gel retardation and methylation interference assays. However, using the intact GR or progestin receptor, binding to the entire GRE was seen and it was concluded that the intact receptors exist as stable dimers already in solution (Tsai et al., 1988). It would seem that other domains of the receptor, apart from the DNA-binding domain, are of importance for stabilizing the dimer. Similar results have also been reported for the estrogen receptor (Kumar and Chambon, 1988).

6. Closing Remarks

The ability to purify GR in an active form has been shown to be a powerful tool. Not only has it enabled some structural studies of the protein, particularly with regard to its various functions, but it has also enabled studies on the mechanism of gene regulation. These results are not only of interest for an understanding of the mechanism of action of steroid hormones, but also are of more general interest for an understanding of gene regulation in eukaryotic cells. The steroid receptors, and in particular GR, are among the more readily available transcriptional factors today. The ability to produce large quantities of functional domains of these proteins should prove of great use in these studies. The recombinant proteins should also enable further structural studies of the receptor proteins using, for instance, X-ray crystallography or 2-D NMR. Such studies using DBD$_r$ are at present in progress.

Acknowledgments

The work described in this chapter was supported by grants from the Swedish Medical Research Council (grant 2819) and the Swedish National Board for Technical Development (grant 404).

References

Akerblom, I. E., Slater, E. P. Beato, M., Baxter, J. D., and Mellon, P. L. (1988) *Science* **241,** 350–353.

Alberts, B. and Herrick, G. (1971) *Methods Enzymol.* **21,** 198–217.

Atger, M. and Milgrom, E. (1976) *Biochemistry* **241,** 4298–4304.

Baxter, J. D., Rousseau, G. G., Benson, M. C., Garcea, R. L., Ito, J., and Tomkins, G. M. (1972) *Acad. Sci. USA* **69,** 1892–1896.

Beato, M., Arnemann, J., Chalepakis, G., Slater, E., and Willmann, T. (1987) *J. Steroid Biochem.* **27,** 9–14.

Becker, P. B., Gloss, B., Schmid, W., Strähle, U., and Schütz, G. (1986) *Nature* **324,** 686–688.

Berg, J. M. (1986) *Science* **232,** 485–487.

Bonifer, C., Dahlman, K., Lanzer, M., Bujard, H., Flock, J.-I., and Gustafsson, J.-Å. (1988) in *E. coli. Proc. UCLA Symp. on Steroid Hormone Action* **75,** 201–210.

Carlstedt-Duke, J., Gustafsson, J.-Å., and Wrange, Ö. (1977) *Biochim. Biophys. Acta* **497,** 507–524.

Carlstedt-Duke, J., Wrange, Ö., Dahlberg, E., Gustafsson, J.-Å., and Högberg, B. (1979) *J. Biol. Chem.* **254,** 1537–1539.

Carlstedt-Duke, J., Okret, S., Wrange, Ö., and Gustafsson, J.-Å. (1982) *Proc. Natl. Acad. Sci. USA* **79,** 4260–4264.

Carlstedt-Duke, J., Strömstedt, P.-E., Wrange, Ö., Bergman, T., Gustafsson, J.-Å., and Jörnvall, H. (1987) *Proc. Natl. Acad. Sci. USA* **84,** 4437–4440.

Carlstedt-Duke, J., Strömstedt, P.-E., Persson, B., Cederlund, E., Gustafsson, J.-Å., and Jörnvall, H. (1988) *J. Biol. Chem.* **263,** 6842–6846.

Carlstedt-Duke, J. Strömstedt, P.-E., Dahlman, K., Rae, C., Berkenstam, A., Hapgood, J., Jörnvall, H., and Gustafsson, J.-Å. (1989) in *The Steroid/Thyroid Hormone Receptor Family and Gene Regulation* (Carlstedt-Duke, J., Eriksson, H., and Gustafsson, J.-Å., eds.), Birkhäuser, Basel, pp. 93–108.

Dahlman, K., Strömstedt, P.-E., Rae, C., Jörnvall, H., Flock, J.-I., Carlstedt-Duke, J., and Gustafsson, J.-Å. (1989) *J. Biol. Chem.* **264,** 804–809.

Dellweg, H.-G., Hotz, A., Mugele, K., and Gehring, U. (1982) *EMBO J.* **1,** 285–289.

Denis, M., Poellinger, L., Wikström, A.-C., and Gustafsson, J.-Å. (1988) *Nature* **333,** 686–688.

Evans, R. M. (1988) *Science* **240,** 889–895.

Freedman, L. P., Luisi, B. F., Korszun, Z. R., Basavappa, R., Sigler, P. B., and Yamamoto, K. R. (1988) *Nature* **334,** 543–546.

Geisse, S., Scheidereit, C., Westphal, H. M., Hynes, N. E., Croner, B., and Beato, M. (1982) *EMBO J.* **1,** 1613–1619.

Giguère, V., Hollenberg, S. M., Rosenfeld, M. G., and Evans, R. M. (1986) *Cell* **46,** 645–652.

Godowski, P. J., Rusconi, S., Miesfeld, R., and Yamamoto, K. R. (1987) *Nature* **325,** 365–368.

Govindan, M. V., Spiess, E., and Majors, J. (1982) *Proc. Natl. Acad. Sci. USA* **79,** 5157–5161.

Grippo, J. F., Tienrungroj, W., Boborodea, V., Pratt, S. E., Carlson, M. D., and Pratt, W. B. (1986) in *Thioredoxin and Glutaredoxin Systems: Structure and Function* (Holmgren, A., Bränden, C.-I., Jörnvall, H., and Sjöberg, B.-M., eds.), Raven, New York, pp. 377–390.

Higgins, J. J., Rousseau, G. G., Baxter, J. D., and Tomkins, G. M. (1973) *J. Biol. Chem.* **248,** 5866–5872.

Hollenberg, S. M., Weinberger, C., Ong, E. S., Cerelli, G., Oro, A., Lebo, R., Thompson, E. B., Rosenfeld, M. G., and Evans, R. M. (1985) *Nature* **318,** 635–641.

Hollenberg, S. M., Giguère, V., Segui, P., and Evans, R. M. (1987) *Cell* **49,** 39–46.

Jantzen, H.-M., Strähle, U., Gloss, B., Stewart, F., Schmid, W., Boshart, M., Miksicek, R., and Schütz, G. (1987) *Cell* **49,** 29–38.

Kadonaga, J. T. and Tjian, R. (1986) *Proc. Natl. Acad. Sci. USA* **83,** 5889–5893.

Kalimi, M., Colman, P., and Feigelson, P. (1975) *J Biol. Chem.* **250,** 1080–1086.

Kumar, V. and Chambon, P. (1988) *Cell* **55,** 145–156.

Mainwaring, W. I. P. and Irving, R. (1973) *Biochem. J.* **134,** 113–127.

Marston, F. A. O. (1986) *Biochem. J.* **240,** 1–12.

Miesfeld, R., Okret, S., Wikström, A.-C., Wrange, Ö., Gustafsson, J.-Å., and Yamamoto, K. R. (1984) *Nature* **312,** 779–781.

Miesfeld, R., Rusconi, S., Godowski, P. J., Maler, B. A., Okret, S., Wikström, A.-C., Gustafsson, J.-Å., and Yamamoto, K. R. (1986) *Cell* **46,** 389–399.

Miesfeld, R., Godowski, P. J., Maler, B. A., and Yamamoto, K. R. (1987) *Science* **236,** 423–427.

Miller, J., McLachlan, A. D., and Krug, A. (1985) *EMBO J.* **4,** 1609–1614.

Nilsson, B., Abrahamsén, L., and Uhlén, M. (1985) *EMBO J.* **4,** 1075–1080.

Okret, S., Carlstedt-Duke, J., Wrange, Ö., Carlström, K., and Gustafsson, J.-Å. (1981) *Biochim. Biophys. Acta* **677,** 205–219.

Okret, S., Wikström, A.-C., Wrange, Ö., Andersson, B., and Gustafsson, J.-Å. (1984) *Proc. Natl. Acad. Sci. USA* **81,** 1609–1613.

Payvar, F., Wrange, Ö., Carlstedt-Duke, J., Okret, S., Gustafsson, J.-Å., and Yamamoto, K. R. (1981) *Proc. Natl. Acad. Sci. USA* **78,** 6628–6632.

Payvar, F., Firestone, G. L., Ross, S. R., Chandler, V. L., Wrange, Ö., Carlstedt-Duke, J., Gustafsson, J.-Å., and Yamamoto, K. R. (1982) *J. Cell. Biochem.* **19,** 241–247.

Payvar, F., DeFranco, D., Firestone, G. L., Edgar, B., Wrange, Ö., Okret, S., Gustafsson, J.-Å., and Yamamoto, K. R. (1983) *Cell* **35,** 381–392.

Pfahl, H. (1982) *Cell* **31,** 475–482.

Potuzak, H. and Dean, P. D. G. (1978) *FEBS Lett.* **88,** 161–166.

Rousseau, G. G., Higgins, S. H., Baxter, J. D., Gelfand, D., and Tomkins, G. M. (1975) *J. Biol. Chem.* **250,** 6015–6021.

Rusconi, S. and Yamamoto, K. R. (1987) *EMBO J.* **6,** 1309–1315.

Sakai, D. D., Helms, S., Carlstedt-Duke, J., Gustafsson, J.-Å., Rottman, F. M., and Yamamoto, K. R. (1980) *Genes Delevop.* **2,** 1144–1154.

Scheidereit, C. and Beato, M. (1984) *Proc. Natl. Acad. Sci. USA* **81,** 3029–3033.

Strähle, U., Schmid, W., and Schütz, G. (1988) *EMBO J.* **7,** 3389–3395.

Tsai, S. Y., Carlstedt-Duke, J., Weigel, N., Dahlman, K., Gustafsson, J.-Å., Tsai, M.-J., and O'Malley, B. W. (1988) *Cell* **55,** 361–369.

Willmann, T. and Beato, M. (1986) *Nature* **324,** 688–691.

Wrange, Ö. and Gustafsson, J.-Å. (1978) *J. Biol. Chem.* **253,** 856–865.

Wrange, Ö., Carlstedt-Duke, J., and Gustafsson, J.-Å. (1979) *J. Biol. Chem.* **254,** 9284–9290.

Wrange, Ö., Okret, S., Radojcic, M., Carlstedt-Duke, J., and Gustafsson, J.-Å. (1984) *J. Biol. Chem.* **259,** 4534–4541.

Wrange, Ö., Carlstedt-Duke, J., and Gustafsson, J.-Å. (1986) *J. Biol. Chem.* **261,** 11770–11778.

Yamamoto, K. R. (1985) *Ann. Rev. Genet.* **19,** 209–252.

Glucocorticoid Receptor Purification

Patrick Lustenberger
and Pierre Formstecher

1. Introduction

1.1. Properties of the Glucocorticoid Receptor

Glucocorticoids, like other steroid hormones, exert their biological actions after binding to a specific intracellular soluble receptor (Yamamoto and Alberts, 1976). This protein is found in cytosol prepared from ruptured tissues or cells, in a non-DNA-binding form, called nontransformed receptor (or nonactivated receptor). In target cells, binding of the steroid agonist to its receptor implies a modification in the receptor's ability to interact with DNA (Rousseau et al., 1973; Munck and Foley, 1979; Markovic and Litwack, 1980; Miyabe and Harrison, 1983). This latter form of the glucocorticoid receptor is commonly named transformed-receptor (or activated-receptor).

Both forms differ in numerous aspects, but not in steroid binding. Nontransformed cytosolic glucocorticoid receptor of target cells has a sedimentation coefficient of $\cong 9\,S$ and a Stokes radius of 7–8 nm, consistent with an oligomeric structure having a mol wt of $\cong 300,000\,kDa$ under nondenaturing conditions (Sherman and Stevens, 1984; Holbrook et al., 1985). This large 9 S complex can be stabilized in the presence of molybdate, or other transition metal oxyanions from the group VI-A (Leach et al., 1979; Dahmer et al., 1984). The nuclear binding ability (transformation or acti-

Receptor Purification, vol. 2   ©1990 The Humana Press

vation) can be acquired in vitro under a variety of conditions (e.g., elevation of tem-perature or ionic strength of the cytosol, dilution of cytosol, or alteration in pH). The above treatments of the cytosol are thought to cause a subunit dissociation (Vedeckis, 1983) and the loss of a 90kDa nonsteroid binding phosphoprotein identified as the 90kDa heat shock protein (Mendel et al., 1986; Denis et al., 1988b). Physicochemical analysis of the transformed complexes demonstrates a M_r of 100,000 Da (sedimentation coefficient of $\cong 4\,S$ and Stokes radius of 5–6 nm). Analysis under denaturing conditions reveal that the steroid-binding entity, covalently labeled with either the affinity label dexamethasone mesylate (Simons and Thompson, 1981) or the photoaffinity labels triamcinolone acetonide (Dellweg et al., 1982; Gehring and Hotz, 1983) or promegestone (Nordeen et al., 1981) is a M_r 94,000 Da protein in rat liver (Carlstedt-Duke et al., 1987), in human Hela S 3 cells (Brönnegard et al., 1987) and in mammalian cells in general (Gronemeyer and Govindan, 1986).

Numerous results that clarify the structure of both forms (transformed and nontransformed) of the glucocorticoid receptor and the mechanism of transformation have been published and have led to progress in the purification of the receptor by conventional techniques. Moreover, receptor purification also has improved with affinity chromatography, and, more recently, with high performance liquid chromatography. The present review is focused on various methods of glucocorticoid receptor purification. This will include partial purification and more complete purification of both forms of the glucocorticoid receptor.

1.2. Approaches to Glucocorticoid Receptor Purification

It is clear that the glucocorticoid receptor may be found in three distinct forms in crude cytosol in accordance with the following scheme:

$$R \underset{\longleftarrow}{\overset{\longrightarrow}{}} RH \longrightarrow R'H$$

Approaches to receptor purification can focus on (i) purifying only one form of receptor, this can be achieved whether by maintaining a single form throughout purification or by selecting one of the forms in the last step; (ii) obtaining the appropriate molecular form in view of the experiments envisioned. It is difficult to reverse the conversions depicted. Once the receptor is bound to the steroid, dissociation to produce free receptor or steroid exchange is not easily accomplished; likewise, reversion of the transformed form to the heteromeric 8–9 S nontransformed form does not appear to be feasible at present.

In a previous review on this subject, Santi et al. (1979) described the difficulties precluding glucocorticoid receptor purification. Among these, the limited sources of tissue abundant in receptor, the necessity of adrenalectomy, the presence of other steroid binding proteins than receptor, and the high degree of purification required to obtain homogenous preparations still remain unresolved. The only improvement is relevant to the stability of the glucocorticoid receptor. In fact, addition of molybdate allows stabilization of all forms of the receptor and blocking of the transformation process (Dahmer et al., 1984). This is of particular importance, especially in view of purification. Several purifications of the glucocorticoid receptor have been reported. Our main contribution in this field was the careful design of an affinity adsorbent suitable for this purification (Lustenberger et al., 1981; Idziorek et al., 1985; Lustenberger et al., 1985). In this paper, we will summarize procedures and new approaches that appear to be reproducible and generally applicable.

2. Partial Purification of the Receptor

The use of partially purified preparations of glucocorticoid receptor is advantageous because it should reduce some complications inherent in using crude cytosol. These complications are mainly attributed to nonspecific binding proteins or to proteases.

The crude cytosol is normally rather dilute, contains a considerable number of different proteins and has a large volume. The fractionation step chosen must be able to handle the amount of material present and, if possible, concentrate the sample for further processing.

We have investigated both precipitation and adsorption methods, with bound or unbound glucocorticoid receptor.

2.1. Precipitation

2.1.1. Ammonium Sulfate Precipitation

This method was used as the initial step in the first attempts of glucocorticoid receptor (Hackney and Pratt, 1971; Rousseau et al., 1975) and progesterone receptor purification (Kuhn et al., 1975). According to their results, the precipitation of steroid bound receptor was obtained at 25–33% saturation of ammonium sulfate, bringing a 10–40-fold purification with 70–80% yield. The method can be used to precipitate either steroid-receptor complexes or uncomplexed receptors.

In the case of glucocorticoid receptor, our procedure is as follows. The crude liver cytosol is brought to 40% saturation ammonium sulfate by dropwise addition of saturated ammonium sulfate. After 30 min, the precipitate is collected by centrifugation and redissolved in $0.05M$ borate buffer pH 7.4 (with $0.01M$ molybdate, $0.02M$ β mercaptoethanol, and 20% v/v glycerol). At this concentration of ammonium sulfate, corticosteroid-binding globulin (CBG) is partly precipitated. The recovery is about 60–70% for a two- to threefold purification.

2.1.2. Protamine Sulfate Precipitation

Originally described by Rousseau et al. (1975) the procedure was also used by Govindan and Manz (1978). After elimination of nucleic acids by streptomycin, the crude cytosolic receptor was purified five- to tenfold in about 75–90% yield by precipitation with protamine sulfate. Protamine was removed by ion exchange chromatography on carboxymethyl cellulose.

Protamine sulfate (0.75%) in water is added dropwise to stirred cytosol (prepared in hypotonic buffer) in ice until the final solution is 0.075% in protamine. After 30 min, the precipitate is collected by centrifugation at 35,000 g for 15 min. The receptor pellets are redissolved in $0.16M$ phosphate buffer pH 7.4 (including $0.01M$ molybdate, $0.02M$ β mercaptoethanol, and 20% v/v glycerol).

Various alternatives have been tested as substitute for the high molarity phosphate buffer. Results are presented in Fig. 1. In the case of phosphate buffers, the optimal molarity is $0.16M$. When Tris buffers are used, only the addition of $0.2M$ molybdate allows the resolubilization of the receptor. Another way that avoids a high salt concentration is a heparin solution. A concentration of 1 mg/mL heparin in $0.02M$ Tris buffer is satisfactory for receptor resolubilization in good yield. Whatever the procedure chosen, the purification does not exceed tenfold.

The protamine sulfate extracts can be frozen at –80°C and stored for weeks with no detectable change in hormone binding activity. Moreover, it must be noted that CBG is not precipitated by protamine.

2.1.3. Other Precipitation Methods

Combining protamine sulfate precipitation and ammonium sulfate precipitation is possible but does not spectacularly improve the purification of glucocorticoid receptor, essentialy because of the poor yield. The polyethylenimine precipitation described for the 1,25-dihydroxycholecalciferol receptor (Pike and Haussler, 1979) was unsuccessful when applied to the glucocorticoid receptor.

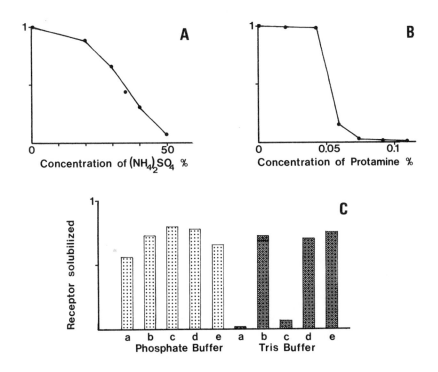

Fig. 1. Precipitation of the glucocorticoid receptor. Aliquots of rabbit liver cytosol prepared in 0.02M phosphate buffer were added with **A** saturated ammonium sulfate or **B** 0.75% protamine sulfate solution, until the desired final concentrations. After 30 min, the supernatant fractions were recovered by centrifugation. Specific binding was measured after a 16 h incubation at 0–4°C in the presence of $2 \times 10^{-8}M$ [^{3}H]dexamethasone. Results are expressed as ratio of binding in the supernatants (B) to binding in the initial cytosol (B$_o$). **C** Unbound glucocorticoid receptor was precipitated in aliquots (2 mL) of cytosol by 0.075% protamine sulfate. After 30 min, the precipitates were recovered by centrifugation and redissolved in 0.4 mL buffer. Specific binding was measured as previously. Results are expressed as ratio of receptor recovered to receptor in the initial cytosol.

Phosphate buffer	Tris buffer
a. 0.12M phosphate buffer	a. 0.02M Tris buffer
b. 0.14M phosphate buffer	b. 0.02M Tris buffer + 0.2 M molybdate
c. 0.16M phosphate buffer	c. 0.02M Tris buffer + 0.5 mg/mL heparin
d. 0.18M phosphate buffer	d. 0.02M Tris buffer + 1 mg/mL heparin
e. 0.20M phosphate buffer	e. 0.02M Tris buffer + 2 mg/mL heparin

All buffers were adjusted to pH 7.4 at 4°C and contained 0.02M β-mercaptoethanol, 0.01M molybdate, and 20% (v/v) glycerol.

Finally, we have no experience with polyethylene glycol precipitation of the glucocorticoid receptor. When applied to the progesterone receptor, this procedure allows a twofold purification with 80% yield of the unliganded form (Maggi et al., 1981).

2.2. Chromatographic Procedures

2.2.1. Ion Exchange Chromatography

An alternative to precipitation is adsorption. Ion exchange chromatography is a useful choice for this stage. Glucocorticoid receptor, which is about 5.1–5.4 pI (Ben-Or and Chrambach, 1983; Idziorek et al., 1985; Lustenberger et al., 1985) is negatively charged at physiological pH, so an anion exchanger is normally appropriate. Therefore, DEAE exchangers have been largely used. DEAE purifications were usually performed in column with a linear salt (NaCl or KCl) gradient elution. In these conditions, purification of molybdate stabilized steroid bound glucocorticoid receptor is about four- to eightfold. Yields are typically 80% or more. In case of unbound receptor, recovery is lowered to 50–60%. Rather than a column procedure, the batch adsorption can be used as a rapid alternative. The procedure for batch adsorption is to mix the cytosol (2.5 vol) with DEAE-trisacryl (1.0 vol). After 15 min stirring at 4°C, the gel is then washed three times with 3 vol buffer (0.02M phosphate buffer, pH 7.4 with 0.05M NaCl, 0.01M molybdate and 20% glycerol). The receptor is desorbed at high ionic strength (0.16M phosphate buffer, 2 × 1 vol). This procedure has proved to be very efficient especially for unbound receptor, allowing a 20-fold purification with 60% yield. Synthetic ion exchangers like DEAE-Trisacryl show exceptional chromatographic properties and give the best results up to now. Nevertheless, new silica based supports like Acell® (Waters) or Spherosil® (IBF) require to be evaluated.

Nontransformed glucocorticoid receptor does not interact with polyanionic exchangers like carboxymethylcellulose, phosphocellulose, or DNA cellulose. This kind of supports has proved very efficient for the purification of the transformed receptor (*see* Section 3.2.). When used as prepurification step, the treatment with polyanionic exchangers leads to a minimal 1.2–1.5-fold purification with 90–100% yield. But as stated by Moudgil et al. (1985), this treatment removes many proteolytic enzyme activities that adsorb to phosphocellulose and thus mimimizes further degradation of the receptor. Moreover, it also removes a significant part of DNA-binding proteins, an interesting feature when the purified receptor is then transformed and used for binding experiments to specific DNA fragments. In order to avoid transformation of the receptor to the DNA binding form (Atger and Milgrom, 1976), the flowrate must be high, or molybdate must be added during the procedure.

2.2.2. Hydroxylapatite Chromatography

Hydroxylapatite is widely used in preparative biochemistry. Because of its chemical structure, hydroxylapatite interacts with negatively charged moieties on the protein surface as well as with positively charged protein groups. The ability of hydroxylapatite to adsorb most proteins, including the glucocorticoid receptor, from solutions of low ionic strength, and to desorb them at higher salt concentrations, is the basis for a method for receptor purification. Partially purified receptor can be eluted stepwise from hydroxylapatite in excellent yield using $0.3M$ phosphate buffer. A two- threefold purification is achieved by this procedure.

2.2.3. Heparin Sepharose
Chromatography and Apparented Procedures

Heparin is principally composed of glucosamine and glucuronic acid, whose certain hydroxylic groups are esterified with sulfuric acid or sulfated. As a result of its composition and biochemical role, heparin has the property of combining with a number of proteins, enzymes, and polycationic organic compounds in general, the interactions being either of specific or of electrostatic nature. Because of the above-mentioned particularities, heparin immobilized on insoluble carriers has been exploited in the field of affinity chromatography.

Initially described for the prepurification of estradiol receptor (Molinari et al., 1977; Secco et al., 1979; Sica and Bresciani, 1979; Ratajczak and Hähnel, 1980), this procedure allowed to remove nonspecific esterases present in the cytosol. Heparin immobilized on agarose can also be used to isolate glucocorticoid receptor (Blanchardie et al., 1984; Weisz et al., 1984).

Both forms (transformed and untransformed) of cytosolic steroid bound glucocorticoid receptor have been purified. In presence of $0.05M$ Mg^{2+} ions, the purification factor is 70-fold with 80% yield for the nontransformed form. Figure 2 shows the result of a typical experiment of purification. Elution is obtained with a linear gradient of NaCl ($0-0.5M$) or heparin ($0-0.5$ mg/mL). Stepwise elution is also practicable leading to identical purification (Blanchardie et al., 1984). The transformed receptor has a higher affinity for immobilized heparin, and, therefore is eluted at $0.37M$ NaCl (vs $0.17M$ for the nontransformed form). This results in a slightly better purification factor. When the binding of ligand-free receptor was tested, poor results were obtained, probably as a result of the

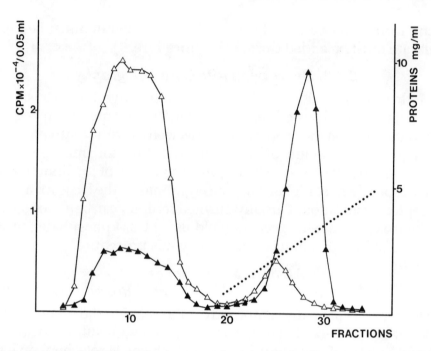

Fig. 2. Heparin Sepharose chromatography. A column (5 × 1.1 cm) was filled with heparin-Ultrogel (IBF) and equilibrated in 0.02M phosphate buffer (pH 7.4, 0.02M phosphate, 0.05M MgCl$_2$, 0.02M β-mercaptoethanol, 0.01M molybdate and 20% (v/v) glycerol). [³H]dexamethasone labeled rat liver cytosol (8.5 mL) was treated with dextran-charcoal and applied at 6.0 mL/h flowrate. After washing with 5 mL buffer, elution was obtained using a 10 mL NaCl gradient (0–0.3M in phosphate buffer). Fractions collected were 1.0 mL each. Radioactivity (▲) was measured for each fraction (0.05 mL) and proteins (△) were assayed using the Coomassie Brilliant Blue procedure. Salt gradient is depicted by the dotted line. The initial receptor concentration in cytosol was 0.40 pmol/mg protein. In the pooled fractions (24–30) the concentration was 28.2 pmol/mg protein.

inactivation of the receptor by heparin (McBlain and Shyamala, 1984; Hubbard and Kalimi, 1983). Another major drawback is represented by the limited capacity and hydrodynamic properties of agarose that make the procedure time consuming. The immobilization of heparin on more rigid particles would yield a stationary of higher capacity and allow the high flowrates precluded by the gel compression typically observed with conventional agarose based media.

Beside heparin, other "specific ion exchangers" have been tested. Schrader (1975) report the chromatography of the progesterone receptor on protamine-Sepharose. The receptor is eluted at high salt molarity (0.8M KCl) with 90% yield corresponding to a 66-fold purification. Recently, the same bioselective matrice has been shown to interact with the glucocorticoid receptor (Nemoto et al., 1987). The nontransformed glucocorticoid receptor was bound by protamine Sepharose and arginine Sepharose via the nonhormone binding component corresponding to the 90kDa heat shock protein. No information were available about the performance of the purification. It could be attractive to further test this possibility.

2.2.4. Dye-Ligand Chromatography

Since the introduction of Cibacron Blue for the purification of nucleotides cofactor enzymes, dye-ligand chromatography has greatly expanded and its numerous applications include the purification of nearly two hundred enzymes and proteins (Dean and Watson, 1979; Subramanian, 1984). Mellon (1984) has pioneered in dye-ligand interactions with the 1,25-dihydroxycholecalciferol receptor. He demonstrated that steroid receptor complexes are retained on several immobilized triazinyl dyes: Cibacron Blue F3GA, Procion Red HE3B, and Green A dye. Beside this work, triazinyl dyes were first used as inhibitors of receptor transformation, particularly in the cases of estrogen, androgen, and progesterone receptors, as reviewed by Grody et al. (1982). Cibacron Blue-Sepharose chromatography was used for removing albumin from the cytosol but resulted in low recovery of estradiol receptor (Ratajczak and Hähnel, 1980). More recently, affinity immobilization on Cibacron Blue Sepharose was proposed by Iqbal et al. (1985) for the microassay of estrogen or androgen receptors.

Several dye-ligand have been examined for purifying glucocorticoid receptor complexes from rabbit liver in our laboratory. Only Procion Red HE3B was found to be useful in this purification. Cibacron Blue F3GA, Green A, Orange A and Blue B dyes that display different selectivity and binding properties were found ineffective.

As can be seen in Fig. 3, the [^{3}H]dexamethasone-receptor complexes formed at 0–4°C are retained by immobilized Procion Red. Addition of 0.05M divalent cation Mg^{2+} enhanced the binding fivefold. Elution with a salt gradient (0–1M KCl) resulted in a peak of specifically bound radioactivity eluting maximally at 0.25M KCl. We have obtained a four- sixfold purification with 70% yield for the nontransformed glucocorticoid receptor. Biospecific elution with nucleotides (AMP, GMP, ADP, ATP, NAD at 0.025M) was unsuccessful. We have not attempted to further optimize the conditions of purification or to study the behavior of unbound receptor.

2.2.5. Hydrophobic-Interaction Chromatography

The use of *n*-alkyl agarose was proposed by Santi et al. (1979). Using hydrocarbon coated Sepharose of varying chain lengths, it has been shown that the ethyl and *n*-butyl agarose columns retain virtually all of the recep-

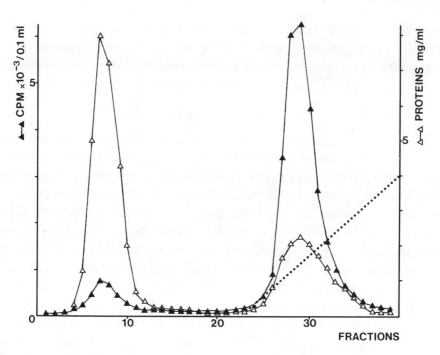

Fig. 3. Dye-ligand chromatography. A 4 mL aliquot of labeled rabbit liver cytosol (treated with dextran-charcoal) was applied to a Red A column (5 × 1.1 cm; 5 mL gel). Chromatography was initiated by washing the column with 15 mL equilibrating buffer (0.02M phosphate, 0.05M $MgCl_2$, 0.02M β-mercaptoethanol, 0.01M molybdate and 20% (v/v) glycerol) at 6.0 mL/h flowrate. A linear KCl gradient (0–1M) of 10 mL was used to elute the column. Fractions (1 mL) were collected and 0.1 mL aliquots were removed to determine the radioactivity (▲). The proteins (△) were assayed using the Coomassie Brilliant Blue procedure. Salt gradient is depicted by the dotted line. The initial parameters of cytosol were: 0.18 pmol receptor/mg proteins, and 2.5 pmol/mL cytosol. In the pooled fractions (25–32), the concentration of receptor was 0.5 pmol/mg proteins and 0.87 pmol/mL.

tor. More hydrophobic columns have the same performance, except that nonspecific binding is higher. Elution is obtained with increasing the ionic strength of the buffer. The performances, seven- tenfold purification with 60–70% recovery, appear satisfactory. The only drawback comes from the limited capacity of the n-alkyl agarose. In the case of estrogen receptor, a similar gel, i.e., octyl-agarose was found to denature the receptor likely during the elution with a 1% sodium caprylate solution (Ratajczak and Hähnel, 1980). Fractionation of bound protein by decreasing the polarity of the eluent (e.g., with ethylene glycol) have not been tested. Phenyl-Sepharose chromatography has been used to study the hydrophobic properties of the glucocorticoid receptor. Because of the steroid-receptor complex dissociation induced by the high concentration of ammonium sulfate, the procedure can be applied only to covalently labeled receptor. The purification abilities of phenyl-Sepharose chromatography have not been yet evaluated (Sablonnière et al., submitted).

2.2.6. Gel Filtration Chromatography

Because of its limited capacity, gel filtration chromatography was not often retained as a first fractionation step. Nevertheless, Santi et al. (1979) claimed that it is one of the most successful procedure allowing a five-eightfold purification with 80–90% yield. The major drawbacks of conventional gel filtration (poor resolution and low flowrate) have been removed by the development of high performance size exclusion chromatography procedures. In view of preparative use, the preparative large TSK G 4000 column designed for high performance size exclusion chromatography afforded good separations. Up to 2 mL of cytosol or biological samples could be injected. The recoveries were in the range 70–90% (Sablonnière et al., 1987).

Other separative techniques, such as electrophoresis or sucrose gradient ultracentrifugation, remain useful analytical tools. Because of the limited size of the sample, application to prepurification remains exceptional.

Although the purification afforded by all these methods is thus far relatively low, they provide rapid methods of removing 90% of extraneous proteins in cytosol. This might be useful in reducing complications caused by nonspecific binding proteins in experiments dealing with receptor structure and function, as well as in subsequent purification steps.

3. Global Purification of Glucocorticoid Receptor

Partial purification procedures can be applied to all three forms of receptor. On the other hand, only steroid-bound receptor can be purified to near homogeneity.

3.1. Purification of Nontransformed Glucocorticoid Receptor by Affinity Chromatography

Because of the reversibility of the binding of steroid to the glucocorticoid receptor, affinity chromatography appears as a suitable procedure for the specific separation of receptor from other proteins. First attempts were unsuccessful mainly because of the structure of the affinity matrixes, the ligand being immobilized through a labile ester linkage. Convincing results were obtained only when efficient and stable affinity gels were proposed.

In a recent review (Formstecher and Lustenberger, 1987), we have described the various steps of the design of specific adsorbents. Therefore, we will only briefly summarize these results.

3.1.1. Design of the Affinity Matrix

Our experience led us to propose the following tentative rules for the design of specific affinity matrixes (Lustenberger et al., 1981):

1. Select a ligand specific for the glucocorticoid receptor. Dexamethasone, a synthetic fluorinated glucocorticoid, was preferred to cortisol or corticosterone, the natural ligand of the receptor, but also of plasma transcortin, to which dexamethasone does not bind.
2. Select the linkage point of the spacer arm in such a way that the specificity is not radically modified. The 17 β-position of dexamethasone allowed to preserve the features (1–2 insaturation; 9 α-fluor, 16 α-methyl- and 11 β-hydroxyl-substitutions) that promote high affinity for the receptor and prevent binding to transcortin (Rousseau et al., 1979).
3. Introduce at this position a carboxylic function that will easily react with an amino spacer arm. The resulting amide bond is very stable and not susceptible to hydrolysis. This is easily accomplished by periodic oxydation of dexamethasone leading to the 17 β-carboxylic acid, which can be then connected to primary amines using classical peptide synthesis reagents. Dexamethasone spacer derivatives were prepared by using various primary monoamines.
4. Use a long hydrophobic spacer arm, the simplest being a linear saturated aliphatic chain (at least nine carbons). This kind of spacer arm does not drastically change the affinity for the receptor, since the binding site is globally hydrophobic. Moreover, aliphatic diamines are commercially available in a wide range of lengths, which limit the requirement of chemical synthesis.
5. Check that the immobilized ligand will allow adsorption and elution of the receptor. This can be easily tested with dexamethasone derivatives by competitive binding experiments and exchange experiments between this compound and [^{3}H]dexamethasone. In the case of the 17 β-carboxamide derivative substituted with nonylamine, the affinity is suitable, with a $K \cong 0.4 \ \mu M$, and this compound is easily exchangeable with a $t1/2 \cong 8$ h.

All of the above is related to the ligand and the spacer arm. Nevertheless, efficient affinity adsorbents can only be obtained if precautions are taken during the preparation of the gels.

1. Select a chromatographic media with good properties. Agarose is well suited for affinity chromatography purpose. Chemically stabilized crosslinked agarose (Sepharose CL4B, Pharmacia) is of particular in-

terest because it displays good stability in organic solvent, which is the media required for the coupling of the ligand.

2. Select a low degree of ligand substitution. Because of the high affinity and specificity of the immobilized dexamethasone derivative, a high specific saturation of the derivatized beads can be achieved (i.e., a high percentage of available ligand sites effectively occupied by a receptor molecule). This will be beneficial for the purification efficiency, by increasing the ratio of specific vs nonspecific binding and allowing easier elution. Moreover, elution in a concentrated form can be obtained.

3. Prepare the affinity matrix with careful monitoring of each step (CNBr activation and assay of activated species; coupling of the spacer arm and assay of free amino groups; coupling of the ligand and determination of the bound steroid). It is the only way to obtain an affinity matrix with the desired degree of substitution and devoid of ionic exchange or hydrophobic properties that are related to persistance of free spacer arm or to excessive activation.

The careful design of all the tools and steps involved in the preparation of affinity matrixes tailor-made for the purification of the glucocorticoid receptor have to be done to obtain the most efficient gels (Fig. 4).

3.1.2. Purification Procedure

The purification procedure that we developed, is described in details in previous publication (Idziorek et al., 1985; Lustenberger et al., 1985). Briefly, the three step procedure is as follows:

1. Prepurification by protamine sulfate precipitation. The usefulness of this step does not lie in the degree of purification afforded, but rather in the concentration of the biological material and in the elimination of extraneous protein to a great extent. At this stage, the glucocorticoid receptor has been extracted from cytosol and some of the major contaminants have been removed. Moreover, the protamine sulfate extract is rather stable when stored at –70°C. The overall performance of this step as compared to others are described in Section 2.1.

2. The next stage is represented by the affinity chromatography step. The adsorption of the receptor is performed using the batch procedure with a ratio of 1 vol gel for 50 vol biological material. This procedure was preferred to the column one because of the high affinity of the ligand and the relatively poor stability of the unliganded receptor. The various parameters, i.e., steroid concentration of the gel, length time of the batch, ratio of cytosol/gel volumes were carefully studied (Lustenberger et al., 1981; Formstecher and Lustenberger, 1987). The finality lies in the necessity to elute the receptor in a concentrated form that could be more easily handled for subsequent experiments.

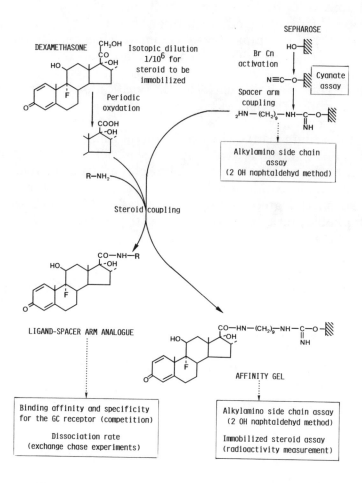

Fig. 4. Design of a specific affinity matrix. Controls and experiments necessary appear in the framed areas.

Washing of the loaded gel before elution is of critical importance for the purity of the final product. The foreign extraneous proteins entrapped or nonspecifically bound to the gel might represent the main part of the eluted proteins. Thus, rapid and efficient washing procedures displaying the maximum stringency possible without loss or damage of the receptor have been developed. This included a sequence of buffer with low and high ionic strengths at 4° or 25°C. The entire procedure was performed in column and lasted less than 1 h.

Biospecific elution is performed with tritiated dexamethasone or triamcinolone acetonide (concentration 2–5 µMol in 1–4 gel volumes) at 0°C for the rat liver receptor or 20°C for the rabbit liver receptor. Because of the low concentration of the immobilized ligand, the reassociation of the receptor with immobilized ligand sites is nearly excluded. Therefore, it is not necessary to resort to chaotropic agents or organic solvents to help dissociation. Purifications of 300- to 2000-fold with 20–40% recovery were usually obtained.

Two kinds of affinity matrixes have been developed for the glucocorticoid receptor purification. On the one hand, desoxycorticosterone gels described by Failla et al. (1975) and by Grandics et al. (1984) and marketed as Sterogel A® have been used by several groups (Weisz et al., 1984; Webb et al., 1985; Housley et al., 1983,1985; Singh and Moudgil, 1985). On the other hand, we first described dexamethasone matrixes (Lustenberger et al., 1981) that were since widely used (Lustenberger et al., 1985; Idziorek et al., 1985; Govindan and Gronemeyer, 1984; Rousseau et al., 1987; Lombes et al., 1987; Hapgood and Von Holt, 1987). The main differences between these two affinity procedures lie essentially in the following.

1. In the specific saturation obtained after the adsorption step: 6.6–13.2% for dexamethasone gels vs 0.007–0.25% for DOC gels (expressed as ratio of the sites occupied by receptor to total sites available);
2. In the concentration of receptor in the eluate: 60–260 pmol/mL vs 5–25 pmol/mL for dexamethasone and DOC gels, respectively;
3. In the performance of the purification: 400- 2000-fold vs 100- 300-fold, respectively. This relies on the high specificity of dexamethasone, as compared to DOC.

In our opinion, the results of receptor purification have to report not only the purification factor and the yield but also the actual concentration of the purified protein. This latter parameter is of great importance, especially for the experiments envisioned with purified material.

In a recent work, Hapgood and Von Holt (1987) discussed several technical points relating to adsorption or elution conditions that do not modify the procedure we have already described (Lustenberger et al., 1985; Idziorek et al., 1985).

3.1.3. Final Purification Step

Owing to the high value of the purification factor necessary to obtain glucocorticoid receptor in an homogeneous state, the affinity chromatography step, even in the best case, was unable to produce a totally purified protein. With our procedure and affinity gel, the affinity eluates were

10–30% pure on the basis of the molecular mass of the native receptor. Thus, a further purification step was required, even if it was hazardous because of the well-known instability of purified steroid-receptors and the difficulty of handling very dilute protein solutions without significant loss.

Various procedures were tested, such as

1. Hydroxylapatite chromatography;
2. DEAE-trisacryl chromatography;
3. High performance size exclusion chromatography on TSK 3000 column.

Comparison of the results obtained is depicted on Table 1. Both the efficacy and recovery of size exclusion chromatography are the most satisfactory. With an analytical column, because of the limited size of the sample (0.5 mL) it is necessary to proceed by iterative injections after concentration by adsorption on DEAE with stepwise elution. Recent experiments demonstrate that high performance ion-exchange chromatography on TSK-DEAE or mono Q give also very satisfactory results for analytical or preparative work on affinity eluates. The implementation of high performance size exclusion chromatography is possible only because the receptor adsorbed on a small vol of gel could then be eluted in a concentrated form.

Conventional techniques such as gel filtration chromatography (Failla et al., 1975) or DEAE chromatography (Weisz et al., 1984; Grandics et al., 1984; Hapgood and Von Holt, 1987) were usually used as final steps. However, high performance techniques resulted in clearly better separations than conventional chromatography and undoubtly afforded a significant improvment in our previous purification procedures. This is especially well illustrated by the results depicted in Fig. 5. The two experiments were carried out virtually in the same conditions, with purified glucocorticoid receptor. The separation between the two forms, transformed and nontransformed, appears better with high performance ion exchange chromatography.

3.1.4. Conclusion

The design and synthesis of a suitable affinity matrix seem to us to be the rational procedure. This can be carried out advantageously through experiments with modified ligands in solution and also with matrix bound ligands. Further efforts need to be spent for the optimization of the affinity chromatography procedure, and especially of its adsorption and washing steps, the care afforded largely determines the quality

Table 1
Comparative Study of Various Procedures
for the Final Purification of the Affinity Eluate[a]

	Hydroxylapatite chromatography	DEAE trisacryl chromatography	High performance size exclusion chromatography
Purification			
Specific activity cpm/mg proteins × 10⁻⁶	50.5	64.5	121.7
Fold	2.4	3.1	5.9
Recovery Yield%	33	62	86

Hydroxylapatite chromatography

 1.5 mL of affinity eluate was applied on a hydroxylapatite (BIORAD) column: 1.5 × 1.1 cm. The flowrate was 36 mL/h. The elution involved a 20 min washing with $0.02M$ phosphate buffer followed by a 10 mL linear phosphate gradient $(0.02–0.4M)$.

DEAE chromatography

 1.5 mL of affinity eluate was applied on a DEAE Trisacryl (IBF) column: 1.5 × 1.1 cm. The flowrate was 36 mL/h. The elution involved a 20 min washing with $0.02M$ phosphate buffer followed by a 10 mL linear KCl gradient $(0–0.4M)$.

High performance size exclusion chromatography

 0.5 mL of affinity eluate was applied on a TSK G 3000 (LKB) column: 60 × 0.7 cm). The flowrate was 48 mL/hr with the following eluent: $0.01M$ Tris buffer.

Buffers

 Phosphate buffer: $0.02M$ phosphate, $0.02M$ β-mercaptoethanol, $0.01M$ molybdate, 20% (v/v) glycerol, pH 7.4.
 Tris buffer: $0.01M$ Tris, $0.01M$ EDTA, $0.4M$ KCl, pH 7.0.

[a]The specific activity of the affinity eluate was 20.8×10^6 cpm/mg proteins. The purification procedures were carried out at 0–4°C.

of the receptor purification. The higher the specific saturation of a good matrix, the more efficient the washing step, and the better are the results. Finally resorting to high performance chromatography techniques as final step represent a decisive progress, owing to their excellent resolution and speed.

 This rational and systematic approach necessary to obtain optimal affinity matrixes has proved to be efficient. Indeed, we were able to design other affinity matrixes suitable for the purification of the progesterone receptor (Renoir et al., 1982) and the mineralocorticoid receptor (Lombes et al., 1987).

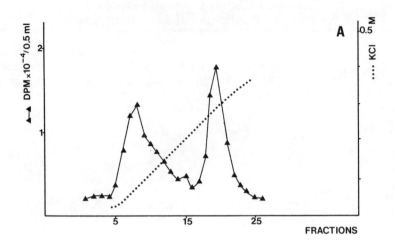

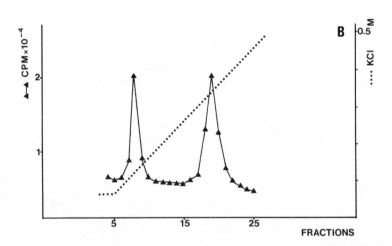

Fig. 5. DEAE chromatography of affinity eluates. **A** Conventional DEAE-cellulose (Schmidt et al., 1985). After removing the molybdate by gel filtration, an aliquot (1 mL) of affinity eluate was transformed and applied on a DEAE cellulose column (3 mL) equilibrated with $0.05M$ phosphate buffer pH 7.0. After washing with $0.05M$ phosphate buffer, the column was eluted with a linear KCl gradient (0–$0.4M$). Fractions (1 mL) were collected and assayed for radioactivity. **B** High performance ion exchange chromatography (Idziorek et al., 1985). An aliquot (0.5 mL) of affinity eluate was applied on a TSK DEAE 545 column (1.25 mL) equilibrated with $0.05M$ phosphate buffer pH 7.4. After a 20 min washing, elution was carried out with a linear 0–$0.5M$ KCl gradient. Fractions (1 mL) were collected and 0.05 mL aliquots were assayed for radioactivity.

3.2. Purification
of Transformed Glucocorticoid Receptor

The classical procedure, i.e., repeated chromatography on phosphocellulose or DNA-cellulose with a transformation step introduced midway through the protocol, was first described by Atger and Milgrom (1976) and Eisen and Glinsmann (1978). A definitive procedure was proposed by Wrange et al. (1979). The principle of the purification lies in the differential behavior of transformed and nontransformed glucocorticoid receptor, toward polyanionic exchangers. The transformed form is more acidic than the untransformed form.

Using a similar approach, we developed a one-step purification using high performance ion exchange chromatography (Denis et al., *J. Chromatog.*, accepted for publication).

3.2.1. Analysis
of the Glucocorticoid Receptor on Mono Q Column

In the presence of 10 mM molybdate, both liganded and unliganded glucocorticoid receptors were eluted as a single and sharp peak at 0.32M NaCl. In the absence of molybdate and after heat transformation, another peak of specifically bound [^{3}H]dexamethasone was eluted with 0.08M NaCl. These two molecular forms of the liganded receptor were further characterized. The most acidic (eluted with 0.08M) had a Stokes radius of 5.1 nm and a sedimentation coefficient of 4.6 S leading to a calculated M_r $\cong$ 100,000. The second peak when similarly analyzed gave the following results: Stokes radius = 7.3 nm and S_{20} = 9.0 S, calculated M_r $\cong$ 280,000. Analysis of both forms with minicolumns of DNA-Ultrogel, DEAE-Trisacryl, and hydroxylapatite (HA-Ultrogel) confirmed the identity of the two peaks with the transformed and the nontransformed glucocorticoid receptor complexes, respectively.

As first suggested by Parchman and Litwack (1977), the activated species can be generated by the conditions of the mono Q procedure itself providing molybdate is omitted in the eluent buffer (Denis et al., 1988a). This prompted us to study the conditions and kinetics of transformation on the column. Figure 6 shows the results of such experiments. Labeled samples of cytosol prepared in presence or absence of molybdate were loaded onto the mono Q column. Before initiating the linear salt gradient, washing with buffer without molybdate was performed in order to remove free dexamethasone. The longer the washing, the greater the relative abundance of the first peak. Heating the column at 20°C during the washing period considerably accelerated this phenomenon. The addition of molybdate (0.01M) in the buffer entirely prevented transformation on the column.

Because of the important difference between the elution molarities of the two forms and of the generation of the transformation by the column itself, it was of interest to develop a purification procedure. Labeled cytosol prepared in buffer supplemented with molybdate was loaded on the mono Q column. A 15-min washing with Tris-buffer (0.01M supplemented with molybdate 0.01M and NaCl 0.25M) allowed to eliminate the free steroid and a large amount of proteins. At this stage, are retained on the column all proteins eluted above 0.25M including the [³H]dexamethasone nontransformed receptor. Then washing was pursued with Tris buffer (0.01M) during 35 min. Because of the absence of molybdate, this allowed dissociation of the heat shock protein, the transformed receptor remaining bound to the support.

A linear salt gradient (0–0.25M) was applied. The transformed receptor which is eluted at 0.08M is the only protein whose affinity toward the support had changed.

The overall purification needs 60 min, and its performance is depicted in Table 2. The calculated receptor homogeneity, assuming an M_r of 90,000 per steroid binding unit, was in the range 10–20%. The physicochemical and immunological characteristics of the preparation demonstrated unequivocally the presence of transformed receptor (Denis et al., *J. Chromatog.*, accepted for publication). The procedure can be applied to the glucocorticoid receptor of various sources (rat liver, rabbit liver) with only slight modifications in the molarity of buffers during the washing and elution steps. Adaptation to preparative colunms allows the processing of larger volumes of biological materials, up to 100 mL cytosol (Formstecher, unpublished results).

High performance ion exchange chromatography appears very suitable both in efficiency and quickness for purification of transformed glucocorticoid receptor.

4. General Comments

4.1. Partial Purification vs Global Purification

The global purification of the glucocorticoid receptor has been attempted only by a few groups and convincing results have been obtained with both the transformed and the nontransformed forms. However, this purification remains difficult and has lost much of its interest, since some

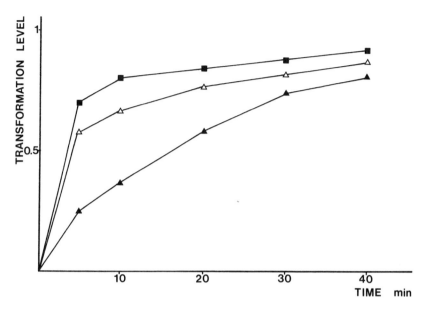

Fig. 6. Kinetics of the transformation of the glucocorticoid receptor induced by the anion exchanger. Cytosol was prepared in phosphate buffer (0.02M phosphate, 0.02M β-mercaptoethanol, 20,% (v/v) glycerol, pH 7.4) with (▲■) or without (△) molybdate (0.01M). Labeled samples (0.5 mL) deleted of unbound steroid by charcoal treatment, were loaded onto a Mono Q column. After washing with Tris buffer (0.02M Tris, 0.02 M β-mercaptoethanol, 2% (v/v) acetonitrile), elution was carried out with a linear NaCl gradient (0–0.5M). For each duration of washing, the transformed form and the non-transformed form were collected and the level of transformation was calculated.

Cytosol prepared without MoO₄ △} --- washing at 4°C
Cytosol prepared with MoO₄ ▲ }
Cytosol prepared with MoO₄ ■ } --- washing at 4°C

of its main goals have been already reached, i.e., obtaining specific antire-ceptor antibodies and cloning the receptor gene. On the other hand, be-cause of the limited sources of receptor, physicochemical and structural studies on purified receptor cannot be envisioned at the present time. Large quantities of receptor would be available using in vitro expression of a transfected receptor cDNA. Even in this case, however, purification will be necessary. Choosing a purification procedure is only dependent on the molecular form of receptor to be purified. Nontransformed molybdate stabilized glucocorticoid receptor can be obtained by affinity techniques. Ion exchange chromatography, whether using open-bore columns or high pressure chromatography equipment, is undoubtedly the most useful gen-eral procedure for the purification of the transformed form of the glucocor-ticoid receptor.

Table 2
One-Step Purification
of Transformed Glucocorticoid Receptor from Rabbit Liver

Step	Total protein, mg	Total activity, pmol	Specific activity, pmol/mg prot.	Purification, –fold	Yield, %
Cytosol	292	90.5	0.31	–	100
HPIEC	0.033	35	1,060	3,421	38.7

Labeled cytosolic preparation, prepared in phosphate buffer, was applied to the Mono Q column. All steps were carried out at 0–4°C with a flowrate of 1 mL/mn. The following sequence of washing and elution step was controlled by the gradient programmer: buffer A supplemented with 0.275M NaCl 10 mL, buffer B 40 mL; gradient 0–100% buffer A supplemented with 0.250 mM NaCl 10 mL. Fractions (1 mL) were collected and assayed for radioactivity and proteins.

Buffer A: 0.01M Tris HCl, 0.01M molybdate, 0.02M β-mercaptoethanol, 0.02M β-mercaptoethanol, acetonitrile 2% (v/v), pH 7.4.

Buffer B: 0.01M Tris HCl, 0.02M β-mercaptoethanol, acetonitrile 2% (v/v), pH 7.4.

On the other hand, partial purification is often needed for many studies. The finality is quite different. The purification appears necessary either to remove some contaminant present or added in cytosol or to select exclusively one molecular form of the receptor. The choice of a procedure must be made according to the study envisioned. The separation of molecular forms in order to carry out binding or interaction experiments is better achieved using high performance ion exchange techniques. A simple enrichment is preferably obtained with precipitation procedures. The obtention of molecular forms differing in their composition can be carried out with hydrophobic or gel filtration procedures.

4.2. Checking the Purity

After every purification step, even in case of prepurification, it is very important to check the physicochemical characteristics and functional properties of the receptor. This is carried out with usual criteria: determination of Stokes radius by gel filtration, measurement of the sedimentation coefficient, analysis of ionic parameters (pI), and so on. Recent progress in methodologies like high pressure liquid chromatography or ultracentrifugation in vertical rotor, appear very beneficial for the rapidity and efficiency of these analyses. Functional properties include binding parameters (specificity, association, and dissociation kinetics), transformation study, and interaction with DNA. The minicolumn procedure described by Holbrook et al. (1983) represents a simple biological assay to rapidly verify the molecular form of the receptor.

Other criteria that are relatively simple to perform experimentally may be applied. For example, glucocorticoid receptor can be covalently labeled with [³H]dexamethasone mesylate allowing to perform electrophoretic analysis under denaturing conditions. Immunologic criteria can also be used either by immunoprecipitation or immunoblotting. Whatever, the purification implemented, criteria other than the simple assay of macromolecular bound [³H]steroid should always be performed in order to check the quality of receptor–steroid complexes and avoid artifacts.

4.3. Future Directions

The need for purified glucocorticoid receptor preparations is still present. Only by studying such preparations can reliable conclusions be drawn as to the nature of the receptor subunits, their phosphorylation and their requirement for receptor function.

Answers are already given and purification of associated component of receptor was recently described (Nemoto et al., 1987; Denis, 1988). Similarly, interaction with DNA fragments can be tested with the help of specific affinity matrix primarily designed for the purification. This new DNA receptor binding test introduced by Rousseau et al. (1987) compares very favorably with the classical DNA-cellulose binding test. Another field of development is represented by the interaction between receptor and antiglucocorticoid derivatives. Besides the classical study based upon binding studies of the tritiated antiglucocorticoids (RU 486 or progesterone for example) that we have already performed (Lefebvre et al., 1988), a more direct approach could be developed using RU 486 derivatized matrix.

Finally, the new technical development of HPLAC must not be forgotten. This technique combines the high speed characteristics of HPLC with the specificity of affinity chromatography. The application of this new technique is under evaluation in our laboratory. The rapid purification of antidexamethasone antibodies by such a procedure with dexamethasone derivatized beads allows to give an account of the possibilities of such affinity matrixes (Formstecher et al., 1986). Work is currently in progress to adapt this procedure to the purification of glucocorticoid receptor.

Acknowledgments

First the authors would like to express their gratitude and appreciation to those colleagues, particularly P. Blanchardie, M. Denis (Nantes); T. Idziorek, B. Sablonniére, and P. Lefebvre (Lille), whose research efforts have contributed significantly to the experiments discussed in this chapter.

The studies presented in this chapter were supported by the University of Nantes and the University of Lille II and by grants from INSERM (CRL contract no. 844013 and 854008), ARC (contract no. 62.92) and the EEC (contract no. STZ-0075-1-B). We are also indebted to Mrs. Combalot for her excellent secretarial assistance.

References

Atger, M. and Milgrom, E. (1976) *Biochemistry* **15**, 4298–4304.
Ben-Or, S. and Chrambach, A. (1983) *Arch. Biochem. Biophys.* **221**, 343–353.
Blanchardie, P., Lustenberger, P., Orsonneau, J. L., and Bernard, S. (1984) *Biochimie* **66**, 505–511.
Brönnegard, M., Poellinger, L., Okret S., Wikstrom, A. C., Bakke, O., and Gustafsson, J.-A. (1987) *Biochemistry* **26**, 1697–1704.
Carlstedt-Duke, J., Stromstedt, P.-E., Wrange, O., Bergman, T., Gustafsson, J.-A., and Jornvall, H. (1987) *Proc. Natl. Acad. Sci USA* **84**, 4437–4440.
Dahmer, M. K., Housley, P. R., and Pratt, W. B. (1984) *Ann. Rev. Physiol.* **46**, 67–81.
Dean, P. D. G., and Watson, D. H. (1979) *J. Chromatog.* **165**, 301–319.
Dellweg, H.-G., Hotz, A., Mugele, K., and Gehring, V. (1982) *EMBO J.* **1**, 285–289.
Denis, M., Guendouz, F., Blanchardie, P., Bernard, S., and Lustenberger, P. (1988a) *J. Steroid Biochem.* **30**, 281–285.
Denis, M. (1988) Two-step purification and *N*-terminal aminoacid sequence analysis of the rat M_r 90,000 heat shock protein. *Anal. Biochem.* **173**, in press.
Denis, M., Poellinger, L., Wikstom, A. C., and Gustafsson, J. A. (1988b) *Nature* **333**, 686–688.
Eisen, H. J. and Glinsmann, W. (1978) *Biochem. J.* **171**, 177–183.
Failla, D., Tomkins, G. M., and Santi, D. V. (1975) *Proc. Natl. Acad. Sci USA* **72**, 3849–3852.
Formstecher, P., Hammadi, H., Bouzerna, N., and Dautrevaux, M. (1986) *J. Chromatog.* **369**, 379–390.
Formstecher, P. and Lustenberger, P. (1987) in *Recent Advances in Steroid Hormone Action* (Moudgil, V. K., ed.), Walter de Gruyter, Berlin, 499–535.
Gehring, V. and Hotz, A. (1983) *Biochemistry* **22**, 4013–4018.
Govindan, M. V. and Manz, B. (1978) *Eur. J. Biochem.* **108**, 47–53.
Govindan, M. V. and Gronemeyer, H. (1984) *J. Biol. Chem.* **259**, 12915–12924.
Grandics, P., Miller, A., Schmidt, T. J., Mittman, N. D., and Litwack, G. (1984) *J. Biol. Chem.* **259**, 3173–3180.
Grody, W. W., Schrader, W. T., and O'Malley, B. W. (1982) *Endocrol. Rev.* **3**, 141–163.
Gronemeyer, H. and Govindan, M. V. (1986) *Mol. Cell. Endocrinol.* **46**, 1–19.
Hackney, J. F. and Pratt, W. B. (1971) *Biochemistry* **10**, 3002–3008.
Hapgood, J. P. and Von Holt, C. (1987) *J. Steroid Biochem.* **28**, 769–777.
Holbrook, N. J., Bodwell, J. E., Jeffries, M., and Munck, A. (1983) *J. Biol. Chem.* **258**, 6477–6485.
Holbrook, N. J., Bodwell, J. E., Mendel, D. B., and Munck, A. (1985) in *Molecular Mechanism of Steroid Hormone Action* (Moudgil, V. K., ed.), Walter de Gruyter, Berlin.
Housley, P. R. and Pratt, W. B. (1983) *J. Biol. Chem.* **258**, 4630–4635.
Housley, P. R., Sanchez, E. R., Westphal, H. M., Beato, M., and Pratt, W. B. (1985) *J. Biol. Chem.* **260**, 13810–13817.

Hubbard, J. R. and Kalimi, M. (1983) *Biochim. Biophys. Acta* **755**, 363–368.

Idziorek, T., Formstecher, P., Danze, P.-M., Sablonniére, B., Lustenberger, P., Richard, C., Dumur, V., and Dautrevaux, M. (1985) *Eur. J. Biochem.* **153**, 65–74.

Iqbal, M. J., Corbishley, T. P., Wilkinson, M. L., and Williams, R. (1985) *Anal. Biochem.* **144**, 79–85.

Kuhn, R. W., Schrader, W. T., Smith, R. G., and O'Malley, B. W. (1975) *J. Biol. Chem.* **250**, 4220–4228.

Leach, K. L., Dahmer, M. K., Hammond, M. D., Sando, J. J., and Pratt, W. B. (1979) *J. Biol. Chem.* **254**, 11884–11890.

Lefebvre, P., Formstecher, P., Richard, C., and Dautrevaux, M. (1988) *Biochem. Biophys. Res. Commun.* **150**, 1221–1229.

Lombes, M., Claire, M., Lustenberger, P., Michaud, A., and Rafestin-Oblin, M. E. (1987) *J. Biol. Chem.* **262**, 8121–8127.

Lustenberger, P., Formstecher, P., and Dautrevaux, M. (1981) *J. Steroid. Biochem.* **14**, 697–703.

Lustenberger, P., Blanchardie, P., Denis, M., Formstecher, P., Orsonneau, J. L., and Bernard, S. (1985) *Biochimie* **67**, 1267–1278.

Maggi, A., Compton, J. G., Fahnestock, M., Schrader, W. T., and O'Malley, B. W. (1981) *J. Steroid Biochem.* **15**, 63–68.

Markovic, R. D. and Litwack, G. (1980) *Arch. Biochem. Biophys.* **202**, 374–379.

McBlain, W. A. and Shyamala, G. (1984) *J. Steroid Biochem.* **20**, 1211–1220.

Mellon, W. S. (1984) *Molec. Pharmacol.* **25**, 79–85.

Mendel, D. B., Bodwell, J. E., Gametchu, B., Harrison, R. W., and Munck, A. (1986) *J. Biol. Chem.* **261**, 3758–3763.

Miyabe, S. and Harrison, R. W. (1983) *Endocrinology* **112**, 2174–2180.

Molinari, A. M., Medici, M., Moncharmont, B., and Puca, G. A. (1977) *Proc. Natl. Acad. Sci. USA* **74**, 4886–4890.

Moudgil, V. K., Healy, S. P., and Singh, V. B. (1985) *J. Biochem.* **98**, 963–973.

Munck, A. and Foley, R. (1979) *Nature* **278**, 752–754.

Nemoto, T., Ohara-Nemoto, Y., and Ota, M. (1987) *J. Biochem.* **102**, 513–523.

Nordeen, S. K., Lan, N. C., Showers, M. O., and Baxter, J. D. (1981) *J. Biol. Chem.* **256**, 10503–10508.

Parchman, L. G. and Litwack, G. (1977) *Arch. Biochem. Biophys.* **183**, 374–382.

Pike, J. W. and Haussler, M. R. (1979) *Proc. Natl. Acad. Sci. USA* **76**, 5485–5489.

Ratajczak, T. and Hähnel, R. (1980) *J. Steroid Biochem.* **13**, 439–444.

Renoir, M., Yang, C. R., Formstecher, P., Lustenberger, P., Wolfson, A., Redeuilch, G., Mester, J., Richard-Foy, H., and Baulieu, E. E. (1982) *Eur. J. Biochem.* **127**, 71–79.

Rousseau, G. G., Baxter, J. D., Higgins, S. J., and Tomkins, G. M. (1973) *J. Mol. Biol.* **79**, 539–554.

Rousseau, G. G., Higgins, S. J., Baxter, J. D., Gelfand, D., and Tomkins, G. M. (1975) *J. Biol. Chem.* **250**, 6015–6021.

Rousseau, G. G., Kirchhoff, J., Formstecher, P., and Lustenberger, P. (1979) *Nature* **279**, 158–160.

Rousseau, G. G., Eliard, P. H., Barlow, J. W., Lemaigre, F. P., Lafontaine, D. A., DeNayer, P., Economidis, I. V., Formstecher, P., Idziorek, T., Mathy-Hartert, M., Voz, M. L. J., Belayew, A., and Martial, J. A. (1987) *J. Steroid Biochem.* **27**, 149–158.

Sablonniére, B., Lefebvre, P., Formstecher, P., and Dautrevaux, M. (1987) *J. Chromatog.* **403**, 183–186.

Santi, D. V., Wastien, W., and Pogolotti, A. L. (1979) in *Monographs on Endocrinology: Glu-*

cocorticoid hormone action, vol XII (Baxter, J. D. and Rousseau, G. G., eds.), Springer, Heidelberg, pp. 109–122.

Schmidt, T. J., Miller-Diener, A., Webb, M. L., and Litwack, G. (1985) *J. Biol. Chem.* **260**, 16255–16262.

Schrader, W. T. (1975) in *Methods in Enzymology*, vol 36 (O'Malley, B. W. and Hardman, J. G., eds.), Academic, NY, pp. 187–210.

Secco, C., Redeuilh, G., Radanyi, C., Baulieu, E. E., and Richard-Foy, H. (1979) *C. R. Acad. Sci. Paris* **289**, 907–910.

Sherman, M. R. and Stevens, J. (1984) *Ann. Rev. Physiol.* **46**, 83–105.

Sica, V. and Bresciani, F. (1979) *Biochemistry* **18**, 2369–2378.

Simons, S. S. and Thompson, E. B. (1981) *Proc. Natl. Acad. Sci. USA* **78**, 3541–3545.

Singh, V. B. and Moudgil, V. K. (1985) *J. Biol. Chem.* **260**, 3684–3690.

Subramanian, S. (1984) *CRC Crit. Rev. Biochem.* **16**, 169–205.

Vedeckis, W. (1983) *Biochemistry* **22**, 1983–1989.

Webb, M. L., Miller-Diener, A. S., and Litwack, G. (1985) *Biochemistry* **24**, 1946–1952.

Weisz, A., Puca, G. A., Masucci, M. T., Masi, C., Pagnotta, R., Petrillo, A., and Sica, V. (1984) *Biochemistry* **23**, 5393–5397.

Weisz, A., Baxter, J. D., and Lan, N. C. (1984) *J. Steroid Biochem.* **20**, 289–293.

Wrange, O., Carlstedt-Duke, J., and Gustafsson, J. A. (1979) *J. Biol. Chem.* **254**, 9284–9290.

Yamamoto, K. R. and Alberts, B. M. (1976) *Ann. Rev. Biochem.* **45**, 721–746.

Glucocorticoid Receptor Purification and Characterization Using Tissue Culture Cells

Wayne V. Vedeckis

1. Introduction

Purification of steroid receptors from tissue culture cells requires certain specific conditions for the implementation of this goal. In this chapter I will discuss some of the unique problems and advantages of utilizing cell cultures for the purification and characterization of the glucocorticoid receptor protein. Although our laboratory has not carried out extensive purification of this receptor from cell cultures, our studies have led to certain conclusions about receptor purification. I emphasize some unpublished observations that may have significant impact on the purification of steroid receptor proteins from established cell lines. In many cases our observations have not been followed up by detailed experimentation. However, these "impressions" and "routine observations" have led us to certain methods of procedure and conclusions that we feel are important in the purification of steroid receptors. The actual physicochemical behavior of the receptor using various purification protocols has been published previously, and these references will be cited when appropriate. Although I will cite, primarily, the work performed in our laboratory on the glucocorticoid receptor from the mouse AtT-20 pituitary tumor cell line, numerous other investigations in various cells and tissues have contributed to our understanding of glucocorticoid receptor structure and function (*reviewed*

Receptor Purification, vol. 2   © 1990 The Humana Press

in Vedeckis, 1985). However, I wish to use this chapter to concentrate on the everyday mechanics of steroid receptor purification that we use, and that do not often appear in the scientific literature.

2. Structure of the Glucocorticoid Receptor

Numerous studies have contributed to our current concept of glucocorticoid receptor structure (Vedeckis, 1985). Our current concept of glucocorticoid receptor (GR) structure is presented in Fig. 1. In the unliganded state, the GR is untransformed, that is, it is incapable of binding to the specific DNA sequence (glucocorticoid response element, GRE) that is necessary for the stimulation of specific gene transcription. A reasonable amount of information has now been accumulated on steroid receptor transformation, i.e., the conversion of the receptor from a non-DNA-binding protein to a form (the transformed receptor) that is able to bind to DNA. It is widely accepted that glucocorticoid receptor transformation results from subunit dissociation (Vedeckis, 1983b). Furthermore, it is also generally held that the untransformed receptor contains one or more (probably two) molecules of the heat shock protein, hsp90 (*reviewed in* Pratt, 1987).

Therefore, the current molecule model for glucocorticoid receptor transformation can be stated as follows (Fig. 1). The untransformed (non-DNA-binding) GR is a heterooligomeric complex of the GR protein ($M_r \sim 86,000$; based on the amino acid sequence) and two molecules of hsp90. Not excluded from this model is the presence of one or more undefined components (receptor binding factors, RBFs), which also interact in a specific manner with the GR. Dephosphorylation of some component in the cytosol appears to be important for GR transformation (Reker et al., 1987). However, recent studies by Munck and colleagues (Mendel et al., 1987) and Pratt and coworkers (Tienrungroj et al., 1987) indicate that dephosphorylation of the GR itself is not important for GR transformation.

The initial work of Litwack and coworkers suggested that the dissociation of a low mol wt component from the untransformed receptor complex is also necessary for GR transformation (Goidl et al., 1977), and the structure of this molecule has been proposed to be an ether aminophosphoglyceride (Bodine and Litwack, 1988a,b). The normal physiological effector of GR transformation is the hormone itself. Thus, it is likely that hormone binding in the intact cell causes some alteration of structure in the untransformed GR. This, along with dephosphorylation of some molecule and the dissociation of the low molecular inhibitor from the complex, results in a destabilization of the heterooligomeric

GR STRUCTURE AND TRANSFORMATION

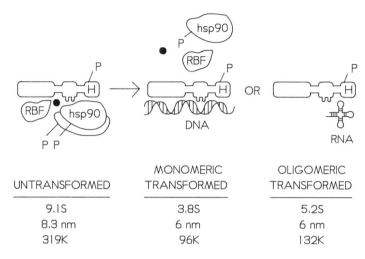

UNTRANSFORMED	MONOMERIC TRANSFORMED	OLIGOMERIC TRANSFORMED
9.1S	3.8S	5.2S
8.3 nm	6 nm	6 nm
319K	96K	132K

Fig. 1. Model of the structure and mode of transformation of the glucocorticoid receptor. The untransformed GR is a heterooligomeric complex which contains at least one hormone-binding subunit. This protein consists of three domains, separated by protease-sensitive regions (represented by constricted areas). The amino terminal domain is the immunogenic domain, while the middle and the caboxyl terminal domains bind to DNA and hormone (H), respectively. Other components of the untransformed GR may include hsp90, unidentified macromolecular receptor binding factors (RBF), and an endogenous, low mol wt inhibitor of transformation (●). The interaction of hsp90 with the hormone-binding subunit may occlude the DNA-binding site, which contains two cysteine-zinc fingers. Hormone-binding, dephosphorylation of some molecule (but not the hormone-binding subunit), and dissociation of the low mol wt inhibitor may all contribute to destabilization of the subunit inter-actions. Subunit dissociation (transformation) liberates the GR monomer, with its DNA-binding site now exposed. The monomer can then bind to DNA and regulate gene expression. The monomer can also bind to a small RNA molecule (such as tRNA) to form an oligomeric transformed complex. The role of this RNA-binding activity has not yet been determined. Further details of this model are presented in the text.

complex. Subunit dissociation occurs with the liberation of the transformed, DNA-binding, GR monomer. In vitro, the GR monomer can further associate with a small RNA molecule, to form a transformed GR containing bound RNA (Kovacic-Milivojevic et al., 1985; Ali and Vedeckis, 1987a,b,c). The role of this RNA-binding activity is currently unknown, and is being studied by a number of laboratories.

Although not thought to be a normal physiological occurrence, the monomeric GR can be proteolyzed into two relatively discrete hormone-binding forms. This is based on the fact that the native monomer is comprised of three protein domains separated by protease-sensitive regions (*reviewed in* Vedeckis, 1985). The partially proteolyzed receptor contains both the DNA- and hormone-binding domains, whereas the mero-receptor contains only the hormone-binding domain. The physicochemical properties of the different GR forms we have studied in our laboratory are presented in Table 1.

3. Use of Cell Lines

3.1. Advantages

There are a number of significant advantages in using tissue culture cells for the purification of the GR. First, the cell population is homogeneous, so that the protein contaminants will be similar in all cells. Second, the medium in which the cells are grown is defined (with the exception of serum). The biggest advantage of this is the absence of endogenous glucocorticoids in the medium (except in the serum), as opposed to the situation in intact (nonadrenalectomized) animals. This means that all of the GR hormone-binding sites are vacant, and this allows efficient, high-level labeling of the GR by added tritiated steroid. Third, cell cultures are easily manipulated, so that rapid timecourse experiments (e.g., in studying receptor transformation) are possible. Fourth, except for that present in the serum, there is no other source of nonreceptor, glucocorticoid-binding macromolecules. Therefore, many of the problems that plague steroid receptor purification and studies in intact animals and tissues are eliminated by the use of cell cultures.

3.2. Serum Requirements

In most cases, optimal cell growth in culture requires the use of serum, probably because it is a rich source of growth factors and other undefined nutrients that promote cell vitality. Although it is possible to use either a completely defined medium or commercial serum substitutes to maintain cell growth in culture, this is still an uncommon practice. In our laboratory, we have succeeded in using completely defined medium to grow cells, but the growth kinetics were not comparable to those obtained using serum. In addition, extended subculture resulted in a loss of cell viability and in noticeable alterations in cellular morphology.

Most of the disadvantages in using tissue culture cells stem from the use of serum. A number of these problems can be avoided by merely washing cells free of serum (e.g., with Tris/saline, 0–4°C, three times),

Table 1
Properties of the AtT-20 Cell Glucocorticoid Receptor Forms and Proteolytic Fragments

Receptor form	Sedimentation coefficient, S	Stokes radius, nm	M_r, g/mole	f/f_o	Axial ratio	DNA-binding activity
Untransformed	9.1	8.3	319,000	1.69	13	No
Oligomeric transformed	5.2	6	132,000	1.64	12	?[b]
Monomeric transformed	3.8	6	96,000[a]	1.82	16	Yes
Partially proteolyzed	3.2	3.9	53,000	1.45	8	Yes
Mero-receptor	2.4	2.4	24,000	1.15	3	No

[a]The actual M_r of the monomeric GR (calculated from the derived amino acid sequence of the cloned gene) is 86,000.
[b]Although this form can bind polyanions (e.g., RNA) it is not known if it can bind to both DNA and RNA simultaneously.

prior to extracting the GR. This eliminates contaminating serum proteins (an initial purification step) and other steroid-binding proteins that could complicate the purification procedure, such as corticosteroid-binding globulin (CBG). There is a finite amount of glucocorticoids present in serum, and most of this is bound to CBG. Although free steroid, and some bound steroid, can be removed by dialysis of the serum against saline, the most effective way of removing the endogenous steroid is by charcoal-stripping of the serum. This is of greatest importance in studies using steroid-sensitive T-lymphoid cells, which are killed by glucocorticoids. Because the glucocorticoid levels in different serum lots can vary, it is possible that enough hormone will be present to either kill the sensitive cells or suppress their growth rate. In both cases, the result would be a selected cell population that contains no GR, lowered GR levels, or mutant GR species. Any of these situations would be counterproductive for GR purification. In addition, some cell lines that do grow in the presence of hormone demonstrate a down-regulation in GR levels when the steroid is present (Svec and Rudis, 1981; Cidlowski and Cidlowski, 1981), again, an unfavorable situation for receptor protein purification.

We have made two modifications in our culture of mouse AtT-20 pituitary tumor cells to deal with these problems. First, we use fetal calf serum exclusively, because it is likely that the level of hormone will be lower than in newborn calf or horse serum. (We have also found that AtT-20 cells grow very poorly in newborn calf serum.) Second, we have found that 5% serum supports the growth of AtT-20 cells as well as does 10% serum, whereas 2% serum is suboptimal. Besides being cost effective, this reduces the level of endogenous glucocorticoids by one-half. The use of 5% dialyzed fetal bovine serum, therefore, is optimal for growing this cell line. It is probably wise for the investigator to spend some time analyzing the serum requirements before attempting GR purification from a specific cell line. A several-fold increase in the intracellular GR level (and, therefore, a "several-fold initial purification" of the receptor) could well be obtained.

3.3. Saturation Density

It is always best to determine the optimal conditions for cell growth with each individual cell line to be used for GR purification. This will include the seeding density, media change schedule, and maximum saturation density to be used for the cell culture. This is particularly important because the synthesis of the GR may occur in a cell-cycle specific manner.

For example, it has been reported that the GR is synthesized in S phase of the cell cycle in HeLa S_3 cells (Cidlowski and Michaels, 1977). Thus, if one were to attempt receptor purification from a stationary phase (high saturation density) culture, it is possible that the GR levels may be lower (on a per cell basis) than for midlog phase cells. Although we have not made a systematic study of this, we have noticed that high density cultures of AtT-20 cells have a lower level of GR than do log phase cells. Also, for bulk cultures we normally seed AtT-20 cells (a suspension cell line) in a spinner flask and do not change the medium before the cells are harvested. It is possible that replacement with fresh medium (e.g., 24 h before harvesting) could elevate the intracellular GR levels. In any event, it would be worthwhile to take these culture conditions into account before attempting to purify the GR from a certain cell line. A substantial increase in the amount of starting GR protein is feasible if these points are considered.

3.4. GR Levels in Various Cells

Another potentially easy way of increasing the yield and purity of the GR protein is to choose a cell line that contains high levels of GR or can be induced to produce high GR levels. The average number of GR molecules in a diploid cell is about 30,000 (Bourgeois and Newby, 1977). For example, WEHI-7 lymphoma and L929 fibroblast cells contain about these many receptors. On the other hand, the mouse S49 lymphoma line contains about 15,000 hormone-binding (wild type) GR molecules and an approx equal quantity of a mutant, nonhormone-binding receptor species (Northrop et al., 1985). Therefore, this cell line would be a poor choice for the purification and characterization of the native, wild-type GR. The mouse AtT-20 cell contains 75,000 GR molecules per cell (Svec and Harrison, 1979), that is, more than twice the amount present in most other cell lines. Because the total amount of cytoplasmic protein is not similarly increased, the choice of this cell line results in a "twofold purification" over the GR found in other cell types. In a recent study (Allen and Vedeckis, unpublished observations), we have found that the amount of GR mRNA present in two lymphoid cell lines (mouse S49, human CEM C7) was increased two- to threefold by treatment with glucocorticoids. Assuming a similar increase in GR protein levels, a brief treatment of these cells with hormone could result in a substantial augmentation of the starting GR protein levels. Once suitable overexpression systems using cloned GR gene sequences become available (*see below*), it may be possible to even more dramatically increase the GR level to enhance the yield and purity during protein purification.

4. Preparation of Cytosol

Although isolated nuclei can be used as a source for purification of in vivo-transformed receptor, the starting point for GR purification is typically a soluble cytoplasmic extract (cytosol) from the cells. There are a number of factors that will influence the yield of the GR at this initial step.

4.1. Buffer Requirements

In general, any common buffer within the normal physiological pH range can be used to make cytosol. In our laboratory, we use 20 mM Tris-HCl, pH 7.4 at 25°C. EDTA (~1 mM) is commonly added to the buffer. This is a "traditional" practice, based upon the desire to inhibit any metalloenzymes (especially hydrolases) present in the extract. There is no compelling data for the inclusion of this agent in the buffer. For certain tissues, other serine and cysteine protease inhibitors (PMSF, leupeptin) can be added to prevent receptor cleavage. Although we have not performed extensive studies, it appears that tissue culture cell extracts may not contain as much protease activity as certain intact tissue extracts (e.g., liver). Because transition metal oxyanions (molybdate, tungstate, vanadate) are potent inhibitors of receptor transformation (*see* Vedeckis, 1985), 20 mM sodium molybdate is often included in the buffer, especially when one wishes to preserve the structure of the untransformed receptor during purification.

Perhaps the most important buffer component to be included is a reducing agent. It has been shown that reduced cysteine residues are required in the hormone-binding site of steroid receptors (*see* Vedeckis, 1985). Therefore, if there is no reducing agent in the buffer a dramatic loss of hormone-binding activity can occur, presumably due to cysteine oxidation. For example, our laboratory uses 1-thioglycerol in our homogenization buffer. We have found that if 1-thioglycerol is not added from a concentrated stock solution *just prior* to the preparation of cytosol, we can lose up to 90% of the hormone-binding activity in the cytosol. This is because 1-thioglycerol is relatively unstable in dilute aqueous solution. A stock buffer solution containing this reducing agent rapidly loses its protective effect on prolonged storage. We have also used β-mercaptoethanol and dithiothreitol (DTT) as reducing agents, both of which will work at appropriate concentrations. However, if the concentration is too high, the reducing agent can promote receptor transformation, and even cause a *loss* of hormone-binding activity (perhaps by disrupting the overall protein structure; Vedeckis, 1983b). Again, although detailed studies have not been carried out in our laboratory, it is likely that DTT is the best reducing agent to use in purification of steroid receptors. Once hormone binding has occurred, the presence of a reducing agent in the buffer is less crucial, presumably because the hormone in the binding site protects cysteine from oxidation.

4.2. Cytosol Preparation

There are some minor points about cytosol preparation from tissue culture cells that may improve the yield and stability of the GR. First, we allow the cells to swell for about 30 min in the homogenization buffer. This may improve lysis and the extractability of the GR, although solid data on this point has not been obtained. We prefer to use about three mL of buffer for every mL of packed cells. This gives us a good extraction of the GR, plus it keeps the total protein concentration high enough to stabilize the GR protein. We have found that a stainless steel Dounce homogenizer (Kontes) works best for lysing the cells while giving minimum surface denaturation of the protein (provided that the vigor of extraction is not excessive). One big advantage of using tissue culture cells is that only one centrifugation needs to be performed. Because of the large amount of cellular debris and connective tissue found in intact tissues, a low speed spin is usually performed, followed by a high speed ($190,000\ g_{av}$) centrifugation of the supernatant. In liver cytosol, there is also a large floating lipid layer that must be removed after each centrifugation. With tissue culture cells, the homogenate can be centrifuged immediately at high speed, and the lipid layer obtained is quite small. To preserve the structure of the GR, all of the above procedures (and those that follow) should be carried out at 0–4°C.

5. GR Labeling with Hormone

5.1. In Vivo Labeling

One of the great advantages of using tissue culture cells for steroid receptor studies is that the hormonal environment can be easily and rapidly controlled by the addition and removal of hormone from the medium. This allows one to perform experiments that are difficult or impossible to carry out in intact animals.

For receptor purification studies, it is sometimes advantageous to label the receptor with tritiated hormone while it is present in the intact cell. For example, if studies wish to be done on the effects of sulfhydryl groups on DNA-binding by the receptor, reducing agents must be absent. As mentioned above, if cytosol is prepared in the absence of reducing agent, hormone-binding activity is often lost. However, if the receptor is labeled in the intact cell, and cytosol is then prepared in the absence of reductant, hormone binding is stable. Another instance in which in vivo labeling of the GR is advantageous is with the use of the covalent affinity ligand, dexamethasone 21-mesylate (Simons and Thompson, 1981). This ligand can be used to covalently radioactively label the GR. This can be valuable in GR purification, because the hormone-binding activity of the GR is irreversibly lost as the receptor becomes purer. We and others have found that the amount of nonspecific labeling of cellular proteins by this electrophilic affinity ligand is greatly reduced when performed in the intact cell, when compared to in vitro affinity-labeling of the receptor. The reason for this is as yet unknown.

In vivo GR labeling is most conveniently carried out as followed. Cells are harvested and washed with either cold Tris/saline or serum-free medium. If a large number of cells are to be used, they can be resuspended in about one-tenth to one-twentieth the original volume of culture medium. Tritiated hormone is added at a concentration that is at least tenfold higher (>10 nM) than the binding affinity of the GR for the ligand. [An exception to this is found in the labeling of the GR with dexamethasone 21-mesylate. In this case, the concentration of ligand is kept fairly low (0.5–2 nM). This is to reduce the amount of nonspecific labeling of cellular proteins.] A rapid labeling (~15–30 min) of the GR can be achieved by incubation at 25–37°C. However, this can result in GR transformation with resultant tight nuclear binding (which can lower the yield of GR in the cytosol). It is best to label the GR in intact cells at 0°C for about 2 h. This results in nearly complete labeling of the hormone-binding sites, with the receptor being in the most nearly physiological structure possible. Because the GR is labeled at low temperature, it will be in the untransformed, heterooligomeric state. Cytosol can then be prepared and receptor purification begun. One problem with in vivo labeling of the GR is that subsequent steroid affinity chromatography purification is not possible. One other point that needs to be considered is the metabolic status of the cells during the in vivo labeling. That is, if the culture volume is reduced to one-tenth to one-twentieth volume to conserve tritiated hormone, the pH of the medium can drop dramatically because of the production of large quantities of lactic acid in a small volume. This is most important when the concentrated cells cultures are in vivo-labeled with hormone at 25 or 37°C. If the culture medium turns yellow or orange during the labeling procedure, doubts are raised about the physiological relevance of the results obtained.

5.2. *In Vitro Labeling*

The labeling of the GR in cytosol merely requires the addition of radiolabeled hormone and subsequent incubation at 0–4°C. Again, sufficient saturation of the hormone-binding sites requires that the hormone is added to a level of at least ten times the K_d of binding of the GR. The only other variables in labeling are the temperature and time. Again, 0–4°C is preferred, as this does not cause receptor transformation. Although hormone binding is usually complete within 2–4 h, we find it convenient to label the GR overnight (12–16 h). This is only feasible if the cytosol preparation contains a low level of endogenous protease activity.

6. Purification of the Untransformed Receptor

As mentioned in Section 2, the untransformed GR is apparently a heterooligomeric complex of uncertain composition. Thus, it would be of considerable importance to be able to purify large quantities of this GR form to homogeneity. With the discovery of the stabilizing effect of sodium molybdate on untransformed steroid receptors, this goal has been partially accomplished.

6.1. *[³²P] Labeling*

One of the great advantages of using cell cultures is the ease of metabolically labeling intracellular, macromolecular components (nucleic acids, proteins). The fact that both the GR protein and hsp90 are phosphoproteins has been exploited by in vivo labeling of these molecules during incubation of cells with [³²P]labeled orthophosphate. Also, we have used this approach to analyze for the presence of RNA in the untransformed complex (Kovacic-Milivojevic and Vedeckis, 1986). After growing cells to an appropriate density, the medium is removed and replaced with phosphate-free medium containing dialyzed serum. At this point, the volume of the cell culture can be reduced to conserve the amount of isotope. We have reduced the culture volume to one-sixth of the original volume. After labeling the cells for 3–4 h at 37°C, the volume of the culture medium was increased to 1.5 times the original level and the incubation continued for another 14–16 h. This protocol allowed a rapid equilibration of the radiolabeled phosphate with intracellular pools, and permitted a long-term labeling of the intracellular components with [³²P]labeled phosphate. Also, the pH of the medium did not drop substantially during the initial incubation of the concentrated cell culture.

6.2. Three-Step Purification

By far the most useful purification protocol for the untransformed GR is that of Grandics et al. (1984). The value of this procedure is that a highly purified untransformed GR preparation, in reasonable yield (30–50%), can be obtained rapidly and with a minimum of steps. The fact that the untransformed GR is stabilized by sodium molybdate is of crucial importance. The purification scheme involves only three steps, *all of which* are performed in the presence of 10–20 mM molybdate.

After the preparation of cytosol in the presence of molybdate, it is incubated with a steroid affinity resin, Sterogel A (Sterogene Bioseparations, Inc., Arcadia, CA). It has been our experience that (with the possible exception of attempts to obtain steroid receptor antibodies) more frustrating investigator-years have been spent on affinity chromatography of receptors than on any other laboratory procedures. The Sterogel A resin is a reliable product and its properties are very good. We have successfully used this resin to purify the untransformed GR from AtT-20 cells (Kovacic-Milivojevic and Vedeckis, 1986). In spite of this, other individuals in the laboratory, using the same batch of affinity resin, have been unsuccessful in using this procedure as an effective purification step. We are at a loss to explain why affinity chromatography is such a difficult procedure, except for the fact that there are a substantial number of variables at each of the individual steps in the procedure. When it works, affinity chromatography results in a substantial purification (up to 500-fold) with good yield (~70%). More typical purification factors using affinity chromatography are 100–200-fold (Grandics et al., 1984).

The affinity eluate is then applied to a Bio-Gel A-1.5m agarose column which has been equilibrated in molybdate-containing buffer. This step achieves about a fourfold purification of the GR (Grandics et al., 1984). The peak fractions, which contain the 8.3 nm, untransformed GR, are then pooled. The final step in the purification is DEAE-cellulose chromatography, with elution of the GR using a linear salt gradient. This step results in good purification (~eightfold; Grandics et al., 1984) and concentrates the receptor protein. The presence of molybdate in the buffer is sufficient to overcome the dissociating (transforming) effect of the moderate salt concentration (~0.2 M) at which the untransformed GR elutes from DEAE-cellulose. The purified, untransformed GR is then pooled and used for further analysis.

As is typical with any protein purification, one must balance the purity of the final product with the yield of the protein. That is, higher purity with lower yield is obtained if only the top few fractions of the GR peaks are pooled from the gel filtration and DEAE-cellulose columns. We typically pool about 80% of the hormone-binding peak from each column. Because three separate physical parameters (steroid-binding, charge, and size) are used for each step, this allows both an excellent purification and good overall yield (~30%) using this procedure. The only other effective method for purifying the untransformed GR involves immunoaffinity chromatography. The disadvantage of this approach is that elution of the purified GR from the immunoaffinity resin involves denaturation of the protein. Thus, one cannot study the properties of the native, untransformed GR when this purification protocol is used.

6.3. Characterization of the Purified Receptor

Using the three-step purification protocol described above, it is possible to analyze the subunit composition and mechanism of transformation of the untransformed GR. These analyses have led to some interesting, but as yet equivocal, results. Our laboratory has grown AtT-20 cells in [^{32}P] orthophosphate for 14–20 h in order to covalently label both phospho-proteins and ribonucleic acid (Kovacic-Milivojevic and Vedeckis, 1986). The untransformed GR was then purified using the three-step procedure described above. The GR obtained was untransformed, as evidenced by its sedimentation coefficient (~9S) on sucrose gradients. One curious finding was that there was no evidence of hsp90 in the purified complex, as analyzed by either silver-staining of the gels or by performing autoradiography. The most likely explanation for this is that the denaturing gel system used (10% total acrylamide) was not optimal for distinguishing between the GR (96K) and hsp90 (90K) bands. However, as far as this author is aware, hsp90 has never been demonstrated to be present in untransformed GR preparations obtained by the three-step purification procedure. The reason for this needs to be determined.

The main goal of our study on the purified, untransformed GR was to determine if RNA is an essential subunit of this receptor form. Two [^{32}P]labeled bands were obtained in the final purified preparation, with mol wts of 96 and 24K. Both of these were protease-, but not ribonuclease-sensitive. Amino acid analysis demonstrated the presence of phospho-serine in both radiolabeled bands. Finally, kinetic analyses showed the direct conversion of the 9S, untransformed GR to the 4S monomer, without production of the 6S, RNA-containing, oligomeric GR form. It was, therefore, concluded that RNA was not present in the untransformed GR (Kovacic-Milivojevic and Vedeckis, 1986).

On the contrary, other laboratories have published detailed studies that suggest that RNA is present in the untransformed GR complex (Webb et al., 1986; Schmidt et al., 1986; Sablonnière et al., 1988; Unger et al., 1988). We do not yet understand the discrepancies in these studies. Our laboratory used AtT-20 cells that were in vivo-labeled with [^{32}P]orthophosphate, followed by the three-step purification protocol. The other groups used rat liver, in vivo or in vitro-labeling procedures, fractions from the final DEAE-cellulose column that were not coincident with the GR, and/or somewhat different purification protocols. One possibility is that RNA is present in the untransformed GR, but that it turns over so slowly that it was not labeled in the 14–20 h period used in our study. The other possibility, although dealt with by the other groups, is that the RNA they found was adventitiously associated with the untransformed GR complex and not an integral subunit. Clearly, further studies are required to resolve this important question.

7. Purification of the Transformed Monomer

A large number of procedures have been developed to purify the monomeric GR receptor. Our laboratory has not tried very many purifications of the monomeric GR. Our limited experience has been that the procedures of Gustafsson and his colleagues (Wrange et al., 1979,1984) appear to give the most satisfactory and reproducible results. I would briefly like to mention some general facts and specific hints that should be kept in mind when purifications of the monomeric GR are attempted. Because the protein contaminants present in different cell lines and tissue sources will vary, I will purposely present only general features of the techniques described. It is essential for each investigator to individually determine the optimal conditions for the purification of the GR in their own laboratory and from their specific tissue or cell source.

7.1. Prefiltration on Anionic Resins

The most striking difference between the untransformed and transformed GR species is that the latter is a DNA-binding protein. This difference has been exploited to allow purification of various monomeric, transformed steroid receptors. Basically, after labeling the GR with hormone, the cytosol is passed over an anionic resin. This can be either phosphocellulose, DNA-cellulose, or both. Because cytosol contains untransformed GR, the receptor protein does not adsorb, but rather, elutes in the flow-through fraction. However, any other DNA-binding proteins (or basic proteins) will bind to this resin. At some subsequent step in the purification, the GR is transformed (i.e., converted to its DNA-binding species). Theoretically, only proteins that have acquired DNA-binding activity subsequent to the first anionic column chromatography should be now capable of binding to an anionic resin, and this should be a very small number of proteins. In principle, a second chromatography on an anionic column should result in an enormous purification of the steroid receptor. In addition, because many mechanisms for causing steroid receptor transformation require that the receptor be liganded, only the specific steroid receptor that has been labeled with hormone should bind to the second anionic resin. This approach (prefiltration on an anionic resin, receptor transformation, filtration on a second anionic resin) results in excellent (many thousand-fold) receptor purification (Wrange et al., 1979,1984).

Four factors appear to be most important to achieve the best purification and yield when using this approach. First, the flowrate of the initial anionic column must be slow enough to allow adsorption of those proteins originally present in cytosol that can bind to the resin (Wrange et al., 1979). Although it is generally assumed that ion-exchange occurs almost immediately upon applying a protein to the resin, this appears to not always be the case. Alternatively, it is possible that non-DNA-binding protein complexes can be converted into DNA-binding forms upon prolonged exposure to the polyanion. If this rate is substantially faster than the spontaneous transformation of the receptor under the same conditions, these contaminating proteins (which may adsorb to the second anionic resin) can be eliminated with this first anionic column. The second important point is that the GR itself can be transformed by extended exposure to the first anionic column. Thus, if the filtration rate is too slow, substantial losses in yield can occur at this step. One way to minimize this effect to is filter *unlabeled* cytosol over the first anionic column (because GR transformation does not readily occur in the absence of hormone). The flow through fraction can then be labeled with hormone.

The third important consideration is the method used to elute the bound GR from the second anionic column. A salt (KCl, NaCl) gradient affords a good purification of the bound GR. It is best to run a fairly shallow gradient to obtain the best separation of the GR from contaminant proteins also bound to the column. However, one must, again, balance the purity of the GR with the yield obtained. If the gradient is too shallow, the GR preparation will be dilute, and this could render the hormone-binding activity of the receptor unstable. Besides monovalent salts, two other agents can be used to elute the GR from DNA-cellulose columns. Milli-molar concentrations of $MgCl_2$ or pyridoxal 5'-phosphate are effective in eluting the GR in a good yield and state of purity (Wrange et al., 1979,1984). Normally, a step elution of the receptor with either of these agents is used. We have not tried gradient elutions using these substances. Although one would assume that a gradient elution would result in improved purity of the product, we do not know if a discrete, concentrated peak of GR would be obtained by gradient elution. This possible alteration in methodology deserves further study.

The fourth important consideration is that steroid receptors tend to get "stickier" as they are purified. Therefore, all glassware and plasticware should be siliconized before use in the purification of the transformed GR.

7.2. Receptor Transformation

At some point in the purification of the monomeric GR, the untransformed GR must be converted into the transformed, monomeric species. This is usually not difficult—many of the commonly used purification procedures, in themselves, promote receptor transformation. However, it is advisable to control this process and allow it to occur at a point that will contribute most to the purification protocol. Most commonly-used methods of GR transformation afford no purification. That is, methods that disrupt subunit interactions (heating, salt-treatment) and those that remove the low mol wt endogenous inhibitor of transformation (dilution, dialysis) result in no purification of the GR. One exception to this is gel filtration, which both removes the low molecular inhibitor and can purify the GR. One method that continues to be ignored is salting out using ammonium sulfate. We have utilized this in the past, and found that a concentration of 35–40% saturated ammonium sulfate causes transformation and precipitation of almost all of the GR. In addition, this step results in a moderate purification of the GR. We have also found that almost no GR is precipitated at 20% saturated ammonium sulfate. Therefore, the flow through fraction from an anionic prefiltration column can be labeled with hormone and then brought to a concentration of 20% saturated ammonium sulfate. After pelleting by centrifugation, the supernatant could be adjusted to 40% saturated ammonium sulfate, and the insoluble, transformed GR pelleted by centrifugation. The receptor can then be dissolved in buffer. Besides obtaining purified, transformed GR, one can dissolve the precipitated GR in a small volume of buffer, thereby concentrating it as well. This is a neglected possible step in GR purification that, together with some further refinement, could be a very useful addition to standard receptor purification protocols. Also, because the principle of this technique is not based solely on the charge or size properties of the protein, it may provide a further purification even if ion-exchange chromatography and gel filtration are also used in subsequent purification steps.

7.3. Ion-Exchange Chromatography

Because the behavior of the GR on various ion-exchange resins has been recently reviewed (Vedeckis, 1985), a detailed description of these results will not be repeated here. Both cationic and anionic resins have been used to purify the GR protein.

I have already mentioned that the untransformed GR elutes at about 0.2 M KCl from DEAE-cellulose. DEAE-Sephadex or DEAE-Agarose can also be used with similar results (although the untransformed GR elutes at a slightly higher salt concentration from these resins). After receptor transformation, the GR elutes at a lower salt concentration from these cationic columns (e.g., at about 0.08 M KCl in the case of DEAE-cellulose). It is curious that this differential chromatographic behavior of these two GR forms has not been used extensively to purify the transformed GR (as have the anionic resins mentioned earlier).

The use of anionic resins (phosphocellulose, DNA-cellulose) to purify the transformed GR has been mentioned earlier. The transformed GR elutes at a somewhat higher salt concentration from phosphocellulose (0.17 M KCl) than from DNA-cellulose (0.14 M KCl).

7.4. Gel Filtration

Gel filtration is the most widely used method for purifying the GR based upon size. The transformed GR has a Stokes radius of about 6 nm. We have found that a Bio-Gel A-1.5m agarose column is ideal for the purification of the transformed GR. More recently, we have used high performance size exclusion chromatography to study this receptor form (LaPointe et al., 1986). With further application of this technique on a preparative scale (discussed in Vedeckis et al., 1987), this may be the method of choice for GR receptor size fractionation in the future.

7.5. Lesser-Used Methods

A fair number of alternative methods of GR purification have been attempted. It is beyond the scope of this chapter to review them all. However, I will briefly mention a few that have been used in our laboratory, with varying degrees of success.

Hydoxylapatite (HAP) chromatography is a potentially useful method for GR purification. We have found that the GR (untransformed or transformed) adsorbs to HAP and elutes at a phosphate concentration of about 0.11M (Vedeckis, 1981). The amount of purification obtained by this technique has not been critically analyzed. One problem with this matrix is that it contains calcium. Therefore, if the cell or tissue source contains a large amount of calcium-activated protease, the GR can be cleaved into various fragments (Vedeckis, 1983a). However, we have found that some cell lines (e.g., AtT-20 cells) contain low enough quantities of endogenous proteases as to make HAP chromatography a potentially useful purification step (Vedeckis, 1983b). HAP chromatography or batch adsorption can also be used to concentrate dilute GR preparations.

Three additional methods could potentially be of use in receptor purification, although problems exist with each. The first is preparative sucrose gradients. The resolution obtained, amount of sample that can be purified, and length of time necessary for centrifugation have limited the use of this technique for GR purification. However, the utility of vertical tube rotor sucrose gradients (Eastman-Reks et al., 1984; Reker et al., 1985) may warrant a reexamination of this method for receptor purification. The last two techniques of potential use are native gel electrophoresis and isoelectric focusing. The main problem with these techniques (in our laboratory) has been a dramatic loss of hormone-binding activity when these approaches were attempted. However, with the use of a covalent ligand, such as dexamethasone 21-mesylate, these two methods should be quite useful in purification of the GR.

7.6. *Possible Future Approaches*

The possibility of obtaining an improved method for the purification of the transformed GR should not be overlooked. With the advent of very rapid techniques (high performance gel exclusion and ion-exchange chromatography; vertical tube rotor sucrose gradients) it may be possible to carry out rapid gel filtration, anion and cation exchange chromatography, and sedimentation fractionation of the untransformed GR. After GR receptor transformation, it may then be possible to purify the transformed GR (which differs in its behavior from the untransformed GR) with these same techniques. The optimum sequence for these steps would need to be critically analyzed, so as to maximize both the purity and yield of the final purified GR preparation.

8. Proteolytic Receptor Forms

Purification of the two hormone-binding fragments of the GR has been largely ignored. There are substantial charge and size differences between these fragments and the native GR species (*reviewed in* Vedeckis,

1985). It is likely that serious attempts to purify these fragments will be made in the near future. This will be necessary when investigators attempt X-ray crystallography of the GR, particularly with the goal of defining the three-dimensional structure of the receptor domains involved in the binding of the hormone and to DNA. The day when the atomic interactions between the amino acids in the receptor and the hormonal ligand, and between the receptor amino acids and DNA, can be visualized is eagerly awaited. It is likely that these studies will require purified GR domains, as opposed to the large intact monomeric protein.

9. Future Directions

There appear to be two areas in which the purification of the GR can progress at a dramatic rate. The first of these will involve immunoaffinity chromatography of the receptor. There are now a significant battery of receptor antibodies that have been developed by a number of laboratories. The major challenge that remains is to devise approaches to elute the GR from the affinity resin in a native, as opposed to denatured, state. This will then allow the GR to be used for various functional studies (e.g., DNA-binding, stimulation, and inhibition of gene expression, and so on). We have obtained a GR antibody by injecting rabbits with a synthetic peptide corresponding to the immunogenic domain of the mouse GR (Ali and Vedeckis, unpublished observations). We have noted that this antibody is much more reactive to the synthetic peptide than to the native GR. Therefore, one possibility would be to prepare an immunoaffinity column, adsorb the native GR protein, and then compete the bound GR from the affinity column by adding the synthetic peptide. An alternative approach would be to develop a renaturation protocol that could be used with the GR after it is eluted from the affinity column with a denaturant. With the larger number of GR antibodies now available, results from these approaches should be forthcoming.

The second improvement in the purification of the GR will undoubtedly come from the area of recombinant DNA technology. It is only a matter of time before overexpression vector systems are developed for the cloned GR gene. These could be developed in either bacterial, yeast, or mammalian expression systems. Because the GR is posttranslationally modified (e.g., by phosphorylation) and because it may associate with other intracellular components (hsp90), a mammalian expression system would be preferred for functional studies. A problem could arise if overexpression of the GR protein were deleterious (e.g., toxic) to cellular physiology. In this case, it may be necessary to place the cloned GR gene under the control of an inducible promoter (mouse mammary tumor virus long terminal repeat, metallothionein promoter, heat-shock promoter, and so on). In this way, the cells could grow normally, and overexpression of the GR could be induced just prior to receptor purification. In any event, once one is capable of producing large quantities of GR in cells, the subsequent purification of the protein and further structural and functional studies will be greatly facilitated. This will undoubtedly result in a renewed interest in the efficient purification of the GR protein.

Acknowledgments

I would like to sincerely thank all of the past and present members of my laboratory who contributed to our understanding of the structure and function of the AtT-20 cell glucocorticoid receptor. I am also grateful for the support of this research by the National Institutes of Health and the American Cancer Society.

References

Ali, M. and Vedeckis, W. V. (1987a) *Science* **235,** 467–470.

Ali, M. and Vedeckis, W. V. (1987b) *J. Biol. Chem.* **262,** 6771–6777.

Ali, M. and Vedeckis, W. V. (1987c) *J. Biol. Chem.* **262,** 6771– 6777.

Bodine, P. V. and Litwack, G. (1988a) *Proc. Natl. Acad. Sci. USA* **85,** 1462–1466.

Bodine, P. V. and Litwack, G. (1988b) *J. Biol. Chem.* **263,** 3501–3512.

Bourgeois, S. and Newby, R. F. (1977) *Cell* **11,** 423–430.

Cidlowski, J. A. and Cidlowski, N. B. (1981) *Endocrinology* **109,** 1975–1982.

Cidlowski, J. A. and Michaels, G. A. (1977) *Nature* **266,** 643–645.

Eastman-Reks, S. B., Reker, C. E., and Vedeckis, W. V. (1984) *Arch. Biochem. Biophys.* **230,** 274–284.

Goidl, J. A., Cake, M. H., Dolan, K. P., Parchman, L. G., and Litwack, G. (1977) *Biochemistry* **16,** 2125–2130.

Grandics, P., Miller, A., Schmidt, T. J., Mittman, D., and Litwack, G. (1984) *J. Biol. Chem.* **259,** 3173–3180.

Kovacic-Milivojevic, B. and Vedeckis, W. V. (1986) *Biochemistry* **25,** 8266–8273.

Kovacic-Milivojevic, B., LaPointe, M. C., Reker, C. E., and Vedeckis, W. V. (1985) *Biochemistry* **24,** 7357–7366.

LaPointe, M. C., Chang, C.-H., and Vedeckis, W. V. (1986) *Biochemistry* **25,** 2094–2101.

Mendel, D. B., Bodwell, J. E., and Munck, A. (1987) *J. Biol . Chem.* **262,** 5644–5648.

Northrop, J. P., Gametchu, B., Harrison, R. W., and Ringold, G. M. (1985) *J. Biol. Chem.* **260,** 6398–6403.

Pratt, W. B. (1987) *J. Cell. Biochem.* **35,** 51–68.

Reker, C. E., Kovacic-Milivojevic, B., Eastman-Reks, S. B., and Vedeckis, W. V. (1985) *Biochemistry* **24,** 196–204.

Reker, C. E., LaPointe, M. C., Kovacic-Milivojevic, B., Chiou, W. J. H., and Vedeckis, W. V. (1987) *J. Steroid Biochem.* **26,** 653–665.

Sablonnière, B., Economidis, I. V., Lefebvre, P., Place, M., Richard, C., Formstecher, P., Rousseau, G. G., and Dautrevaux, M. (1988) *Eur. J. Biochem.* **177,** 371–382.

Schmidt, T. J., Diehl, E. E., Davidson, C. J., Puk, M. J., Webb, M. L., and Litwack, G. (1986) *Biochemistry* **25,** 5955–5961.

Simons, S. S., Jr. and Thompson, E. B. (1981) *Proc. Natl. Acad. Sci. USA* **78,** 3541–3545.

Svec, F. and Harrison, R. W. (1979) *Endocrinology* **104,** 1563–1568.

Svec, F. and Rudis, M. (1981) *J. Biol. Chem.* **256,** 5984–5987.

Tienrungroj, W., Sanchez, E. R., Housley, P. R., Harrison, R. W., and Pratt, W. B. (1987) *J. Biol. Chem.* **262,** 17342–17349.

Unger, A. L., Uppaluri, R., Ahern, S., Colby, J. L., and Tymoczko, J. L. (1988) *Mol. Endocrinol.* **2,** 952–958.

Vedeckis, W. V. (1981) *Biochemistry* **20,** 7237–7245.

Vedeckis, W. V. (1983a) *Biochemistry* **22,** 1975–1983.

Vedeckis, W. V. (1983b) *Biochemistry* **22,** 1983–1989.

Vedeckis, W. V. (1985) in *Hormonally Responsive Tumors* (Hollander, V. P., ed.) Academic Press, New York, pp. 3–61.

Vedeckis, W. V., LaPointe, M. C., and Kovacic-Milivojevic, B. (1987) *BioChromatography* **2,** 178–185.

Webb, M. L., Schmidt, T. J., Robertson, N. M., and Litwack, G. (1986) *Biochem. Biophys. Res. Commun.* **140,** 204–211.

Wrange, O., Carlstedt-Duke, J., and Gustafsson, J.-A. (1979) *J. Biol. Chem.* **254,** 9284–9290.

Wrange, O., Okret, S., Radojcic, M., Carlstedt-Duke, J., and Gustafsson, J.-A. (1984) *J. Biol. Chem.* **259,** 4534–4541.

The Glucocorticoid Receptor

Purification and Characterization

Manjapra V. Govindan

1. Introduction

A number of purification methods have been described for the gluco-corticoid receptor (GR), ranging in molecular masses between 37–150 kDa. Do these discrepancies in mol wt result from proteolytic degradation or are they contained in different mRNA species or encoded within several receptor genes? We will be able to answer these questions because of the recent development in the molecular biology of hormone receptors.

Glucocorticoids have been shown to regulate among others, the expression of mouse mammary tumor virus (Lee et al., 1981), mouse metallo-thionein (Hager et al., 1980) and growth hormone genes (Evans et al., 1982). Recognition of specific nucleotide sequences by rat liver GR in the cloned MMTV (Scheidereit et al., 1983), human metallothionein (Karin et al., 1984), chicken lyzozyme (Renkawitz et al., 1984), and rat tyrosine amino-transferase (Becker et al., 1986) genes supports the view that the GR–hormone complex functions as a eukaryotic transacting transcriptional regulatory factor. Recent cloning of the complementary DNA for GR (Miesfeld et al., 1984 Hollenberg et al., 1985; Govindan et al., 1985; Danielsen et al., 1986), estradiol receptor (Walter et al., 1985; Green et al., 1986; Maxwell et al., 1987), progesterone receptor (Loosfelt et al., 1986), androgen receptor (Govindan et al., 1987; Lubahn et al., 1988), mineralo-corticoid receptor (Arriza et al., 1987), vitamin D receptor (McDonnell et al., 1987), thyroid hormone receptor (Sap et al., 1986), and elucidation of their amino acid sequence deduced from the cDNA sequencing, reveal the interrelationship between these receptor proteins and their evolutionary origin from verbA protein (Jannson et al., 1983).

2. Results and Discussion

2.1. Purification of Rat Liver GR

Affinity chromatography using matrix-bound steroid has therefore been exploited for the isolation and purification of steroid receptors (Failla et al., 1975). Failla et al. (1975) covalently linked a corticosterone derivative to agarose and used this as an affinity ligand for partially purifying the glucocorticoid receptor from hepatoma (HTC) cells. In 1975, Sweet and Adair introduced an affinity matrix in which the steroid molecule was linked through a disulfide linkage (Sweet and Adair, 1975). For the purification of glucocorticoid receptor from rat liver, we synthesized suitable steroid derivatives that were linked with the matrix through the disulfide bridge using cystaminium-dichloride.

2.1.1. Synthesis of Affinity Columns and Applications

^{3}H/deoxycorticosterone-21-hemisuccinate with a specific activity of 1.578 mCi/mmol was reacted with cystaminiumdichloride (Govindan and Sekeris, 1976) in the presence of l-ethyl-3-(3-dimethylaminopropyl) carbodiimide hydrochloride (EDC). The cystaminium monoamide derivative was coupled to CH-Sepharose 4B at pH 4.5 using EDC. The paper on partial purification of rat liver glucocorticoid binding proteins by affinity chromatography (Govindan and Sekeris, 1976) documents the utilization of the above-affinity matrix.

The steroid bound to the matrix by an ester bond was found to be labile during the affinity chromatography of rat liver cytosol. Since rat liver cytosol contains esterases, we observed a rapid depletion of the steroid ligand from the column during the affinity chromatography procedure.

2.1.2. ll-DOC-2l-Cystaminiumdiamide-Sepharose 4B

Because of the above-mentioned reasons, we chose to attach the steroid ligand on the matrix by a C–N bond, resistant to the hydrolytic enzymes or cytosol. We synthesized sH-deoxycorticosteroid-21-chloride (Counsell et al., 1968) and coupled of this ligand to CH-Sepharose.

To obtain optimal purification by affinity chromatography, many of the material that interacts with the matrix nonspecifically had to be removed by a prepurification step. For this, we concentrated the glucocorticoid-receptor from rat liver cytosol by protamine sulfate precipitation

(Rousseau et al., 1975). The receptor protein can be concentrated as much as 8 times with practically no loss of binding activity. In this concentration procedure, it was possible to remove the serum cortisol binding globulin (CBG), thus separating the dexamethasone binder quantitatively.

Using a combination of the above procedure and affinity chromatography, a 12,000-fold purification of the glucocorticoid-receptor was obtained. Purification of two dexamethasone-binding proteins from rat liver cytosol (Govindan and Sekeris, 1978) documents this purification procedure.

The affinity-purified receptor was chromatographed on a DEAE-cellulose column. On the basis of their elution characteristics from the DEAE-column, the presence of three dexamethasone binding components in rat liver cytosol was demonstrated (Schmid et al., 1976). They were named DE-1, present in the flowthrough; DE-2, eluting at 0.1 mM NaCl; and DE-3, eluting at 0.2 mM NaCl. On the basis of the above elution characteristics, we obtained two receptor fractions, DE-2 and DE-3, after chromatography of affinity purified material on DEAE-cellulose. We were unable to detect DE-1 in the flow through of the purified receptor. It may be possible that the loss or inactivation of this component occurred during the concentration step by protamine sulfate precipitation or during extraction with 0.16M phosphate buffer.

Analytical SDS-polyacrylamide gel electrophoresis (Laemmli, 1970) of the fractions containing radioactivity after acid precipitation showed that DE-2 has a mol wt of 45,000, and DE-3 consisted of a protein of mol wt 90,000. DE-2 (fractions 18–21) and DE-3 (fractions 23–25) were obtained in approx 95% pure form (Govindan and Sekeris, 1978).

2.1.3. DEX-20-Cystaminiumdiamide Sepharose 4B and DEX-21 Deoxy-21- Aminocystaminium Monoamide Sepharose 4B

[^{3}H]dexamethasone-17-β carboxylic acid (Mondero, 1937) and [^{3}H] dexamethasone-21-methanosulfonate (Govindan and Manz, 1980) were synthesized with a specific activity of 2.5 μCi/mmol. These derivatives were coupled with cystaminiumdichloride using EDC. The monoamide products were characterized (Govindan and Manz, 1980) and used in synthesizing affinity matrixes to purify the glucocorticoid receptor (Govindan et al., 1982). The 17β-carboxamides (Rousseau et al., 1979) formed by coupling DEX-17β-COOH to amines are well characterized derivatives with almost the same affinity to receptor as glucocorticoids. The affinity matrixes contained 0.1–0.08 μmoles of ^{3}H-dexamethasone bound per gram of gel.

2.2. Purification Procedures

These affinity matrixes were used in purifying glucocorticoid receptor from rat liver cytosol by direct incubation. The affinity matrixes were used to bind glucocorticoid-receptor after an overnight incubation. After affinity elution, the eluted material was submitted to DNA-cellulose chromatography (Milgrom et al., 1973). The apparently native cytoplasmic form of the receptor, the 6.1 nm complex, was purified by Wrange et al. (1979) using a method based on the differential affinity of the nonactivated and activated forms of the receptor complex for DNA. The stability of the steroid–receptor complex was improved by affinity eluting the complex from DNA-cellulose with pyridoxal 5'-phosphate (Cake et al., 1978). Pyridoxal phosphate has also been shown to extract the glucocorticoid receptor from the cell nucleus, in which case the extracted complex is much smaller than the cytoplasmic complex (Cidlowski et al., 1978). Characterization of the purified receptor by gel filtration on Sephadex G-200 yielded a mol wt of 85,000 as calculated according to Siegel and Monty (Siegel and Monty, 1966). By SDS-electrophoresis analysis, a mol wt of 89,000 was observed (Wrange et al., 1979).

The purified receptor by affinity chromatography on DOC-affinity column showed two receptor bands. The major form had a mol wt of 90,000 (DE-3) and the other 45,000 (DE-2). The 90,000 d component isolated by affinity chromatography seemed to be the same as the 89,000 d glucocorticoid receptor subunit purified by Wrange et al. (1979).

By employing DNA-cellulose chromatography after steroid affinity chromatography, the native cytosol receptor could be purified to homogeneity. The second band enriched from rat liver cytosol by dexamethasone affinity column with mol wt of approximately 35,000, did not bind to DNA-cellulose and was found in the flow through fraction. The nature of this 35,000 d polypeptide, which is a degradation product of GR binding dexamethasone, has been recently verified. The unactivated, molybdate-stabilized glucocorticoid receptor has been purified using the affinity chromatography approach (Grandics et al., 1984).

The major mol wt component present in the purified glucocorticoid receptor preparation (Govindan et al., 1982) and the receptor component purified by DNA-cellulose affinity chromatography (Wrange et al., 1979; Govindan and Gronemeyer, 1984) has virtually identical mobility on the same gel system. The component with the mol wt 90,000 is the predominant one present in the purified receptor preparation.

The nature of 45,000 and 30,000 d polypeptides with regard to the cytosol receptor has been discussed by Wrange et al. (1984). They demonstrated proteolysis of glucocorticoid receptor with lysosomal enzymes (Carlstedt-Duke et al., 1977) and observed a fragment of 3.6 nm. It was suggested that the 45,000 d is identical to the 3.6 nm. Climent et al. (1977) have reported a partial purification of glucocorticoid receptor with a mol wt of 33,500. The lower mol wt component purified from rat liver cytosol by dexamethasone affinity column is probably identical to the receptor component partially purified by Climent et al. (1977). This 35,000 d polypeptide is also a proteolytic fragment of the glucocorticoid receptor, as discussed by Wrange et al. (1979).

2.3. Affinity Labeling of Rat Liver Glucocorticoid Receptor

The 9a-fluoro-16a-methyl-2l-diazo-2l-deoxy prednisolone synthesized (Manz and Govindan, 1980) in an overall yield of 30% has about 300-fold less affinity for the glucocorticoid receptor. Affinity constant was ~$1 \times 10^6 M^{-1}$. This derivative inactivated dexamethasone binding in crude cytosol to 20%. The receptor preparations obtained after affinity labeling and subsequent DNA-cellulose chromatography show 90,000 d as the predominant polypeptide band. Fluorography of the gel containing receptor enriched from the affinity labeled material by the second DNA-cellulose chromatography show that there is only one polypeptide band labeled, namely the 90,000 d.

In order to follow the radiolabeling of the second DNA-cellulose excluded material, the flowthrough was 35% ammonium sulfate precipitated. Following dialysis to remove ammonium sulfate against Buffer C (Govindan and Gronemeyer, 1984), the dialysate was chromatographed on a DEAE-Sepharose 6B (50 mL, flowrate 2.5 mL/min). The bound material was eluted with a linear gradient of 50–500 mM NaCl. Fractions of 10 mL were collected and individual fractions were precipitated with 10% TCA and SDS-polyacrylamide gel electrophoresis was performed. The gel was stained with Coomassie Blue and fluorographed to identify the labeled bands. Fractions 8–10 and 24–30 were radioactive.

The affinity labeling provided direct evidence that the 90,000 d polypeptide is one of the subunits of glucocorticoid receptor if not the intact glucocorticoid receptor. The other labeled product identified after ammonium sulfate precipitation, and DEAE chromatography has a mol wt of 35,000. This may be a degraded product of glucocorticoid receptor. From the labeling to the last step of purification, it took three days, and we assumed that the intact receptors were proteolytically degraded by then.

Further characterization of the rat liver GR was performed by purifying the GR by different techniques and analyzing the purified product by photoaffinity labeling with unmodified ligand as well as purifying the affinity labeled receptor with dexamethasone-21-mesylate prior to purification.

2.4. Comparison of Rat Liver Glucocorticoid Receptor Purified by DNA Affinity Column (Method A) and Ligand Affinity Chromatography

The principle of the method is based on the differential binding of nonactivated and thermally activated receptor to DNA-cellulose (Wrange et al., 1979; Payvar et al., 1983).

After incubation with [^{3}H]triamcinolone acetonide, the cytosol from 16 rat livers was filtered over a 400-mL phosphocellulose column to remove hormone-binding serum proteins and a major portion of DNA-binding proteins. The phosphocellulose flowthrough adjusted to pH 7.8 was applied onto the first DNA-cellulose column. As soon as one-half of the flowthrough was obtained, it was immediately thermally activated, cooled to 0°C, and applied onto a second DNA-cellulose column. During this time, the rest of DNA-cellulose column flowthrough was thermally activated, cooled to 0°C, and applied onto the second DNA-cellulose column.

The first DNA-cellulose column was washed with 3 volumes of buffer C, 3 volumes of 0.5× TBE, and eluted with 0.5M NaCl in buffer C. Approximately 5% of the purified hormone-binding activity was eluted and immediately irradiated (Govindan and Gronemeyer, 1984). In a manner similar to the first DNA-cellulose flowthrough, the flow through obtained from the second DNA-cellulose column was collected in three fractions that were independently incubated for 15 min at 20°C. This mode of fractionation was convenient because relatively low flow rates are necessary for the quantitative binding of activated receptor to DNA-cellulose (Wrange et al., 1979). After cooling to 0°C, each fraction was directly applied onto a third DNA-cellulose column. DNA-cellulose columns II and III were washed and the eluates were irradiated as described for the first DNA-cellulose column (Gronemeyer et al., 1985; Govindan and Gronemeyer, 1984). Using the method of photoaffinity and electrophilic crosslinking, we investigated the nature of the hormone-binding activities eluted from the three columns.

Scanning of the fluorograph obtained after photoaffinity labeling shows that more than 95% of the binding activity present in the three eluates represents the intact glucocorticoid receptor polypeptide. Several distinct minor labeled bands can also be detected on the fluorographs. These bands are not a result of artifactual crosslinking, e.g., occurring during irradiation, as the were present with identical mobilities in both affinity labeling reactions. Most probably they result from receptor breakdown products with intact hormone and DNA-binding domains that are copurified with the glucocorticoid receptor.

The radioactivity eluted from the first DNA-cellulose column represents binding to the intact receptor, which may have been activated either endogenously or during the initial stages of purification. The majority of the binding activity is recovered from the second DNA-cellulose column and represents all the receptor molecules which have been activated during the first thermal treatment (47.5% of the first DNA-cellulose flowthrough). This is in agreement with other published results (Wrange et al., 1979; Govindan and Gronemeyer, 1984). Attempts to bind the activity of the second DNA-cellulose flowthrough fractions on a further DNA-cellulose column or longer initial heat treatments were unsuccessful. However, when this fraction was again heated (15 min at 20°C) and applied onto a third DNA-cellulose column, no binding activity could be detected in the flow through. Analysis of the eluate obtained from this column by affinity labeling demonstrated that not only the activity present in the flowthrough of the second DNA column could be reactivated, but also that it could be quantitatively isolated as intact receptor. With the inclusion of this step, the overall recovery of activated receptor from DNA-cellulose columns II and III represented 80% of the phosphocellulose flowthrough. Thirty-eight and 28% of the receptor initially present in the cytosol are removed in DNA-cellulose eluates II and III with purifications of approx 2700- and 460-fold, respectively.

From the mol wt determined as described above, the pure rat liver glucocorticoid receptor is 10,870 pmol/mg of protein, assuming one steroid-binding site/92kDa receptor polypeptide. Therefore, the calculated purity of receptor in DNA column II and III eluates is 25 and 4%, respectively. Two main contaminating proteins are present in the eluate of the

DNA-cellulose II column. When necessary, the majority can be easily removed by DEAE chromatography, after removal of ammonium sulfate by extensive dialysis, to give approx 60% pure receptor protein with an overall 15% recovery (Wrange et al., 1984).

3. Purification of Rat Liver Glucocorticoid Receptor by Dexamethasone Affinity Chromatography (Method B)

For affinity chromatographic purification, we used a matrix containing dexamethasone 17β-carboxylic acid bound by an amide bond to Affi-Gel-102. The non[^{3}N]TA-labeled cytosol was filtered through a phosphocellulose column as described for method A, and the flowthrough was applied onto the affinity column at a flow rate of 1.5 mL/min. This flowrate was optimum to enable the maximum interaction of the binding components with the stationary matrix. We chose a column technique because the purity of the material was greater than for a batchwise procedure; it also avoided losses caused by sticking of the matrix to the walls of the incubation container and was more effective for removing unbound material.

After washing the column extensively overnight with buffer A and then 250 mM NaCl in buffer A, the elution was performed by incubating the matrix for 5 h at 0°C in 20 mM Tris-HCl, pH 7.4, 10 mM mercaptoethanol, and 10% glycerol containing 50 mM NaSCN (110) and 1 μM triamcinolone acetonide (20 mL). After elution, the column was washed with an additional 30 mL of elution buffer and the eluate and the washing were combined. A 5-mL aliquot of this fraction was immediately filtered through Sephadex G-50 and the excluded protein peak was used to determine the amount of [^{3}H]TA-binding activity released from the matrix as well as for photoaffinity labeling studies. More than 90% of the binding activity could be eluted from the affinity matrix, being nearly 5900-fold purified. Analysis of this activity by silver staining and photoaffinity labeling shows that the same band is crosslinked as in method A.

For further purification, the affinity eluate was diluted, adjusted to pH 7.8, activated, and applied onto a DNA-cellulose column. The column was washed and eluted, and the eluate was immediately crosslinked as described above. The final purification thus achieved is approx 7800-fold, with a recovery of about 63% (Govindan and Gronemeyer, 1984). Analysis of the DNA-cellulose purified affinity eluate by affinity labeling and silver staining shows that intact glucocorticoid receptor had been purified.

3.1. Preparative SDS-Polyacrylamide Gel Electrophoresis of Glucocorticoid Receptor Purified by Method A

DNA-cellulose column eluates II and III were further purified by preparative SDS-polyacrylamide gel electrophoresis using a continuous flow elution devise (Govindan and Gronemeyer, 1984). Thirty-five μg of DNA-cellulose II and 100 μg of DNA-cellulose III eluates were analyzed by SDS-polyacrylamide gel electrophoresis and fluorography before the preparative gel electrophoresis. The differences observed between the autoradiograph may be a result of proteolysis. Whether this proteolysis takes place during the storage or thawing is not yet clear.

Even though most of the degradation products are also present, initially they have increased during storage. Most of the degradation products are common to both DNA-cellulose II and III eluates. Elution and recovery were followed by measuring crosslinked radioactivity in 10-μg aliquots of each fraction. The preparative gel fractions obtained from DNA-cellulose II and III eluates exhibited only one silver-stained protein band at the height of the major radioactive peak. BII and BIII, these peak fractions contained two different components that may be due to limited proteolysis.

Method A is a modification of the purification procedures described by Eisen and Glinsman (1978) and Wrange et al. (1979). Pure preparations (50–80%) with yields between 20–45% had been achieved previously by specific dissociation of the receptor–hormone complex from DNA-cellulose using either pyridoxal 5'-phosphate (Wrange et al., 1979) or magnesium chloride (Wrange et al., 1984) to avoid quenching of photoaffinity labeling reaction because of pyridoxalphosphate. This reagent was omitted for the dissociation of hormone-receptor from DNA-cellulose. For the isolation and further analysis of all the hormone-binding components bound to DNA-cellulose, we found that quantitative elution can be obtained with NaCl alone.

As has been described by several groups (Simons et al., 1983; Wrange et al., 1979), the efficiency for DNA-cellulose binding of rat liver glucocorticoid receptor after thermal activation is in the order of 30–50%. It has been speculated that this is caused by receptor degradation or a limitation in the activation step. We show here that the hormone-binding activity present in the flow through of the column could be nearly quantitatively bound to and recovered from an additional DNA-cellulose chromatography if a further activation step was employed. Analysis of this material by affinity labeling showed that no degradation had occurred during these two activation steps and that intact receptor was isolated. The binding of receptor to the last DNA-cellulose column was entirely dependent on the second activation step and not caused by insufficient primary activation or limited capacity of the matrix. These possibilities were excluded by increasing the time for the initial activation step, decreasing the flowrate, or doubling the column volume.

Previous studies with dexamethasone 17β-carboxylic acid bound to a disulfide-containing affinity matrix demonstrated the isolation of highly purified receptor (Govindan and Sekeris, 1976). This type of affinity matrix can be used only once, since the hormone–receptor complex is dissociated by cleavage of the disulfide bond between the hormone and the stationary matrix (Green et al., 1986). Previously described affinity matrices (Govindan and Manz, 1980) for the purification of glucocorticoid receptor contain 12–14–CH2-groups as spacers between the ligand and the stationary matrix. By affinity chromatography and subsequent DNA-cellulose chromatography (Govindan and Manz, 1980) or DEAE chromatography (Sakaue and Thompson, 1977), glucocorticoid receptor has been purified 10,548-fold with a 68% recovery. Failla et al. (1975) were able to purify the glucocorticoid receptor 2000-fold using a ll-deoxycorticosterone derivative-bound affinity matrix. Using this matrix, they could absorb 80% of the receptor from crude cytosol and elute 40–50% of the absorbed receptor with $5 \times 10^{-8}M$ (^{3}H)TA. Lustenberger et al. (1981) have described an affinity matrix derived from dexamethasone 17β-carboxylic acid. Using a batchwise procedure, they purified the receptor 7000-fold in two steps with an overall yield of 18%. We prepurified the crude cytosol by phosphocellulose exclusion chromatography and performed the affinity binding by chromatography on a dexamethasone 17β-carboxylic acid-bound matrix. After thorough washing of this column with buffers of different ionic strength, we eluted the hormone–receptor complex quantitatively by in-cubating the affinity matrix with a buffer containing sodium thiocyanate (Green et al., 1980). We further purified the receptor after thermal activation of the affinity eluated by DNA-cellulose chromatography.

The glucocorticoid receptor could be purified approx 7800-fold by employing the hormone-bound affinity matrix with an overall yield of 63%. The affinity matrix absorbed 90% of the receptor present in the phosphocellulose flow through. The elution of the receptor was quantitative using sodium thiocyanate and 1 μM triamcinolone acetonide. This low concentration of chaotropic salt had no negative effect in thermal activation and further purification by DNA-cellulose chromatography. Using one affinity column, this method allows only one purification per week because of time consuming and extensive regeneration of the matrix required before the next isolation. The majority of the receptor in the affinity eluate (~70%) could be thermally activated.

4. Immunological Experiments

Immunization was performed with the eluted proteins from gel slices after preparative gel electrophoresis. Immunization was carried out as described by Vaitukaitis et al. (1971) by injecting subcutaneously. Two rabbits were injected with each antigen. The dose of the antigen used was 5 µg in 1 mL Freund's complete adjuvant. Antibodies were detected 4–5 d after the last injection. Blood was collected by cardiac puncture, 7–8 d after the last injection. The ouchterlony double diffusion experiment with 90,000 and 45,000 d antigens and their respective antibodies against them show that antibodies to 45,000 d reacted with 90,000 d polypeptide, whereas antibodies to 90,000 d polypeptide reacted with 45,000 d polypeptide (Govindan and Sekeris, 1978).

4.1. Immunoprecipitation of Cytosol Glucocorticoid Receptors

In order to determine whether the purified polypeptides are the integral parts of glucocorticoid receptor complex, two experiments were conducted. In the first experiment, cytosol was prepared in tris buffer containing 50 mM KCl and incubated with $5 \times 10^{-8}M$ [^{3}H]-dexamethasone. The unbound hormone was removed from the rat liver cytosol by treatment of the cytosol with dextran-coated charcoal. Aliquots of this labeled cytosol were incubated with various dilutions of antibodies to 45,000 d protein, 90,000 d protein and the preimmune serum as a control for 16 h at 0–4°C. Samples were centrifuged 5000 rev/min for 10 min and the pellet was washed with 0.9% NaCl and radioactivity in the pellet was determined by scintillation counting. Simultaneously, radioactivity remaining in the supernatant was also determined. Both antibody preparations were able to immunoprecipitate the dexamethasone binding activities of rat liver cytosol.

In the second experiment, purified antibodies were coupled to CNBr activated Sepharose and cytosol preparation were applied to the columns. Sepharose bound antibodies to 45,000 mol wt protein and antibodies to 90,000 mol wt protein retained about 72% of rat liver dexamethasone binding activity proving the specificity of these antibodies to the receptor proteins. Antibody bound proteins could later be eluted with 3M sodium thiocyate (Anderson et al., 1975), precipitated with 10% trichloroacetic acid and analytically electrophoresed. It is clear that both antibodies immuno-precipitate the same receptor components from rat liver cytosol. In our experiments with cytosol from adrenalectomized rat livers on the antibody columns, we observe predominantly the 45,000 mol wt component.

4.2. Studies on Translocation by Immunoaffinity Chromatography

In a further experiment, we injected adrenalectomized rats with dexa-methasone, and, after various intervals of hormone treatment, the rats were killed, and cytosol and nuclear extracts were prepared from the rat livers (Chauveau et al., 1956; Govindan, 1980).

Using immunoaffinity chromatography it was possible to show that 90,000 d is the nuclear receptor form. Both cytosol and nuclear form of glucocorticoid receptor are identical in mol wt. The difference between them may be a result of two different confirmation of the receptor after binding with the hormone.

4.3. Immunofluorescence Studies on Translocation

To study receptor translocation, HTC-cells grown on tube slips were processed (Larzarides and Weber, 1974), and all cell suspensions contain-ing the tube slips were incubated with hormone at 4 and at 37°C. To follow the kinetics of translocation, the time intervals of hormone treatment were varied. Control experiments were conducted with preimmune serum. Cells treated with specific antibodies reacted with a strong fluorescence, restricted mainly to the cytoplasm, whereas cells treated with preimmune serum showed only a very weak, diffuse fluorescence in the cytoplasm.

The cells treated with dexamethasone at 4°C for 2, 5, 10, 20, and 30 min differed in no way in the fluorescence distribution pattern, as compared with HTC cells not treated with hormone at 4 or 37°C. By contrast, HTC cells treated with hormone at 37°C showed a change in the fluorescence distribution pattern. As early as 2 min after hormone treatment, the hor-mone receptor complex appeared to migrate from its initial cytoplasmic location. It has been demonstrated that 10 min after hormone treatment, strong fluorescence was no longer observed in the cytoplasm, but most was localized within the nucleus. It should be noted that translocation was almost complete at this time; cells treated with hormone for 20 min showed fluorescence exclusively in the nucleus. At 30 min of incubation at 37°C some fluorescence seemed to be redistributed over the cytoplasm.

Functional characterization of the receptor molecules by raising antibodies against the purified component shows that these antibodies react not only with the purified or partially purified crosslinked and noncrosslinked denatured receptor band but also with native cytosolic receptor as shown by the protein A-Sepharose assay. The antibodies raised against the receptor obtained by affinity chromatography (Govindan and Sekeris, 1978) reacted with the 92,000-Da polypeptide prepared by method A in immunoblotting analysis, and the antibodies raised against the purified glucocorticoid receptor obtained by method A reacted with hormone affinity-purified receptor equally well.

The M_r = 92,000 ± 400 of the unactivated cytosolic receptor protein, observed after protein A immunoadsorbent assay and photoaffinity labeling of the glucocorticoid receptor in the affinity eluate, reported in this chapter, is in close agreement with the size determination of "unactivated" glucocorticoid receptor from rat liver (Grandics et al., 1984).

The proteolytic products purified by preparative SDS-gel electrophoresis were analyzed by Western blotting. The receptor fragments of 45,000 and 64,000 Da reacted with antireceptor antiserum, whereas the 36,000-Da fragment did not react. Thus, the proteolytic fragments of glucocorticoid receptor (45,000 and 64,000 Da) contain not only the hormone-binding and DNA-cellulose-binding domains, but also the immunoreactive domain (Govindan et al., 1985). To analyze whether these antibodies will also react with native nonpurified glucocorticoid receptor molecules, a protein A-Sepharose immunoadsorbent assay was performed. In order to decrease the nonspecific binding of cytosolic proteins to protein A-Sepharose, a preincubation with this matrix was found to be necessary. Thereafter, 1 mL of cytosol containing 50,000 cpm of binding activity was incubated with immune or preimmune serum followed by adsorption to protein A-Sepharose. Totals of 13,500 and 17,000 cpm were eluted from protein A-Sepharose incubated with 1:200 and 1:100 diluted immune serum, respectively. No radioactivity was absorbed on the protein A-Sepharose column when the cytosol was incubated with preimmune serum. Fluorography of the eluates after trichloroacetic acid precipitation and SDS-gel electrophoresis shows that a single band of M_r = 92,000 ± 400 reacted with the antibodies.

5. DNA–Receptor Interaction

Numerous studies have shown that glucocorticoid hormones bind to specific proteins [hormone-receptor complexes (HRCs)] present in their target cells (Yamamoto and Alberts, 1976) and this binding apparently increases the rate of synthesis of specific mRNAs (Martial et al., 1977). DNA binding experiments using crude HRC preparations from cytosol as well as purified glucocorticoid HRC have demonstrated the association of HRCs with DNA (Baxter et al., 1972; Wrange et al., 1979). The major portion of HRC (~10^5 per cell) appears to bind nonspecifically and with relatively low affinity to target cell nuclei in vivo (Williams and Gorski, 1972) and in vitro (Chamness et al., 1974).

Glucocorticoid hormones have been shown to regulate mouse mammary tumor virus (MMTV) production in mammary tumor cell lines (Fine and Arthur, 1976). Induction of MMTV RNA synthesis and accumulation has been associated with a change in the rate of transcription of the proviral genes and may be a primary biochemical response in transcription have provided evidence for the physical linkage of the regulated MMTV gene and the primary DNA sequences that mediate the hormonal response (Hynes et al., 1981).

The availability of cloned viral sequences (Majors and Varmus, 1981) and purified HRC preparations (Govindan et al., 1982) allow characterization of the specificity of the association between HRCs and MMTV DNA. Payvar et al. (1981) used a nitrocellulose filter binding assay to show that HRC binds to MMTV DNA fragments that include either the env gene or the env gene and the long terminal repeat (LTR).

6. Cloning of Human GR cDNA

6.1. Characterization of the Human Glucocorticoid Receptor from MCF-7 Cells

The breast cancer cell line MCF-7 GR was partially purified, photoaffinity labelled with [^{3}H]triamcinolone acetonide (TA), in both the presence and absence of a 100-fold excess of nonlabeled TA, and analyzed by SDS-PAGE. The major labeled band corresponded to a protein of 94kDa that is identical to that observed for the rat liver cytosol GR (Govindan and Gronemeyer, 1984). A minor band of 59kDa was visible on the original fluorograph, which may represent a degradation product of the human GR. Rat liver and MCF-7 GRs were partially purified using method A as previously described (Govindan and Gronemeyer, 1984). Duplicate aliquots of the MCF-7 DNA cellulose I and II eluates and rat liver DNA cellulose I, II, and III eluates were analyzed on Western blots using either a rabbit polyclonal antiserum raised against the rat liver GR or preimmune serum. A single protein of 94kDa was detected for both the rat and human samples using this GR antiserum.

6.2. Cloning of Human Glucocorticoid Receptor cDNAs Expressed in MCF-7 Cells

The rat liver monospecific antibodies, which were capable of detecting as little as 25 pg of purified human GR, were used to screen 500,000 recombinants of an MCF-7 cDNA library (Govindan et al., 1985) in the expression vector λgtll (Young and Davis, 1983), as described previously. Six clones, positive after 3 rounds of screening, were isolated. Three of these clones, which contained small inserts and internal EcoRi sites, were not analyzed further. The two other clones, λHGRl, λHGR9 did not cross-hybridize. The fusion proteins corresponding to these clones displayed apparent mol wt of 119, 122, and 126kDa, which is in good agreement with the cDNA insert sizes. The fusion protein from λGHR16 was studied further by Western blot analysis using the rat liver GR antiserum. This antiserum detected the fusion protein but not wild type (wt) β-galactosidase. The reaction observed with other proteins, in both the λgtll wt and λHGR16 samples, is very likely owing to the presence of antibodies in the rabbit serum directed against *E. coli* proteins. In order to substantiate further that the clones isolated represent the human GR, we used an epitope selection technique (Hall et al., 1984), whereby the purified gtll fusion proteins were used to immunopurify their corresponding antibodies from the rat liver GR antiserum. Both the rat liver and human receptors were recognized by either the rat liver GR antiserum or the HGR16, HGR9, and λHGRl fusion protein-selected antibodies. No proteins were detected using β-galactosidase-selected antibodies. The human GR could not be detected in the MCF-7 cytosol unless it was partially purified.

Rat liver GR was coupled to CNBr-activated Sepharose and used to affinity-purify GR specific antibodies from the rat GR antiserum. These monospecific antibodies immunoabsorbed a [^{3}H]dexamethasone-21-mesylate affinity-labeled protein of 94kDa, following partial purification from either rat liver or MCF-7 cytosols. No crosslinked protein was immunoadsorbed using the preimmune serum. Furthermore, no receptor-bound radioactivity was precipitated with the monospecific antibodies when the human GR was electrophilically labeled in the presence of a 100-fold excess of unlabeled dexamethasone. These data indicate clearly that the rat liver and human GRs have the same size and share common antigenic determinants which are recognized by the rat liver monospecific antibodies described here.

By definition, the GR is a protein which is capable of binding glucocorticoids with both high affinity and specificity. Thus, the fusion protein-selected antibodies, described above, should recognize rat liver and human GRs electrophilically affinity-labeled with [³H]dexamethasone-21-mesylate. As expected, both the rat liver GR monospecific antibodies and each of the fusion protein-selected antibodies precipitated both the rat and human GRs. No receptor-bound radioactivity was seen when either pre-immune serum was used nor when the MCF-7 GR preparation was cross-linked to the mesylate derivative in the presence of a 100-fold excess of unlabelled dexamethasone. Although the three fusion protein-selected antibodies precicipated essentially equivalent amounts of the human GR, the ~HGRl-selected antibodies appeared to be less efficient in precipitating the rat liver GR when compared to the two other selected antibodies.

All of these results support the conclusion that the cDNA sequences of the three isolated clones encode epitopes present in both the human and rat GRs. Cloning of the complete human GR cDNA and revealing the amino acid sequence declined from the cDNA sequenced followed by expression of functional glucocorticoid receptor by gene transfer substantiated these results.

6.3. Northern Blot Analysis
of MCF-7 Poly (A) + RNA Using the Human GR cDNA

MCF-7 cell and rat liver poly(A)+ RNA was analyzed on Northern blots using the cDNA insert of HGR16 as a probe. The human GR cDNA probe cross-hybridized with a rat liver RNA species of about 6.8 kb under high stringency conditions, which indicates a large degree of homology between the two sequences. In comparison, the MCF-7 cell line appears to contain two minor hybridizing poly (A) + RNA species of approx 5 and 6.5 kb, together with a major one of about 7 kb. The same bands were detected when the inserts of either of the other two cDNA clones were used.

From the results presented above, it is evident that the human MCF-7 GR shares common antigenic determinants with, and has the same apparent mol wt (9kDa) as, the rat liver GR. The size of 94kDa is the same as that reported for the GR characterized in two other human cell lines (Harmon et al., 1984). A homogeneous preparation of the rat liver GR allowed the purification of GR clones from a human rgtll cDNA expression library. The fusion proteins of three independent GR clones were used to select their corresponding antibodies from the rat liver GR antiserum. Each of these selected antibodies was capable of recognizing both a rat liver and a human protein of 94kDa by Western blot analysis. Furthermore, partially purified rat liver or human GR, electrophilically affinity-labeled with [³H]dexamethasone-21-mesylate, was immunoadsorbed by each of the three antibody preparations. It is highly likely that each of the three independent noncrosshybridizing cDNA clones selects a different rat liver GR antibody. That each of these antibodies recognizes a protein of 94kDa capable of binding dexamethasone-mesylate, strongly supports the conclusion that these clones correspond to human GR cDNA sequences. Definite proof that we have cloned the human GR is provided by the expression of the complete cDNA sequences encoding the human GR, and the demonstration that the corresponding protein is capable of binding the hormone.

The GR, in common with other steroid hormone receptors, should consist of at least two domains, one responsible for the binding of hormone and the other for specific DNA recognition. Site-directed in vitro mutagenesis together with the comparison to the sequence of the human oestradiol receptor, which has recently been cloned in our laboratory (Govindan et al., 1985), should allow us to locate these functional domains. Expression of GR cDNA in transgenic mice under the regulation of a variety of promoters may help in understanding the role of GR in the control of gene expression during development and in terminally differentiated cells. Taken together, these approaches should allow us to obtain a better understanding of the receptor's mode of action.

7. Summary and Future Perspectives

The glucocorticoid receptor has been purified by a variety of affinity matrices. It has been shown that irrespective of the source of the tissue or species, the GR is a protein of mol wt of 90,000. Antibodies against the purified receptor have been used to characterize the native as well as denatured form of the GR by immunochemical and immunohistochemical techniques. The translocation of the GR in hepatoma tissue culture cells suggests that in the absence of hormone, the GR is predominantly distributed around the nucleus, and hormone facilitates the receptors to tightly bind to the nucleus and chromatin. This finding is contrary to the localization of estradiol or progesterone receptors observed by other groups. The finding that hormone is necessary for highly specific and tight interaction of GR is very strongly supported by the experiments describing the binding of GR with MMTV-LTR. We have been able to show that receptor devoid of hormone has very little if any specific interaction with defined regions of MMTV-LTR. The final evidence for the specific interaction of the receptor–hormone complex with defined DNA sequences is expected. Availability of complete cDNA clones for GR and other receptors will permit to study whether the specific interaction in vitro involves the formation of hormone–receptor complex or the receptor devoid of hormone is efficient or other transacting factors are necessary in specific interaction with DNA.

Affinity labeling using [^{3}H]-21-deoxy-21-diazodexamethasone [^{3}H]-triamcinolone acetonide and [^{3}H]-dexamethason-21-mesylate has been employed in characterizing the hormone-covalently bound GR. By combining the technique of affinity labeling and affinity chromatography the GR purification and characterization is optimized and the proteolytic degradation products are detected. The antibodies prepared against the rat liver GR that crossreacted with the human MCF-7 GR has been used to isolate the cDNA of human GR. The availability of the cloned GR, its expression and construction of chimeric receptor molecules in elucidating biological functions are the recent development of steroid hormone receptor research.

Our aim to isolate the receptor gene and its regulation is well under way. Finally, the experiments already denote the involvement of other transacting factors other than steroid receptors are involved in hormone regulated gene expression. We have developed techniques to characterize and purify such factors and wish to study their interaction with steroid receptors and their regulated genes, in understanding hormone regulated gene expression.

Acknowledgments

This work was partly supported by the Medical Research Council, National Cancer Institute and Kidney Foundation of Canada. I thank C. E. Sekeris, P. Chambon, and F. Labrie for their continued interest and support. I thank also many colleagues who participated during the development of this work and Helen Leclerc for excellent secretarial assistance.

References

Anderson, N. G., Willis, D. D., Holladay, D. W., Caton, J. E., Holleman, J. W., Eveleigh, J. W., Attrill, J. E., Ball, F. L., and Anderson, N. L. (1975) *Anal. Biochem.* **68,** 371–393.
Arriza et al. (1987) *Science* **237,** 268– 274.
Baxter, J. D., Rousseau, G. G., Benson, M. C., Garcea, G. L., Ito, I., and Tomkins, G. M. (1972) *Proc. Natl. Acad. Sci. USA* **69,** 1892–1896.
Beato, M. and Feigelson, P. (1972) *J. Biol. Chem.* **247,** 7890–7896.
Becker, P. B., Gloss, G., Schmid, W., Streaele, U., and Schuetz, G. (1986) *Nature* **324,** 636–688.
Cake, M. H., Di Sorbo, D. M., and Litwack, G. (1978) *J. Biol. Chem.* **253,** 4886–4891.

Carlstedt-Duke, J., Gustafsson, J. A., and Wrange, O. (1977) *Biochim. Biophys. Acta* **497**, 507–524.

Chamness, G. C., Jennings, A. W., and McGuire, W. (1974) *Biochemistry* **13**, 327–331.

Chauveau, J., Moule, Y., and Rouiller, C. (1956) *Exp. Cell Res.* **11**, 317–321.

Cidalowski, J. A. and Thanassi, J. W. (1978) *Biochem. Biophys. Res. Commun.* **82**, 1140–1146.

Climent, F., Doenecke, D., and Beato, M. (1977) *Biochemistry* **16**, 4694–4703.

Counsell, B. H., Hong, B. H., Willette, R. E., and Renade, V. V. (1968) *Steroids* **11**, 817–826.

Cuatrecasas, P. and Anfinsen, C. B. (1971) *Ann. Rev. Biochem.* **40**, 259–278.

Danielsen, M., Northrop, J. P., and Ringold, G. M. (1986) *ENBO J.* **5**, 2513–2522.

Eisen, H. J. and Glinsman, W. H. (1978) *Biochem. J.* **171**, 177–183.

Evans, R. E., Birberg, N. C., and Rosenfeld, M. G. (1982) *Proc. Natl. Acad. Sci. USA* **79**, 7659–7663.

Failla, D., Tomkins, G. M., and Santi, D. V. (1975) *Proc. Natl. Acad. Sci. USA* **72**, 3849–3852.

Fine, D. L., Arthur, L. O., and Young, L. J. T. (1976) *In Vitro* **12**, 693–701.

Govindan, M. V. (1980) *Biochim. Biophys. Acta* **631**, 327–333.

Govindan, M. V. and Sekeris, C. E. (1976) *Steroids* **28**, 499–507.

Govindan, M. V. and Sekeris, C. E. (1978)*Eur. J. Biochem.* **89**, 95–104.

Govindan, M. V. and Manz, B. (1980) *Eur. J. Biochem.* **254**, 1537–1539.

Govindan, M. V. and Gronemeyer, H. (1984) *J. Biol. Chem.* **260**, 12915–12924.

Govindan, M. V., Spiess, E., and Majors, J. (1982) *Proc. Natl Acad. Sci. USA* **79**, 5157–5161.

Govindan, M. V., Devic, M., Green, S., Gronemeyer, H., and Chambon, P. (1985) *Nucl. Acid Reg.* **13**, 8293–8304.

Grandics, P., Miller, A., Schmidt, T. J., Mittman, D., and Litwack, G. (1984) *J. Biol. Chem.* **259**, 3173–3180.

Green, G. L., Nolaru, C., Engler, J. P., and Jensen, E. V. (1980) *Proc. Natl. Acad. Sci. USA* **77**, 5117–5119.

Green, S., Gilna, M., Waterfield, M., Baker, A., Hort, Y., and Shine, J. (1986) *Nature* **320**, 134–139.

Gronemeyer, H., Harry, P., and Chambon, P. (1983) *FEBS Lett.* **156**, 287–292.

Gronemeyer, H., Govindan, M. V., and Chambon, P. (1985) *J. Biol. Chem.* **260**, 6916–6925.

Hager, L. J., McKnight, G. S., and Palmiter, R. D. (1980) *Nature* **28**, 340–342.

Hall, R., Hyde, J. E., Goman, M., Simmons, D. L., Hope, I. A., Mackay, M., Scaife, J., Merkil, B., Richle, R., and Stocker, J. (1984) *Nature (London)* **311**, 378–382.

Harmon, J. M., Brower, S. T., Eisen, H. J., and Thompson, E. B. (1984) *Cancer Res.* **44**, 4540–4547.

Hollenberg, S. M., Weinberger, C., Ong, E. S., Cerelli, G., Oro, A., Lebo, R., Thompson, E. B., Rosenfeld, M. G., and Evans, R. M. (1985)*Nature* **318**, 635–641.

Hynes, N. E., Kenndy, N., Rahmsdorf, U., and Groner, B. (1981) *Proc. Natl. Acad. Sci. USA* **78**, 2038–2042.

Jannson, K. M., Philipson, L. O., and Vennstrom, B. (1983) *EMBO J.* **2**, 561–565.

Karin, M., Haslinger, A., Holtgreve, H., Richards, R. I., Krauter, P., Westphal, H. M., and Beato, M. (1984) *Nature (London)* **308**, 513–519.

Laemmli, U. R. (1970) *Nature* **227**, 680–685.

Larzarides, E. and Weber, K. (1974) *Proc. Natl. Acad. Sci. USA* **71**, 2268.

Lee, F., Mulligan, R., Berg, P., and Ringold, G. (1981) *Nature* **294**, 228–232.

Loosfelt, H., Algev, M., Misrahi, M., Mantel, A. G., Mericl, C., Logeat, F., Benarous, R., and Milgrom, E. (1986) *Proc. Natl. Acad. Sci. USA* **83**, 9045–9049.

Lubahn, D., Joseph, D. R., Sullivan, P. M., Willard, H. F., French, F. S., and Wilson, E. M. (1988) *Science* **240**, 327–330.

Lustenberger, P., Formstecher, P., and Dautrevaux, M. (1981) *J. Steroid Biochem.* **14**, 697–703.

Majors, J. E. and Varmus, H. E. (1981) *Nature (London)* **289**, 253–258.

Manz, B. and Govindan, M. V. (1980) *Noppe Seylers Z Phygiol. Chem.* **361**, 953– 957.

Martial, J. A., Baxter, J. D., Goodman, H. M., and Seeburg, P. H. (1977) *Proc. Natl. Acad. Sci. USA* **74**, 1816–1820.

Maxwell, B. L., McDonnell, D. P., Conneely, O. M., Schulz, T. Z., Green, G. L., and O'Malley, B. W. (1987) *Mol. Endocrinol.* **1**, 25–35.

McDonnell, D. P., Mangelsdorf, D. J., Pilce, W., Haussler, M. R., and O'Malley, B. N. (1987) *Science* **235**, 1214–1217.

Miesfeld, R., Rusconi, S., Godowski, P. J., Maler, B. A., Okret, S., Wilkstrom, A. C., Gustafsson, J. A., and Yamamoto, K. R. (1986) *Cell* **46**, 389– 399.

Miesfeld, R., Okret, S., Wilkstrom, A. C., Wrange, O., Gustafsson, J. A., and Yamamoto, K. R. (1984) *Nature* **312**, 779–787.

Milgrom, E., Atger, A., and Baulieu, E. E. (1973)*Biochemistry* **12**, 5198–5205.

Monder, C. (1971) *Steriods* **18**, 187–194.

Payvar, F., Wrange, O., Carlstedt-Duke, J., Okret, S., Gustafsson, J. A., and Yamamoto, K. R. (1981) *Proc. Natl Acad. Sci. USA* **78**, 6628–6632.

Payvar, F., De Franco, D., Firestone, G. L., Edgar, B., Wrange, O., Okret, S., Gustafsson, J. A., and Yamamoto, K. R. (1983) *Cell* **35**, 381– 392.

Renkawitz, R., Schutz, G., von der Ahe, D., and Beato, M. (1984) *Cell* **37**, 503–510.

Rousseau, G. G., Higgins, S. J., Baxter, J. D., Gelfand, D. and Tomkins, G. M. (1975) *J. Biol. Chem.* **250**, 6015–6021.

Rousseau, G. G., Kirchoff, J., Formstecher, P., and Lustenberger, P. (1979) *Nature* **279**, 158–160.

Sakaue, Y., Thompson, E. B. (1977) *Biochem. Biophys. Reg. Commun.* **77**, 533–541.

Sap, J., Munoz, A., Damm, K., Goldberg, Y., Ghysdael, J., Leulz, A., Beug, H. and Vennstrom, B. (1986) *Nature* **324**, 635–640.

Scheidereit, C., Geisse, S., Westphal, H. M., and Beato, M. (1983) *Nature (london)* **301**, 749–752.

Schmid, W., Grote, H., and Sekeris, C. E. (1976) *Mol. Cell. Endocrinol.* **5**, 223– 241.

Siegel, L. M. and Monty, K. J. (1966) *Biochim. Biophyg. Acta* **112**, 346–362.

Simons, S. S. Jr., Schleenbaker, R. E., and Eisen, H. J. (1983) *J. Biol. Chem.* **258**, 2229–2238.

Sweet, F. and Adair, N. K. (1975) *Biochem. Biophyg. Reg. Commun.* **63**, 99–105.

Tsawdareoglou, G. N., Govindan, M. V., Schmid, W., and Sekeris, C. E. (1981) *Eur. J. Biochem.* **114**, 305–313.

Vaitukaitis, J., Robbins, J. B., Nieschlag, E., and Ross, G.T. (1971) *J. Clin. Endocrinol.* **33**, 988–991.

Walter, P., Green, S., Green, G. L., Krust, A., Bonert, J. M., Jeltsch, J. M., Staub, A., Jensen, E., Scrace, G., Waterfield, M., and Chambon, P. (1985) *Proc. Natl. Acad. Sci. USA* **82**, 7889–7893.

Williams, D. and Gorski, J. (1972) *Proc. Natl Acad. Sci. USA* **69**, 3464–3468.

Wrange, O., Carlstedt-Duke, J., and Gustafsson, J. A. (1979) *J. Biol. Chem.* **254**, 9284–9290.

Wrange, O., Okret, S., Radojcic, M., Carlstedt-Duke, J., and Gustafsson, J. A. (1984) *J. Biol. Chem.* **259**, 4534–4541.

Yamamoto, K. R. and Alberts, B. M. (1976) *Annu. Rev. Biochem.* **45**, 721–746.

Young, R. A. and Davis, R. W. (1983) *Science* **222**, 778–782.

Purification of the Rat Kidney Mineralocorticoid Receptor

Zygmunt S. Krozowski and John W. Funder

1. Introduction

For over a decade following the identification of a mineralocorticoid receptor (MR) the principal ligand thought to occupy this receptor was the mineralocorticoid aldosterone. In addition, the only tissues believed to contain MR were sodium-transporting epithelia such as kidney, colon, and parotid. During the last five years, a considerable body of evidence has demonstrated the existence of indistinguishable binding sites in pituitary, brain, lung, heart, and epididymis (Moguilewsky and Raynaud, 1980; Krozowski and Funder, 1981a,b; Pearce et al., 1986; Pearce and Funder, 1987). Furthermore, studies in the rat hippocampus have established that aldosterone and corticosterone have equivalent affinities for such sites, generically termed Type I receptors. Given the much higher plasma levels of the glucocorticoid, corticosterone is thought to be the principal occupant of these sites (Krozowski and Funder, 1983; Beaumont and Fanestil, 1983). In the kidney, however, Type I receptors are occupied by aldosterone, a phenomenon reflecting, in part, the preferential lowering of free corticosterone by corticosterone-binding globulin (Funder et al., 1973) and the action of the enzyme ll-beta hydroxysteroid dehydrogenase that converts corticosterone and cortisol, but not aldosterone, to their receptor-inactive ll-keto metabolites (Funder et al., 1988).

Corticosterone and cortisol can thus be expected to occupy both Type I receptors and classical, dexamethasone-binding glucocorticoid receptors (GR), depending on the particular tissue and plasma steroid concentration;

Receptor Purification, vol. 2   ©1990 The Humana Press

these latter binding sites are thus sometimes designated Type II glucocorticoid receptors. The capability of Type I receptors to function as modulators of glucocorticoid action in one tissue and of mineralocorticoid action in another strongly suggests that the receptor displays some tissue specific heterogeneity, and an investigation of the purified protein may be required to elucidate any such differences.

The MR has proved exceedingly difficult to purify to homogeneity due to an apparent inherent instability of the binding site and the unavailability of a suitable affinity labeling ligand. Early studies (Agarwal, 1978) were also hampered by the not inconsiderable affinity of ^{3}H-aldosterone for Type II receptors. More recently, the stability of the renal MR in cytosols has been shown to be greatly enhanced by the inclusion of heavy metal complexes such as molybdate into homogenization buffers (Grekin and Sider, 1980). In addition, the availability of highly specific glucocorticoids such as RU26988 (11β-17β-dihydroxy-17α-propynyl-androsta-1,4,6-triene-3-one) and RU28362 (11β,17β-dihydroxy-6-methyl-17-[1-propynyl] androsta-1,4,6-triene-3-one) to block tracer binding to Type II sites has facilitated the unequivocal identification of Type I receptors by the use of ^{3}H-aldosterone.

Physicochemical characterization of renal MR suggests that many of the techniques used to purify other steroid receptors are applicable to the MR. We have used anion exchange, size exclusion, and affinity chromatography to achieve a >18,000-fold purification of the rat renal MR (Krozowski, 1984). Analysis of the purified material by SDS gel electrophoresis showed a mol wt of about 35 kDaltons (kDa) (Krozowski, 1984), which probably represents a breakdown product of the native 108kDa receptor (Arriza et al., 1987). A breakdown product of 39 kDa has also been reported for the closely related purified GR (Wrange et al., 1984). There are a number of other techniques that may further improve prospective MR purification protocols.

2. Chromatography
of Unactivated and Activated Type I Receptors

Eisen and Harmon (1986) have shown that it is possible to activate the renal MR by heating at 25°C for 20 min after the removal of molybdate by gel filtration; 25% of the MR was activated to a form that bound DNA, whereas only 2% of unactivated MR bound. DEAE-cellulose chromatography of heat treated cytosols, however, showed that the Type I binding sites elute over a broad range of salt concentration (40–190 mM NaCl), with about 50% eluting before 60mM NaCl. In contrast, nonactivated Type I receptors eluted as a single sharp peak at 190 m*M* NaCl. These results indicate that the Type I receptor exists as a highly heterogeneous species after 20 min of heat activation. Further studies are needed to ascertain whether heating for a longer period of time will result in a greater proportion of Type I receptor activation or if activation by high salt or by dilution of cytosol will give a homogeneous-activated species. Characterization of Type I receptors by gel filtration on Sephacryl S300 showed that the MR exists as a multimeric complex (MW-450kDa) in the presence of molybdate and as a 70kDa protein in the absence of the heavy metal ion. We have also observed that the partially purified rat renal Type I receptor elutes as a multimeric complex of 450kDa when run on an Ultrogel AcA-34 column (Krozowski, 1984). The capacity of the Type I receptor to exist in various separable physicochemical forms can be readily exploited to further purify the MR. Differential Sephacryl S300 filtration in the presence and absence of molybdate and DNA-cellulose and DEAE-cellulose chromatography, with successive elution of the nonactivated and then the activated forms of the receptor, may allow a considerable degree of further purification.

Colonic MR can be activated to a DNA binding form by treatment with 300 m*M* KCl, raising the pH to 8 or by the addition of EDTA or EGTA (Schulman et al., 1986). However, heating the colonic cytosol did not appear to activate the receptor; gel filtration or addition of ATP, conditions known to activate other steroid receptors, do not activate colonic MR. Activation of the colonic Type I receptor by increasing the pH to 8 resulted in 42% of the Type I sites binding to DNA-cellulose. Characterization of activated cytosol on DEAE-cellulose showed that the receptor was present in two clearly separable forms with the activated species, which comprised 30% of the total Type I sites, eluting at 40 m*M* KCl and the nonactivated form at 190 m*M* KCl. Rat brain Type I receptors, on the other hand, have been shown to be highly unstable to activation in the absence of molybdate (Beaumont and Fanestil, 1986). Differential DNA-cellulose chromatography has been successfully used in the purification of the glucocorticoid receptor, where binding to DNA before and after activation has, in addition, allowed the separation of DNA binding proteins from the receptor (Govindan and Gronemeyer, 1984).

3. Heparin-Sepharose Chromatography

Heparin-Sepharose has been shown to bind steroid receptors and has frequently been used in purification procedures. This matrix has been used in the purification of the molybdate-stabilized estrogen receptor from calf uterus, where as a first step it provided a concentrated receptor preparation with a 40% yield and a 5–10-fold increase in purity (Atrache et al., 1985). Renal Type I sites bind heparin-Sepharose and can be eluted with high salt or heparin-containing buffers (Weisz et al., 1986). When elution was carried out with heparin, 70% of the MR could be recovered either as the steroid-receptor complex or as the unbound receptor. Elution with KCl, however, resulted in half this amount recovered when uncomplexed receptor was used, while the hormone-receptor complex was recovered in 85% yield. Less efficient elution was achieved with total liver RNA, where 2 mg/mL RNA eluted 55% of the Type I binding sites, while dextran sulphate (2 mg/mL) resulted in a recovery of 40%. Little or no elution was obtained with chondroitin sulphate, dextran, or *d*-glucosamine. MR binding to heparin is independent of the state of activation of the receptor and is readily performed in the presence of molybdate.

An additional advantage of using heparin-Sepharose is that corticosterone-binding globulin is not retained by the gel, separating it from MR and simplifying affinity chromatography of the corticosterone-binding globulin rich renal cytosol (Krozowski and Funder, 1983; Stephenson et al. 1984) on deoxycorticosterone(DOC)-agarose matrices. Rat kidney MR was purified 10–fold by a single passage of renal cytosol on heparin-Sepharose. The occupation of corticosterone-binding globulin by the inclusion of cortisol 17-beta acid (Manz et al., 1982) in renal cytosols should further improve affinity purification.

4. Hydrophobic Interaction Chromatography

Another potential method for purifing MR is that of hydrophobic interaction chromatography (HIC), an approach previously used in attempts to purify the rabbit progesterone receptor (Lamb and Bullock 1983). The receptor was found to be increasingly retarded on alkyl-agarose columns as the length of the alkyl chain increased, reflecting the presence of hydrophobic regions on the protein. Absorption did not occur at the steroid binding site and did not require activation to a DNA binding form. Madhok et al. (1987) were able to separate two forms of the estrogen receptor by HIC when fresh cytosol was used; frozen cytosol gave four peaks of estradiol-binding with the most hydrophobic peak corresponding to the largest form of the receptor. This technique has also been used to purify the nuclear androgen receptor in conjunction with micrococcal nuclease digestion of prostatic chromatin (Bruchovsky et al., 1981). Nuclease digestion resulted in the precipitation of 95% of the nuclear receptor; HIC chromatography of the receptor enriched fraction on 5-aminohexyl-agarose gave rise to a 93-fold purification of the binding site with a recovery of 45%. A potential problem with HIC is that the organic solvents used to elute MR may disrupt ligand binding. To examine this possibility, we have investigated the stability of the ^{3}H-aldosterone-MR complex in the presence of various organic solvents. The complex was found to be stable in the presence of up to 20% acetonitrile, 20% isopropanol, or 30% ethylene-glycol (Fig. 1). Rat kidney HR bound to octyl-Sepharose when renal cytosol was made 200 mM in NaCl; elution with 30% ethylene glycol resulted in a 14–fold purification of the receptor with 80% yield (Fig. 2).

5. Ligand Affinity Chromatography

Ligand affinity chromatography has proved to be a powerful tool in the purification of steroid receptors. Rat hepatic glucocorticoid (Type II) receptors have been purified about 4000–fold on deoxycorticosterone-derivatized agarose, gel filtration on Bio-Gel A–1.5m agarose, and DEAE-cellulose chromatography (Grandics, et al. 1984). An affinity matrix consisting of diethylstilbestrol coupled to epoxy-activated agarose has previously been used to purify the estrogen receptor; the bound receptor was eluted with p-sec-amylphenol and sodium thiocyanate (Van Oosbree et al., 1984). We have utilized an affinity resin, comprised of deoxycorticosterone linked by an epoxy bond to agarose to purify the renal Type I receptor using aldosterone to selectively elute the receptor (Krozowski, 1984). This matrix appears resistant to enzymatic degradation, obviating the problems of ligand loss from the gel as formerly reported for deoxycorticosterone-hemisuccinate linked to aminoethyl-agarose (Ludens et al., 1972a,b). When we subjected the renal Type I receptor to anion exchange chromatography on DE–52 cellulose corticosterone-binding globulin eluted in the flowthrough fraction, whereas the receptor was totally retained. The avid binding of deoxycorticosterone by corticosterone-binding globulin necessitates the separation of the plasma binder before affinity chromatography on deoxycorticosterone-agarose gels; inclusion of micromolar RU26988 precludes binding of Type II receptors to the gel. Type I receptors eluting from the anion exchange column were then chromatographed on an Ultrogel AcA-34 exclusion column and the receptor-containing fractions bound to the affinity matrix in batches. Approximately 60% of the receptor bound to the matrix and essentially all of the binding sites could be eluted by incubation overnight at 4°C with 4 micromolar aldosterone. Affinity purification resulted in >100–fold enrichment of the DE–52, Ultrogel–treated receptor yielding >18,000–fold purification and a 29% recovery overall (Krozowski, 1984).

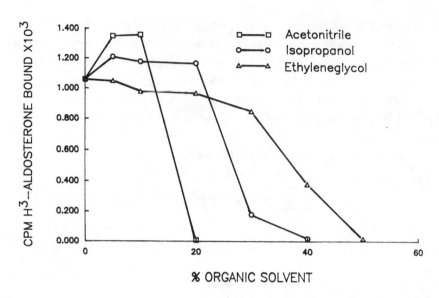

Fig. 1. Stability of the Type I receptor in various organic solvents. Increasing amounts of ice-cold acetonitrile, isopropanol, or ethyleneglycol were added to rat kidney cytosol prepared in TM buffer and the specific binding of ³H-aldosterone measured as previously described (Krozowski and Funder, 1983). Values shown are averages of triplicate determinations.

Lombes et al. (1987) have reported the synthesis of an affinity matrix that binds Type I, but not Type II, receptors. The 3-O-carboxymethyloxime deoxycorticosterone Sepharose gel was used to purify rabbit kidney Type I receptors about 1000–fold with a recovery of 30–40%. When affinity purified Type I receptors were analyzed by size exclusion HPLC, it was found that size exclusion chromatography decreased the Stokes radius of the protein from 6 to about 4 nm; Type II receptors affinity purified on a 17β-dexamethasone matrix and subjected to size exclusion HPLC showed no decrease in size. Since affinity purified Type I and Type II receptors displayed the same molecular size as cytosolic receptors when analyzed by sedimentation on glycerol gradients, it is likely that size exclusion results in the activation or dissociation of the Type I receptor complex when chromatography is performed in the presence of high salt and low protein concentration.

6. Stability of Type I Receptors

Historically, the major obstacle to the purification of MR has been the instability of the binding protein. Despite the substantial stability conferred by the addition of molybdate (Grekin and Sider, 1980), sulfhydryl

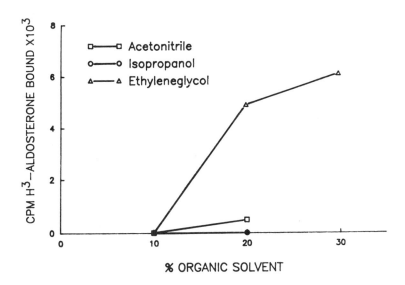

Fig. 2. Elution of renal Type I receptors from octyl-Sepharose by various organic solvents. Cytosol was made 200 m*M* in NaCl and an equal volume of a 50% suspension of octyl-Sepharose was added and the mixture incubated for 30 min at 4°C. The gel was washed with TM buffer and eluted with TM buffer containing various amounts of ethylene glycol, isopropanol, or acetonitrile. Specific binding of ^{3}H-aldosterone was measured in the eluate.

group reagents, glycerol (Emadian et al., 1986), and ligand (Rafestin-Oblin et al., 1977), the MR has not been purified to homogeneity. One reason for this may be that MR, like GR, contains a modulator that is required for ligand-binding (Grippo et al., 1984; Bodine and Litwack, 1986). Successive purification steps may separate receptor from modulator and thus render the ligand binding domain unstable. Since the stability of Type I receptors differs between tissues, it is probable that tissue specific factors such as proteolytic enzymes also play a role in destabilizing the binding sites. Characterization of the Type II receptor in human leukemia cells has shown that the predominant species is the 52kDa form of the receptor, reflecting the action of proteases. The intact Type II receptor (97kDa) was only present when the serine protease inhibitor diisopropylfluorophosphate was present during preparation of the cytosol (Distelhorst and Miesfeld, 1987).

Many of the problems associated with instability could be obviated by the covalent attachment of ligand to MR. Dexamethasone-mesylate interacts covalently with GR (Simons and Thompson, 1981) and DOC-mesylate (DOC-MES) would appear to be a potentially useful compound for binding to MR. To date, we have not succeeded in covalently binding DOC-MES to MR in a specific manner under conditions used by others to bind dexamethasone-mesylate to GR (Simons and Thompson, 1981). Though DOC has modest affinity for Type II glucocorticoid receptors, DOC-MES does not bind GR, suggesting that the mesylate moiety may produce steric hindrance within the binding site (Weisz et al., 1983). Covalent labelng of MR would enable the employment of SDS polyacrylamide gel electrophoresis in the purification of this protein. Though Sakai and Gorski (1984) have reported the reversible denaturation of the estrogen receptor after treatment with SDS and beta-mercaptoethanol, in our hands MR cannot be renatured to a hormone-binding form under identical conditions to those used for the estrogen receptor.

7. Other Sources of Type I Receptors

Attention has focused on the kidney as the source of MR for purification, but the distal colon and hippocampus may prove to be more suitable tissues given the higher abundance of the binding site in these tissues (Moguilewsky and Raynaud, 1980; Pressley and Funder, 1975). The abundance of MR has been shown to increase from proximal to distal parts of the colon (Schulman et al., 1986) where concentrations in excess of 240 fmol/mg of protein have been reported. In addition, messenger RNA for the colonic Type I binding site appears to be the highest of all tissues examined (Reul et al., 1989). Hippocampal cytosol contains about 20 fmol/mg protein in the intact rat, increasing to 120 fmol 24 h after adrenalectomy, with no significant increase in renal Type I sites (Stephenson et al., 1984).

Several groups have studied the physicochemical properties of the brain Type I sites. Beaumont and Fanestil (1986) characterized the molybdate stabilized rat brain Type I receptors by anion exchange chromatography on DEAE-Trisacryl and have shown that the receptor elutes as a single peak at 160 mM NaCl. Similar results have been obtained for rat kidney MR (Krozowski, 1984), whereas multiple peaks were observed in rabbit kidney cytosols (Marver, 1980). Both steroid specificity (Krozowski and Funder, 1983) and isoelectric focusing studies (Wrange and Yu, 1983) have found renal and hippocampal MR to be indistinguishable. Recently, we have compared the rat kidney and hippocampal MR by gel filtration on FPLC and have demonstrated an apparent difference in mol wt between the two species (Doyle et al., 1988). Since these studies were performed in the presence of molybdate, it is possible that the complexing of tissue-specific components to MR gives rise to the differences observed. A common nonhormone-binding component has been reported to be part of several steroid hormone receptor complexes (Joab et al., 1984), and it is possible that other proteins are also bound. Alternatively, a tissue-specific posttranslational modification may also account for the differences observed. It is unlikely that the renal and brain MR differ at the amino acid level, since the mRNA species in the two tissues appear identical in size and there appears to be only one gene, as ascertained by Southern blot analysis (Arriza et al., 1987). Large differences have been observed in the stability of Type I receptors between mouse and rat brain cytosols (Emadian et al., 1986), suggesting that the mouse brain may be a more suitable source of MR.

8. Future Directions

The availability of cDNA's coding for the HR has opened the way to new approaches for the characterization and purification of this protein.

Expression of the MR clone will allow the production of large amounts of the receptor for the generation of antibodies; both polyclonal and monoclonal antibodies have been used to purify other steroid receptors. The progesterone receptor has been purified using an immunoaffinity column in which the anti-receptor antibody was linked to protein A-Sepharose through the Fc fragment (Logeat et al., 1985). We have recently raised polyclonal antibodies to a hexadecapeptide located in the hinge reglon of the human kidney MR (Krozowski et al., 1988). When the antibody was applied to human, rat, and monkey kidney sections, specific immunostaining was observed in cells tentatively identified as being part of the distal convoluted tubule, the connecting piece and cortical collecting duct of the nephron, whereas staining appeared absent in cells thought not to reabsorb sodium in response to steroids. These results are consistent with the known sites of mineralocorticoid action and confirm that the antibody recognizes the physiologically active MR. Since the antibody bound MR in fixed tissue sections, it was anticipated that it would also recognize the denatured form of the protein. Western blot analysis of partially purified renal Type I receptors showed that the antibody recognized an epitope of the expected mol wt. Such immunodetection of denatured MR is particularly important given the difficulties encountered in maintaining the receptor in a hormone-binding form during purification.

Isolation of a modified recombinant MR from transfected eucaryotic cells represents another approach that might provide valuable data. For instance, a recombinant could be constructed whereby a polylysine sequence has been added to the C-terminus of the molecule; such polylysine-containing proteins are strongly retained by anion exchange resins and can be eluted under conditions of high salt. Smith and Johnson (1987) have recently described a glutathione-transferase expression vector containing a thrombin and factor X cleavage site that can be used to generate fusion proteins in which the amino-terminus of the protein of interest is linked to the carboxy-terminus of the enzyme. The resultant recombinant can be readily purified in a single-step procedure using affinity chromatography on glutathione-agarose columns and the purified fusion protein cleaved to yield the protein of interest.

Since the Type I receptor has been cloned and the full length amino acid sequence deduced, it may be pertinent to ask what additional information might be gained by acquiring the native protein in a purified form. In this regard, there are a number of questions that remain unanswered. First, the amino acid sequence derived from the cDNA represents what is thought to be the first translation product; it is possible that the mature or biologically-active Type I receptor may be of a different length and exist in more than one form, analogous to the chicken progesterone receptor which exists as 108,000 dalton and 79,000 dalton proteins. Secondly, though non-receptor mechanisms have been invoked to explain how the Type I receptor acts to induce either glucocorticoid- or mineralocorticoid-responsive genes, tissue-specific post-transcriptional and post-translational modification of the receptor may also play a part in determining which genes are activated. Recently, it has been shown that thyroid hormone receptors exist in various isoforms and that alternative splicing leads to the differential activation of myosin gene transcription (Izumo and Mahdavi, 1988). It is thus important to ascertain whether native Type I steroid receptors exhibit heterogeneity between tissues, and if so, whether such differences can help explain some of the apparently unique characteristics of the Type I mineralocorticoid and glucocorticoid receptors.

References

Agarwal, M. K. (1978) *FEBS Lett.* **85,** 1–8.

Arriza, J. L., Weinberger, C., Cerelli, G., Glaser, T. M., Handelin, B. L., Housman, D. E., and Evans , R. M. (1987) *Science* **237,** 268–275.

Atrache, V., Ratajczak, T., Senafi, S., and Hahnel, R. (1985) *J. Biol. Chem.* **260,** 5936–5941.

Beaumont, K. and Fanestil, D. D. (1983) *Endocrinology* **113,** 2043–2051.

Beaumont, K. and Fanestil, D. D. (1986) *J. Steroid Biochem.* **24,** 513–517.

Bodine, P. V. and Litwack, G. (1986) *J. Biol. Chem.* **263,** 3501–3512.

Bruchovsky, N., Rennie, P. S., and Comeau, T. (1981) *Eur. J. Biohem.* **120,** 399–405.

Distelhorst, C. W. and Miesfeld, R. (1987) *Blood* **69,** 750–756.

Doyle, D., Krozowski, Z. S., Morgan, F. J., and Funder, J. W. (1988) *J. Steroid Biochem.* **29,** 415–421.

Eisen, L. P. and Harmon, J. M. (1986) *Endocrinology* **119,** 1419–1426.

Emadian, S., Luttge, U. G., Densmore, C. L. (1986) *J. Steroid Biochem.* **24,** 953–961.

Funder, J. W., Pearce, P. T., Smith, R., and Smith, A. I. (1988) *Science* **242,** 583–585.

Funder, J. W., Feldman, D., and Edelman, I. S. (1973) *Endocrinology* **92**, 994–1004.

Govindan, M. V. and Gronemeyer, H. (1984) *J. Biol. Chem.* **259**, 12915–12924.

Grandics, P., Miller, A., Schmidt, T. J., Mittman, D., and Litwack, G. (1984) *J. Biol. Chem.* **259**, 3173–3180.

Grekin, R. J. and Sider,R. S. (1980) *J. Steroid Biochem.* **13**, 835–837.

Grippo, J. P., Holmgren, A., and Pratt, W. B. (1984) *J. Biol. Chem.* **260**, 93–97.

Izumo, S. and Mahdavi, V. (1988) *Nature* **334**, 539–542.

Joab, I., Radanyi, C., Renoir, M., Buchou, T., Catelli, M., Binart, N., Mester, J., and Baulieu, E.-E. (1984) *Nature* **308**, 850–853.

Krozowski, Z. S. (1984) in *Handbook of Receptor Research* (Agarwal, M.K. ed.), *Acta Medica,* Rome p. 75–92.

Krozowski, Z. S. and Funder, J. W. (1981a) *Endocrinology* **109**, 1221–1224.

Krozowski, Z. S. and Funder, J. W. (1981b) *Endocrinology* **109**, 1811–1813.

Krozowski, Z. S. and Funder, J. W. (1983) *Proc. Natl. Acad. Sci. USA* **80**, 6056–6060.

Krozowski, Z. S., Rundle, S. E., Wallace, C., Castell, M. J., Shen, J. H., Dowling, J., Funder, J. W., and Smith, A. I. (1988) *Endocrinology* **125**, 192–198.

Lamb, D. J. and Bullock, D. W. (1983) *J. Steroid Biochem.* **19**, 1039–1045.

Logeat, F., Pamphile, R., Loosfelt, H., Jolivet, A., Fournier, A., and Milgrom, E. (1985) *Biochemistry* **24**, 1029–1035.

Lombes, M., Claire, M., Lustenburger, P., Michaud, A., and Rafestin-Oblin, M. E. (1987) *J. Biol. Chem.* **262**, 8121–8127.

Ludens, J. H, DeVries, V. R., and Fanestil, D. D. (1972a) *J. Steroid Biochem.* **3**, 193–200.

Ludens, J. H, DeVries, V. R., and Fanestil, D. D. (1972b) *J. Biol. Chem.* **247**, 7533–7538.

Madhok, T. C., Leung, B. S., and Stout, L. E. (1987) *Horm. Res.* **25**, 29–36.

Manz, B., Grill, H. J., and Pollow, K. (1982) *J. Steroid Biochem.* **17**, 335–342.

Marver, D. (1980) *Endocrinology* **106**, 611–618.

Moguilewsky, M. and Raynaud, J. P. (1980) *J. Steroid Biochem.* **12**, 309–314.

Pearce, P. T. and Funder, J. W. (1987) *Clin. Exp. Pharm. Physiol.* **14**, 859–866.

Pearce, P. T., Lipkevicius, O. R., and Funder, J. W. (1986) *Endocrinology* **118**, 2072–2075.

Pressley, L. and Funder, J. W. (1975) *Endocrinology* **97**, 588–596.

Rafestin-Oblin, M. E., Michaud, A., Clair, M., and Corvol, P. (1977) *J. Steroid Biochem.* **8**, 19–23.

Reul, J., Pearce, P. T., Funder, J. W., and Krozowski, Z. S. (1989) *Mol. Endo.* **3**, 1674–1680.

Sakai, D. and Gorski, J. (1984) *Endocrinology* **115**, 2379–2383.

Schulman, G., Miller-Diener, A., Litwack, G., and Bastl, C. P. (1986) *J. Biol. Chem.* **261**, 12102–12106.

Simmons, S. S. and Thompson, E. B. (1981) *Proc. Natl. Acad. Sci. USA* **78**, 3541–3545.

Smith, D. B. and Johnson, K. S. (1988) *Gene* **68**, 546–563.

Stephenson, G. and Funder, J. U. (1987) *Am. J. Physiol.* **252**, E525–529.

Stephenson, G., Krozowski, Z. S., and Funder, J. W. (1984) *Am. J. Physiol.* **246**, F277–F233.

Van Oosbree, T. R., Kim, U. H., and Mueller, G. C. (1984) *Anal. Biochem.* **136**, 321–327.

Weisz, A., Buzard, R. L., Horn, D., Li, M. P., Dankerton, L. V., and Markland, F. S. (1983) *J. Steroid Biochem.* **18**, 375–382.

Weisz, A., Ciatiello, L., and Bresciani, F. (1986) *J. Steroid Biochem.* **24**, 461–467.

Wrange, O., Okrent, S., Radojscisc, M., Carlstedt-Duke, J., and Gustafsson, J.-A. (1984) *J. Biol. Chem.* **10**, 4534–4541.

Wrange, O. and Yu, Z.-Y. (1983) *Endocrinology* **113**, 243–250.

Purification
of the Androgen Receptor

M. P. Johnson, C. H. Chang, D. R. Rowley, C. Y. F. Young, and D. J. Tindall

1. Introduction

The androgen receptor protein, like many other hormone receptors, is present within its target tissues at a relatively low concentration of approx 0.001% of the total cellular protein (Chan et al., 1989). It is apparent from numerous studies (Wilson and French, 1979; Sherman et al., 1983; Johnson et al., 1987) that the size and steroid binding activity of the androgen receptor are relatively unstable in cellular fractions because the receptor is easily degraded by endogenous proteases in many androgen target tissues. These two characteristics complicate the study of androgen receptor protein structure and of its pivotal role in the mechanism of androgen action. It may seem ironic that proteolysis and low receptor concentration not only provide incentives for purification of the receptor, but also contribute significant difficulty in achieving purification of the intact receptor in large quantities. For this reason, successful protocols for the purification of the androgen receptor require careful consideration of the qualities of the tissue source, the selectivity and limitations of the various enrichment techniques, and verification that the androgen receptor has been purified in its intact form.

Receptor Purification, vol. 2   ©1990 The Humana Press

2. Selection of the Best Tissue Source

The rat ventral prostate gland is a classical androgen target tissue that has been studied more than any other. However, this gland is not an optimal source for androgen receptor purification because of its high content of endogenous proteases, which appear to degrade the receptor immediately upon tissue homogenization. Inclusion of sodium molybdate or protease inhibitors such as leupeptin or diisopropyl-fluorophosphate (DFP) in the homogenization buffer may help stabilize androgen binding activity (Prins and Lee, 1979; Wilson and French, 1979; Chang et al., 1982), but do not totally eliminate cleavage of the receptor molecule during lengthy purification (Johnson et al., 1987). Wilson and French (1979) have reported that rat ventral prostate contains a DFP-insensitive protease that can cleave the androgen receptor. Thus, it is likely that proteolysis accounts for the purification of a small 86kDa androgen receptor from rat ventral prostate by this laboratory (Chang et al., 1983), despite the use of leupeptin.

Androgen receptors have been identified in a number of androgen responsive tissues besides prostate, including androgen sensitive tumors (Bruchovsky and Renie, 1978; Shain et al., 1981; Diamond and Barrack, 1984) and cell lines (Smith et al., 1984; Hasenson et al., 1985), seminal vesicle (Lea et al., 1979; Chang et al., 1982; Schanbacher et al., 1987), epididymis (Hansson and Djoseland, 1972; Blaquier and Calandra, 1973; Tindall et al., 1975; Wilson and French, 1976), testis (Hansson et al., 1974; Tindall et al., 1977; Sanborn et al., 1977; Gulizia et al., 198 3) kidney (Gehring et al., 1971; Isomaa et al., 1982), uterus (Giannopoulos, 1973; Chang and Tindall, 1983), submaxillary gland (Dunn et al., 1973), pituitary (Thieulant and Pelletier, 1979; Schanbacher et al., 1987), pineal (Cardinali et al., 1975), hypothalamus (Pelletier and Caraty, 1981), skin (Eppenberger and Hsia, 1972), muscle (Jung and Baulieu, 1972), aorta, and myocardium (Lin et al., 1981a,b). Of these tissues, the best source of androgen receptor for purification may be the Dunning R3327G rat prostate adenocarcinoma. This tumor subline is easily propagated by subcutaneous transplantation. Although the H subline of the R3327 tumor appears to contain less receptor-degrading protease activity, the G subline possesses a much higher concentration of androgen receptor which is relatively stable (Rowley et al., 1984; Johnson et al., 1987). Another advantage of the G subline is its faster growth rate (Isaacs,1982). This property permits accumulation of sufficient starting material for purification from a small colony of tumor-bearing rats in a short period of time. For example, 25 rats implanted at four sites each will produce approx. 300g of G tumor within six wk. The dissected tissue should be frozen immediately in liquid nitrogen as pieces 1 cm³ or smaller. Some loss of androgen binding activity can be expected during prolonged storage but should be minimal during the first three mo.

3. Development of the Purification Protocol

3.1. Ammonium Sulfate Precipitation

The use of ammonium sulfate precipitation for purification of the androgen receptor has two benefits: receptor enrichment and transformation. Precipitation at approx 40% ammonium sulfate saturation is a common early step for steroid receptor purification (see other chapters in this volume). Simple titration and time-course experiments have demonstrated that slow addition of crystalline ammonium sulfate to 40% saturation, followed by overnight incubation at 4°C, resulted in a fourfold enrichment of the androgen binding activity (Chang et al., 1982; Rowley et al., 1984). Although the overnight period is optimal and convenient, incubation for 1-2 h is usually sufficient. Exposure to the high concentration of ammonium sulfate also transforms the cytosolic androgen receptor to a DNA-binding state.

The term "transformation" is used in an operational sense because the exact molecular changes that occur in androgen receptor structure are poorly understood. Rowley et al. (1986) have described three non-proteolyzed forms of the cytosolic androgen receptor. These include the 9.1 S oligomeric receptor that does not bind DNA, the 4.4 S monomeric receptor that does bind DNA, and the 7.7 S intermediate form that also binds DNA-cellulose. The relationships of these three forms has been characterized to a limited extent. The steroid-binding monomeric subunit is associated with an undefined RNA component (e.g., ribonucleoprotein) as the 7.7 S intermediate form. The oligomeric complex contains the monomer, the 90kDa heat shock protein, and possibly other non-steroid-binding components. Transformation by ammonium sulfate precipitation or other conventional methods is thought to dissociate the oligomeric complex, exposing the monomeric receptor's DNA binding region (Chan et al., 1989).

3.2. Differential DNA Chromatography

The strategy of androgen receptor chromatography on DNA resins is based on the fact that the receptor's ability to bind DNA can be manipulated. When androgen receptor is prepared as a low salt cytosolic extract, the predominant form of the receptor is the nontransformed 9S oligomeric complex (Wilson and French, 1979; Rowley et al., 1986). This large form fails to bind DNA, perhaps because the receptor's DNA-binding domain is occluded by subunit interaction. Therefore, when cytosol is applied to a column of DNA-Sepharose or DNA-cellulose, the nontransformed androgen receptor will be isolated from many DNA-binding proteins that are retained while the receptor flows through. Androgen receptor in this flow-through fraction can be transformed to a DNA-binding state by treatment with high salt (0.5M KCl for 30 min at 4°C), ammonium sulfate precipitation, warming, or prolonged incubation (18 h) at 0°C (Chang et al., 1982; Rowley et al., 1984). Application of the transformed receptor to a second DNA column, after dilution or dialysis to achieve low ionic strength, results in retention of receptor on the DNA resin with consequent isolation from many non-DNA-binding proteins. Thus, significant purification can be achieved by this technique of differential DNA chromatography.

Practical considerations that have a significant effect upon the successful application of this method include the following concerns:

1. *Cytosol preparation.* The first phase of the protocol requires that the cytosol androgen receptor be maintained in the nontransformed 9S complex. To accomplish this, the tissue used should be as fresh as possible. For storage, it should be frozen quickly in liquid nitrogen and stored at –85° C. Binding activity of a given batch will decline over several months and should be monitored closely. Tissue should be thawed quickly in saline at room temperature, then transfered to chilled buffer. Tissue homogenization must be thorough but gentle. Despite these precautions, our laboratory has consistently observed that approximately 20% of the receptor will be transformed in the low salt cytosol preparation if molybdate is not used in the buffer (Rowley et al., 1986). To maintain the other 80% in the nontransformed state, the cytosol must not be allowed to incubate overnight (even at 0°C) prior to application on DNA, and must not be frozen-thawed for later use because these actions will promote transformation. We recommend labeling of receptor with less than 10 mM [3H]-androgen for a short period (2–4 h) at 0° C prior to application onto DNA column I to minimize transformation. Addition of [3H]steroid immediately after homogenization provides an extra hour of labeling time during centrifugation and gives higher yields of binding activity. This apparent stabilizing effect of the steroid ligand may be owing to decreased proteolysis of the occupied receptor (Johnson et al., 1987).

2. *Transformation assays.* In troubleshooting the differential DNA chromatographic technique, it may be necessary to assess the proportion of androgen receptor that is transformed. This can be determined by two assays that are explained in detail in Rowley et al. (1986). The DNA-cellulose minicolumn assay allows determination of the percent receptor which is nontransformed or transformed to a DNA-binding state. Sucrose density gradient analysis using a 90 min vertical tube rotor spin of gradients containing 20 mM molybdate and 0.5M KCl permits assay of the percent receptor that is present as the nontransformed 9 S oligomer or as one of the transformed (7.7 S and 4.4 S) forms.

3. *Molybdate.* Inclusion of sodium molybdate in the homogenization buffer will stabilize the androgen receptor in the nontransformed 9 S oligomeric form (Rowley et al., 1984). Therefore, it might seem to be an ideal reagent for androgen receptor purification by differential DNA chromatography. However, it has not proven to be useful in this manner. The problem is that molybdate binds so tightly to the androgen receptor that it is very difficult to transform the receptor in the drop-through fraction from DNA column I, so that it can bind to DNA column II. Rowley et al. (1986) have demonstrated that the androgen receptor can be transformed despite the presence of 10 mM molybdate if high salt (0.5M KCl) and prolonged incubation (18 h at 0°C) are used. The problem is that the prolonged incubation encourages greater proteolysis. It has been suggested that molybdate prevents transformation by maintaining the integrity of the 9 S oligomer; but once the androgen receptor has been transformed, molybdate (<20 mM) has no effect on binding of receptor to DNA-cellulose (Rowley et al., 1986).

4. *Column I.* The name "differential DNA chromatography" is misleading. This method has not relied on the specific binding of the androgen receptor to a defined sequence of DNA. The binding interaction is ionic, which means that any polyanionic resin like phosphocellulose or nonspecific calf thymus DNA should work well. Experiments in this laboratory (Johnson et al., 1987) have indicated that phosphocellulose efficiently removes endogenous proteases that can degrade androgen receptors. Consequently, use of phosphocellulose as column I or use of a layered phosphocellose/DNA-cellulose column in this drop-through step might be very helpful.

5. *Column II.* Care must be taken to ensure adequate washing of the ammonium sulfate pellet to lower the ionic strength of the resuspended preparation. Failure to do so may prevent binding of the transformed androgen receptor to DNA column II. Similarly, if transformation is achieved by addition of KCl to 0.5M for 30 min at 0°C, then subsequent dilution or dialysis must be sufficient to reduce the ionic strength to less than 50 mM KCl.

3.3. Pyridoxal Phosphate Elution

Pyridoxal phosphate has been used in a number of purification protocols to elute steroid receptors selectively from DNA resins (*see* other chapters in this volume). For the application of this method to purification of the androgen receptor, we found it necessary to make significant modifications of the protocol of Wrange et al. (1979). This standard protocol required equilibration of the DNA II column in 50 m*M* sodium borate buffer followed by elution of receptor using 10 m*M* pyridoxal phosphate in 50 m*M* borate buffer, pH 8.1. However, 50 m*M* borate alone can elute a large amount of androgen receptor from the DNA column even before the application of pyridoxal phosphate (unpublished results). To investigate this phenomenon, a borate elution gradient was performed that indicated negligible elution of androgen receptor at 10 m*M* borate but substantial elution at 30–50 m*M* borate (unpublished observations). Consequently, the modified protocol uses a 10 m*M* borate wash followed by elution with 10 m*M* pyridoxal phosphate in 10 m*M* borate buffer, pH 8.1.

There are three practical concerns that need to be recognized when using pyridoxal phosphate. First, 10 m*M* pyridoxal phosphate will not completely dissolve in 10 m*M* borate unless the pH is adjusted to 8.1. Second, pyridoxal phosphate is a potent quencher during scintillation counting. Accurate quantitation of [^{3}H]-steroid eluted in the pyridoxal phosphate fractions requires accommodation for this quenching effect by the use of internal or external standards. Finally, estimates of protein by the Lowry method are complicated by the presence of pyridoxal phosphate. The simplest remedy for this problem is to use the Bradford protein assay as an alternative method.

Cloning of many steroid receptor genes, including the androgen receptor (Chang et al., 1988; Lubahn et al., 1988; Young et al., 1988), have made it possible to formulate a specific hypothesis regarding the elution of receptor from DNA resins by pyridoxal phosphate. The highly conserved DNA binding domains of steroid receptors are rich in lysine and arginine residues. Furthermore, it is known that pyridoxal phosphate can form a Schiff's base with the epsilon amino group of lysine residues. Accordingly, Grody et al. (1982) have postulated that pyridoxal phosphate might interfere with receptor binding to DNA by interacting with a critical amino group within the receptor's DNA binding region.

3.4. Steroid Affinity Chromatography

One of the important biological functions of the androgen receptor is to bind androgenic steroids with a high degree of affinity and specificity. This property can be used to great advantage for purification of the androgen receptor from crude preparations that are rich in contaminating proteins. For example, the androgenic steroids, testosterone or dihydrotestosterone (DHT), can be covalently linked to a chromatographic matrix to form an affinity resin that will selectively retain androgen receptors, thereby giving an enrichment of several hundred-fold (Chang et al., 1982,1983; Rowley et al., 1984; Johnson et al., 1987).

Purification experiments in this laboratory have typically utilized testosterone 17*b*-hemisuccinate linked to Sepharose 4B via a spacer arm of diaminodipropylamine (Chang et al., 1982). A detailed protocol for synthesis of this affinity resin is given in Chang et al. (1982). In these experiments, cytosol or partially-purified fractions were allowed to bind to the resin using a batch method, rather than applying them to a column of resin. Optimal conditions of time and temperature were determined empirically for each batch of resin to obtain the maximum specific adsorption of receptor. The resin was then washed extensively with buffer containing 0.5*M* NaCl and the androgen receptor was eluted by incubation of the resin in buffer containing 10 μg/mL testosterone. The elution technique is essentially an exchange reaction, for which the time-course was examined to obtain optimal recovery.

One of the problems associated with steroid affinity chromatography is the difficulty in estimating the quantity of receptor eluted. Our laboratory has utilized exchange labeling of the eluted receptor with [^{3}H]androgen followed by separation of bound and free steroid using charcoal adsorption. However, the eluates often contain so much free steroid that the binding capacity of the charcoal can be exceeded, leading to inaccurate quantitiation of receptor. As an alternative procedure, a Sephadex G50 column is recommended to separate free steroid from protein-bound steroid.

Our laboratory has examined other affinity resins and protocols for purification of the androgen receptor. For example, 50 mL of Dunning R3327G tumor cytosol was prepared in TTES buffer (10 m*M* Tes, 12 m*M* monothioglycerol, 2 m*M* EDTA, 250 m*M* sucrose, pH 7.4) and applied to a packed 2 mL column of DHT-17*b*-hemisuccinyl-*w*-propanyl-agarose.

This resin was then washed sequentially with 10 volumes of TTES buffer, 250 volumes of 0.7M KCl in TTES buffer, and 10 volumes of 0.3M NaI in TTES buffer. Androgen receptor was eluted with 50 mM DHT in a solution containing 10% dimethyl formamide, 0.3M NaI, 10 mM Tes, 2 mM EDTA, and 12 mM monothioglycerol. Excess salt and unbound steroid were removed by Sephadex G50 chromatography. The results of these preliminary purification experiments have not been published or extensively quantitated but have served to indicate that variations in length of the spacer arm and use of chaotropic agents during washing and elution can facilitate rapid, efficient purification of the androgen receptor.

3.5. Examples of Purification Results

The following purification tables (*see* Tables 1–3) are presented to demonstrate the quantitative results of each of the enrichment strategies described in previous sections.

4. Verification of the Purified Protein

4.1. The Problem of Proteolysis

The physicochemical and functional properties of the androgen receptor monomer have been characterized in many target tissues, and it appears that androgen receptors from these various sources are very similar. However, for some time there was a discrepancy regarding the molecular weight of the monomeric subunit. For example, reports of the size of the androgen receptor monomer of rat ventral prostate have varied from 28–117kDa (for review, *see* Johnson et al., 1987). In 1979, Wilson and French reported their studies of the effects of proteases and protease inhibitors on the androgen receptor, from which it was concluded that the intact monomeric size was a 4.5–5.0 S, 58 Å (M_r = 117kDa) protein. These authors also described two proteolytic fragments of the monomer: a 3.6 S, 37 Å fragment and a 3.0 S, 23 Å fragment. Similarly, Lea et al. (1979) have reported androgen receptor forms of 5.0 S, 53 Å (115 kDa), 3.6 S, 37 Å (55kDa), and 3.0 S, 22 Å (29kDa). Johnson et al. (1987) have confirmed these observations by reporting that sizes of 4.5–5.0 S, 54 Å (109–12 kDa), 3.5–4.0 S, 33 Å (52–59kDa), and 2.5–3.0 S, 20 Å (22–27kDa) were detected in a variety of target tissues and species. Furthermore, these authors described methods to either stabilize the monomeric form by removal of protease activity using phosphocellulose chromatography, or methods to reproducibly generate a given fragment for de-

Table 1
Purification of Androgen Receptor from Steer Seminal Vesicle[a]

Purification step	Total protein, mg	Receptor, pmol	Spec. act., pmol/mg	Yield, %	X-fold purification
Cytosol	3340	77.6	0.023	100	1.0
DNA I flow through	1960	71.8	0.037	93	1.6
40% ammonium sulfate pellet	925	87.9	0.095	113	4.1
Affinity eluate	3.25	43.2	13.3	56	570
DNA II eluate (pyridoxal phosphate)	0.003	37.4	12500	48	540,000

[a]Reprinted with permission from Chang et al., *Biochemistry* **21,** 4102–4109. Copyright 1982 American Chemical Society.

Table 2
Purification of Androgen Receptor from Rat Prostate[a]

Purification step	Total protein, mg	Receptor, pmol	Spec. act., pmol/mg	Yield, %	X-fold purification
Cytosol	4050	370	0.091	100	1.0
DNA I flow through	3978	290	0.073	78	0.8
40% ammonium sulfate pellet	248	380	1.5	100	16.5
Affinity eluate	2.6	110	42	30	460
DNA II eluate (pyridoxal phosphate)	0.008	87	10900	24	120,000

[a]Reprinted with permission from Chang et al., *Biochemistry* **22,** 6170–6175. Copyright 1983 American Chemical Society.

Table 3
Purification of Androgen Receptor from Dunning H Tumor[a]

Purification step	Total protein, mg	Receptor, pmol	Spec. act., pmol/mg	Yield, %	X-fold purification
Cytosol	1530	81.9	0.054	100	1.0
DNA I flow through	1290	62.7	0.049	77	0.9
40% ammonium sulfate pellet	275	38.5	0.140	47	2.62
Affinity eluate	2.8	31.0	11.1	38	207
DNA II eluate (0.5M KCl)	0.002	9.4	4920	12	91,900

[a]Reprinted from NIH Publication No. 87-2881.

tailed study. From the results of these studies, it has been concluded that the intact androgen receptor monomer has a mol wt of approx 110–120 kDa in cytosol and nuclear extracts of all androgen target tissues. Moreover, the variable appearance of smaller artefactual fragments is the result of tissue-specific differences in protease activity and different laboratory procedures for collecting, storing, and processing tissue.

4.2. Physicochemical Characterization

Verification that the purified protein possesses the same properties as the androgen receptor of crude cytosol has been complicated by the problem of proteolysis, which was described above. In light of this confusion over the mol wt of the monomeric receptor, purification strategies have relied upon the specific high affinity binding of androgens by the purified receptor to validate their protocols. Steroid specificity studies (Chang et al., 1982) and Scatchard analysis (Rowley et al., 1984) have been used with purified receptor preparations for that purpose. Physicochemical analysis of the Stoke's radius and sedimentation coefficient of [3H]-androgen labeled preparations of partially-purified androgen receptor have been conducted using the conventional nondenaturing methods of gel filtration chromatography and sucrose density gradient sedimentation (Chang et al., 1982, 1983; Rowley et al., 1984; Johnson et al., 1987). With the advent of affinity labeling, denaturing methods of analysis could also be used.

The usefulness of affinity labeling techniques for the study of steroid receptors stems from the ability to bind a radiolabeled steroid covalently, so that it cannot be dissociated, even under denaturing conditions. Thus, the molecular weight of the labeled receptor protein can be analyzed by denaturing polyacrylamide gel electrophoresis. Affinity labels fall into two categories, the electrophilic affinity labels and the photoaffinity labels. Electrophilic affinity labels usually possess a good leaving group such as a bromine atom or a methanesulfonyloxy (mesylate) moiety. These leaving groups are displaced, thereby allowing the steroid molecule to attach covalently to the receptor. Photoaffinity labels rely on the double bonds in the steroid molecule, which upon exposure to ultraviolet light of the proper wavelength, will break with concomitant covalent attachment to the receptor protein.

For the androgen receptor protein, 17b-[(Bromoacetyl)oxy]-[3H8]-5-α-androstan-3-one ([3H]DHT-bromoacetate), has proven to be a useful electrophilic affinity label. Chang et al. (1984) and Rowley et al. (1984) have used this compound to covalently label purified androgen receptors from steer seminal vesicle, rat seminal vesicle, rat ventral prostate, and R3327 prostatic tumor. These studies demonstrated that DHT-bromoacetate had a reduced affinity for the androgen receptor (Kd = 4 nM), compared to that of DHT (Kd = 1 nM). However, this was still within a practical range. Studies of steroid specificity suggested that DHT-bromoacetate bound to the androgen receptor in a structure-specific manner. It was also demonstrated that 18% of the receptor-bound affinity label became covalently attached (Chang et al., 1984). Therefore, this technique provided a sensitive method of selectively labeling small amounts of the purified androgen receptor with sufficient radioactivity to enable its molecular weight to be determined by de-naturing gel electrophoresis and fluorography. Accordingly, Rowley et al. (1984) and Johnson et al. (1987) have used [³H]DHT-bromoacetate to identify the ~120kDa monomeric form of the androgen receptor from partially purified R3327H tumor.

In 1982, Chang et al. reported that the synthetic androgen, methyltrienolone ([³H]R1881), was used to affinity label highly purified androgen receptor. This latter compound is a photo-activatable ligand whose affinity for the androgen receptor is similar to that of the natural ligand, dihydrotestosterone, and is substantially greater than that of DHT-bromoacetate. Use of R1881 as a photoaffinity label identified a 60kDa protein in denaturing polyacrylamide gels of androgen receptor from steer seminal vesicle. It is apparent now that this protein was a proteolytic fragment of the receptor monomer (Wilson and French, 1979; Johnson et al., 1987). Nonetheless, R1881 appeared to be a useful photoaffinity ligand for highly purified androgen receptor. A number of subsequent studies have stressed that R1881 was not very efficient or useful as an affinity label for use with crude cytosol preparations (Mainwaring and Randall, 1984; Brinkman et al., 1985, 1986). Brinkman et al. (1985) have reported that a purification of at least 40-fold was essential to use [³H] R 1881 in photoaffinity labeling studies of androgen receptors by SDS polyacrylamide gel electrophoresis.

4.3. Western Immunoblot with Monoclonal Antibody

Monoclonal antibodies specific for the androgen receptor have been isolated from the lymphocytes of patients with prostatic cancer (Young et al., 1988). Of more than 1500 serum samples screened by an immunoprecipitation assay, only five sera demonstrated high affinity autoimmune anti-androgen receptor antibodies. Lymphocytes were obtained from the blood of three of these patients and were subsequently activated with mitogens, transformed with Epstein-Barr virus, and cloned in vitro. The resulting human monoclonal antibodies were shown to cross-react with androgen receptors of several mammalian species but not with other steroid receptors or steroid binding globulins. These antibodies were capable of shifting the sedimentation of the androgen receptor under nondenaturing conditions of sucrose gradient analysis. In addition, one of the monoclonals was used to identify a 116 kDa band corresponding to the androgen receptor monomer using Western immunoblot analysis (Young et al., 1988).

In summary, the methods available to verify that the purified protein is the androgen receptor include: analysis of steroid specificity; demonstration of high affinity for androgens; analysis of [^{3}H]androgen-labeled protein by gel filtration and sucrose gradients; denaturing gel electrophoresis of affinity-labeled protein; and denaturing gel electrophoresis of immunoreactive protein.

5. Concluding Remarks

Despite the many years of effort that have been devoted to developing more efficient protocols for purification of the androgen receptor, the low concentration of receptor in target tissue sources is an obstacle that makes it more feasible to obtain large amounts of receptor protein by an alternative method. Instead of purifying androgen receptor from 'natural' tissues, the recently cloned androgen receptor gene (Chang et al., 1988; Lubahn et al., 1988; Young et al., 1988) will be used to overproduce the receptor in yeast or bacterial systems. The in vitro synthesized receptor will undoubtedly be an abundant protein in the cell lysates, that can be purified by conventional chromatographic methods, and/or the steroid affinity- and differential DNA chromatographic techniques described in this chapter. As larger amounts of purified receptor (antigen) become available, it will not be difficult to obtain sufficient polyclonal and monoclonal antiandrogen receptor antibodies for development of immunoaffinity chromatographic methods. Investigators who are interested in the structure and function of the androgen receptor may integrate these various methods to create new purification strategies that suit their particular needs.

References

Blaquier, J. A. and Calandra, R. S. (1973) *Endocrinology* **93,** 51–60.
Brinkman, A. O., Kuiper, G. G. J. M., de Boer, W., Mulder, E., and van der Molen, H. J. (1985) *Biochem. Biophys. Res. Commun.* **126,** 163–169.

Brinkmann, A. O., Kuiper, G. G. J. M., de Boer, W., Mulder, E., Bolt, J., van Steenbrugge, G. J., and van der Molen, H. J. (1986) *J. Steroid Biochem.* **24,** 245–249.

Bruchovsky, N. and Rennie, P. S. (1978) *Cell* **13,** 273–280.

Cardinali, D. P., Nagle, C. A., and Rosner, J. M. (1975) *Life Sci.* **16,** 93–106.

Chan, L., Johnson, M. P., and Tindall, D. J. (1989) in *Pediatric Endocrinology* (Collu, R., Ducharme, J. R., and Guyda, H., eds.) Raven, New York (in press).

Chang, C. H. and Tindall, D. J. (1983) *Endocrinology* **13,** 1486–1493.

Chang, C. H., Lobl, T. J., Rowley, D. R., and Tindall, D. J. (1984) *Biochemistry* **23,** 2527–2533.

Chang, C. H., Rowley, D. R., and Tindall, D. J. (1982) *Biochemistry* **21,** 4102–4109.

Chang, C. H., Rowley, D. R., and Tindall, D. J. (1983) *Biochemistry* **22,** 6170–6175.

Chang, C., Kokontis, J., and Liao, S. (1988) *Science* **240,** 324–326.

Diamond, D. A. and Barrack, E. R. (1984) *J. Urol.* **132,** 821–827.

Dunn, J. F., Goldstein, J. L., and Wilson J. D. (1973) *J. Biol. Chem.* **248,** 7819–7825.

Eppenberger, U. and Hsia, S. L. (1972) *J. Biol. Chem.* **247,** 5463–5469.

Gehring, U., Tomkins, G. M., and Ohno, S. (1971) *Nature New Biol.* **232,** 106, 107.

Giannopoulos, G. (1973) *J. Biol. Chem.* **248,** 1004–1010.

Grody., W. W., Schrader., W. T., and O'Malley, B. W. (1982) *Endocrine Reviews* **3,** 141–163.

Gulizia, S., D'agata, R., Sanborn, B. M., and Steinberger, E. (1983) *J. Androl.* **4,** 248–252.

Hansson, V. and Djoseland, O. (1972) *Acta Endocrinol.* **71,** 614–624.

Hansson, V., McLean, W. S., Smith, A. A., Tindall, D. J., Weddington, S. C., Nayfeh, S. N., French, F. S., and Ritzen, E. M. (1974) *Steroids* **23,** 823–832.

Hasenson, M., Hartley-Asp., B., Kihlfors, C., Lundin, A., Gustafsson, J.-A., and Pousett, A. (1985) *The Prostate* **7,** 183–194.

Isaacs, J. T. (1982) *Cancer Res.* **42,** 5010–5014.

Isomaa, V., Pajunen, A. E. I., Bardin, C. W., and Janne, O. A. (1982) *Endocrinology* **111,** 833–843.

Johnson, M. P., Young, C. Y. F., Rowley, D. R., and Tindall, D. J. (1987) *Biochemistry* **26,** 3174–3182.

Jung, I. and Baulieu, E. (1972) *Nature New Biol.* **237,** 24, 25.

Lea, O. A., Wilson, E. M., and French, F. S. (1979) *Endocrinology* **105,** 1350–1360.

Lin, A. L., McGill, H. C., Jr., and Shain, S. A. (1981a) *Arteriosclerosis* **1,** 257–264.

Lin, A. L., McGill, H. C., Jr., and Shain, S. A. (1981b) *Circ. Res.* **49,** 1010–1016.

Lubahn, D. B., Joseph, D. R., Sullivan, P. M., Huntington, F. W., French, F. S., and Wilson, E. M. (1988) *Science* **240,** 327–330.

Mainwaring, W. I. P. and Randall, V. A. (1984) *J. Steroid Biochem.* **21,** 209–216.

Pelletier, J. and Caraty, A. (1981) *J. Steroid Biochem.* **14,** 603–611.

Prins, G. S. and Lee, C. (1979) *Steroids* **40,** 189–201.

Rowley, D. R., Chang, C. H., and Tindall, D. J. (1984) *Endocrinology* **114,** 1776–1783.

Rowley, D. R., Premont, R. T., Johnson, M. P., Young, C. Y. F., and Tindall, D. J. (1986) *Biochemistry* **25,** 6988–6995.

Sanborn, B. M., Steinberger, A., Tcholakian, R. K., and Steinberger, E. (1977) *Steroids* **29,** 493–502.

Schanbacher, B. D., Johnson, M. P., and Tindall, D. J. (1987) *Biol. Reprod.* **36,** 340–350.

Shain, S. A., Boesel, R. W., Kalter, S. S, and Heberling, R. L. (1981) *JNCI* **66,** 565–574.

Sherman, M. R., Moran, M. C., Tuazon, Fe. B., and Stevens, Y. W. (1983) *J. Biol. Chem.* **258,** 10366–10377.

Smith, R. G., Syms, A. J., and Norris, J. S. (1984) *J. Steroid Biochem.* **20,** 277–281.

Thieulant, M.-L. and Pelletier, J. (1979) *J. Steroid Biochem.* **10,** 677–687.

Tindall, D. J., Hansson, V., McLean, W. S., Ritzen, E. M., Nayfeh, S. N., and French, F. S. (1975) *Mol. Cell Endocrinol.* **3,** 83– 101.

Tindall, D. J., Miller, D. A., and Means, A. R. (1977) *Endocrinology* **101,** 13–23.

Wilson, E. M. and French, F. S. (1976) *J. Biol. Chem.* **251,** 5620–5629.

Wilson, E. M. and French, F. S. (1979) *J. Biol. Chem.* **254,** 6310–6319.

Wrange, O., Carlstedt-Duke, J., and Gustafsson, J.-A. (1979) *J. Biol. Chem.* **254,** 9284– 9290.

Young, C. Y. F., Murthy, L. R., Prescott, J. L., Johnson, M. P., Rowley, D. R., Cunningham, G. R., Killian, C. S., Scardino, P. T., Von Eschenbach, A., and Tindall, D. J. (1988) *Endocrinology* **123,** 601–610.

Young, C. Y. F., Johnson, M. P., Prescott, J., Heydrick, K., Murthy, L., and Tindall, D. J. (1988) Presented at the 1988 Annual Meeting of the Endocrine Society, New Orleans, LA.

Purification
of the Estrogen Receptor

Francesco Bresciani, Nicola Medici,
Ciro Abbondanza, Bruno Moncharmont,
and Giovanni Alfredo Puca

1. Introduction

The estrogen receptor, like all other steroid receptors, belongs to that highly specialized class of proteins which modulate gene expression in eucaryotes via interaction with DNA regulatory sequences located in the proximity of target genes (Yamamoto, 1985).

The estrogen receptor purified to homogeneity from calf uterus is a protein of about 67kDa endowed with a single estrogen binding site (Molinari et al., 1977; Sica and Bresciani, 1979; Bresciani et al., 1979; Puca et al., 1980). Recently, estrogen receptor cDNA clones were isolated (Walter et al., 1985), sequenced and shown to code for polypeptides of 589 to 595 amino acids, with a calculated molecular weight of about 66kDa (Green et al., 1986). The estrogen receptor, like other steroid receptors, was shown to be endowed with 5 functional domains, one of which is the estrogen binding domain, the E domain, and another the DNA binding domain, the C domain (Kumar et al., 1987). Under native conditions the 67kDa estrogen receptor protein apparently exists as a quaternary structure. When uteri are excised from ovariectomized or prepuberal animals and are homogenized in low-salt buffers, the estrogen receptor activity is associated with a large molecular complex in the cytosol. This complex shows a sedimentation rate of 8–9 S and a Stokes radius of 6.4–7.4 nm; the calculated mol wt is about 240kDa (Puca et al., 1971). On the the other hand, when uteri are removed from estrogen-injected or puberal animal, most of receptor activity is found in the nucleus and can be extracted only by high salt buffers, classically 0.4M KCl (Puca and Bresciani, 1968). This nuclear receptor activity is associated with a molecular complex showing a sedimentation rate of about 5.2 S and a Stokes radius of about 5.7 nm; it has a calculated mol wt of about 120kDa (Puca and Bresciani, 1968; Puca et al., 1970a). The most recent hypothesis (Moncharmont et al., 1984; Redeuilh et al., 1987) suggest that the 8–9 S structure consists of a homo-dimer of the 67kDa protein bound to the 90kDa heat-shock protein (hsp90) (Catelli et al., 1985) and that the 5.2 S estrogen-binding complex is the free homodimer (Moncharmont et al., 1984).

There is a consensus that the 8–9 S complex generates the 5.2 S state as a consequence of binding estrogen. This transformation may also be achieved in vitro upon incubation of crude or purified receptor preparations with physiological concentrations of estrogens at 25–37°C. Contrary to the 8–9 S form that is unable to bind to DNA (therefore called "inactive") the 5.2 S form is endowed with this ability (therefore called "active"). One may note that the homodimeric structure of the "active" form of receptor can be correlated to the dyad symmetry of the consensus sequence of the "hormone-response elements" located near target genes (Jost et al., 1984; Walker et al., 1984).

The ability to bind to DNA explains why, upon homogenization of the tissue the "active" receptor is found in the nuclear fraction, as well as why high salt buffers are required for achieving its extraction from the nuclei. Recently, immunohistochemical analysis (King and Greene, 1984) and sophisticated enucleation techniques (Welshon et al., 1984) have shown that in vivo actually about 70% of the estrogen-free "inactive" receptor activity is also located in the nucleus. That most "inactive" receptor is found in the cytosol when the usual conditions of tissue homogenization are applied, it appears to be a consequence of buffer dilution and low temperature that induce "leaking" of receptor from nuclei to cytosol (Molinari et al., 1985).

Attempts at purification of estrogen receptor started shortly after its discovery (Puca et al., 1970a,b,1971,1975). Conventional methods, however, did not achieve purification to homogeneity especially because of a great tendency toward aggregation of the receptor, which could not be efficiently counteracted without producing dissociation of estrogen from the receptor protein and ensuing loss of the radioactive hormone tag. A great deal of effort was thus put toward developing matrixes for ligand affinity chromatography (Sica et al., 1973). The subsequent discovery of a considerable affinity of the estrogen receptor for heparin (Molinari et al., 1977) as well as the use of chaotropic salts like NaSCN (Parikh et al., 1975; Sica et al., 1976) made it possible to utilize two sequential affinity chromatography steps and thus the first purification to homogeneity of the 67 kDa estrogen receptor was achieved (Sica and Bresciani, 1979; Bresciani et al., 1979; Puca et al.,1980). Later, sodium molybdate was shown to inhibit the transformation of the 8–9 S into the 5.2 S form, and this allowed extensive purification of the 8–9 S form (Redeuilh et al., 1987). Finally, the preparation of monoclonal antibodies to the estrogen receptor (Greene et al., 1980; Moncharmont et al., 1982) made it possible the purification of hormone-free receptor protein by immunoaffinity chromatography (Medici et al., 1988).

In this chapter we will initially describe methodology in general and thereafter present the protocols for purification of the 67 kDa receptor protein and of the 8–9 S molecular complex, both in the estrogen-free or estrogen-bound state.

2. Materials and General Methods

2.1. Materials

[^{3}H]17β-estradiol (specific activity 100 Ci/mmol) (Amersham, UK); Diethylstilbestrol (DES), benzamidine, pepstatin A, antipain, leupepetin, aprotinin, EGTA, guanidine and heparin (Sigma, USA; Superose 6B, Dextran T70 and Protein A-Sepharose (Pharmacia, Sweden); dimethylpimelimidate•2 HCl (Pierce, The Netherlands); Coomassie Brilliant Blue G-250 and urea (Bio-Rad, USA). All other reagents are analytical grade.

2.2. General Methods

2.2.1. Buffers

PM buffer: 50 mM sodium phosphate buffer, pH 7.4 at 20°C, 10 mM sodium molybdate; PMD buffer: PM buffer containing 1 mM dithiothreitol; TM buffer: 50 mM Tris-HCl buffer, pH 7.4 at 20°C, 10 mM sodium molybdate; TMD buffer: TM buffer containing 1 mM dithiothreitol; PMDI, TMI and TMDI buffers: PMD, TM and TMD buffers, respectively, containing the following protease inhibitors: 25 mM benzamidine, 2 mg/mL pepstatin A, 2 mg/mL antipain, 2 mg/mL leupepetin, 1 mg/mL aprotinin, 2 mM EGTA. All the buffers used in chromatographic procedures are fil-

Table 1
Physicochemical Parameters of the Estrogen Receptor[a]

	Native	Dissociated	Transformed
Sedimentation coefficient (S)	*8–9*	4.4	5.2
Stokes radius (nm)	7.4	3.9	5.7
Mol wt by NaDodSO$_4$-PAGE			
(kDa)	*67 + 90*	67	67
Mol wt by gel filtration (kDa)	300	67	129
Mol wt theoretical (kDa)		66.182	
Number of aminoacids		595	
Number of steroid binding sites	2	1	2
K$_d$ for 17β-estradiol (nM)	*0.5*	*0.5*	
Isoelectric point	6.5		6.7

[a]Values in italics were determined with purified receptor preparations. Data from Bresciani et al. (1979) and Redeuilh et al. (1987). The theoretical mol wt was calculated by Green et al. (1986).

tered through 0.22 µm nitrocellulose filters. Dextran-coated Charcoal (DCC) is: 0.5% Norit A, 0.05% Dextran T70, 0.1% gelatin in 5 mM sodium phosphate buffer, pH 7.4.

2.2.2. Assay of Estradiol Binding Activity

Aliquots of cytosol or of purified preparations (20–100 µL) are added to 5 nM [^{3}H]17β-estradiol, in a final volume of 500 µL in presence or absence of a 100-fold molar excess of unlabeled estradiol. At the end of a 4-h incubation at 0°C, an equal volume of DCC suspension is added to the samples which, after a 5-min incubation, are centrifuged in a bench-top microfuge. A 0.5-mL aliquot of the supernatant is counted for radioactivity. The difference between the total and nondisplaceable binding is considered as specific estradiol binding.

2.2.3. Assay of Radioactivity

Ten mL of Instagel (Packard Instrument Co., USA) is added to each sample and the radioactivity measured by a beta scintillation counter.

2.2.4. Protein Assay

Protein concentration is determined by the Coomassie blue method (Bradford, 1976).

2.2.5. Polyacrylamide Gel Electrophoresis (NaDodSO$_4$-PAGE)

β-mercaptoethanol, 5% (v/v), sodium dodecylsulfate, 3% (w/v), glycerol, 10% (v/v) and bromophenol blue, 0.01% (w/v) (Laemmli sample buffer) are added to the samples and after heating for 5 min at 90°C, electrophoresis is performed according to Laemmli (1970) using a 1.5-mm thick

4% stacking gel and 11.5% separating gel (acrylamide/bisacrylamide ratio 37.5/1). At the end of the run the gel is stained with the silver stain method (Morrissey, 1981).

2.2.6. Environmental Temperature

All manipulation and chromatographic procedure reported are performed in an ice-bath or in cold room (2–4°C) unless otherwise indicated.

2.3. Preparation of Superose-Epoxyl-DES

The epoxyl spacer arm is coupled to Superose 6B following a published procedure (Sundberg and Porath, 1974). The coupling of DES to the Superose-epoxyl is performed as follow (according to Van Oosbree et al., 1984). DES (5 mg) is dissolved in 1:1 dioxane:water or dimethylformamide:water and the pH adjusted to 10 with KOH. Superose-epoxyl (5 g) is added and the slurry incubated for 18 h at room temperature. Residual reactive groups are blocked by incubation of the resin for 3 h in 1:1 ethylen glycol:water at pH 10. The resin is washed on a Buchner funnel with 1 L of 1:1 dioxane:water and then packed on a column. The resin is further washed with 4:1 methanol:water at a flowrate of 1 L/d for 2 d. Thereafter, the resin is washed with water and equilibrated with PMD buffer. After use, the resin is sequentially washed on Buchner funnel with 20 vol of PM buffer containing $6M$ guanidine, water, 100 vol of 4:1 methanol: water and stored for further use in 0.05% (w/v) sodium azide at 4°C.

2.4. Preparation of Sepharose-Protein A-JS34/32

Preparation and characterization of the monoclonal antibody to ER JS34/32 has been described elsewhere (Moncharmont et al., 1982, 1984). Protein A-Sepharose (0.75 g dry powder) is resuspended in 50 mL $0.1M$ borate buffer, pH 8.2 in a 50-mL Falcon tube. The beads are sedimented and, after removal of the supernatant, 25 mL of JS34/32 antibody (protein concentration 1 mg/mL) previously purified on a Protein A-Sepharose column according to manufacturer's instructions are added. At the end of a 3-h incubation at room temperature on a roller-shaker, the beads are sequentially washed by centrifugation in the same tube once with 40 mL $0.1M$ borate buffer, pH 8.2, and once with $0.2M$ triethanolamine, pH 8.2. The beads are then resuspended in 45 mL $0.2M$ triethanolamine, 20 mM dimethylpimelimidate: pH 8.2, for 45 min at room temperature on a roller-shaker. The beads are sedimented and resuspended for 15 min at room temperature in 45 mL $0.2M$ triethanolamine, 20 mM ethanolamine, pH 8.2. The beads are finally washed by centrifugation with 40 mL $0.1M$ borate buffer, pH 8.2, twice and packed in 0.5-mL gel bed columns (internal diameter 5 mm). The columns are prewashed with elution buffer and equilibrated with PM buffer just before application of the sample. After use, the resin is washed with 4 mL of 0.4M ammonium hydroxyde, pH 11.7, and stored for further use in 0.05% (w/v) sodium azide at 4°C. If proper care is taken, the column can be reused over 50 times.

2.5. Preparation of Sepharose-Heparin

Sepharose-Heparin is prepared by direct coupling of heparin to CNBr-activated Sepharose as previously described. After the use the column is washed with 5 vol of 2M sodium chloride and 2 vol of 0.1M acetic acid. The column is equilibrated in 0.05% (w/v) sodium azide and stored for further use at 4°C.

2.6. Cytosol Preparation

Frozen calf uteri (50–200 g) are pulverized in liquid nitrogen with an ice-grinder or minced with scissor after thawing. Tissue is homogenized with 4 vol of ice-cold PMD buffer with several 30-s bursts of Ultraturrax (Jane and Kunkel, FRG), until all tissue fragments have disappeared. The homogenate is centrifuged at more than $105,000 \times g$ for 45 min at 2°C. The supernatant is collected and thenceforth referred to as cytosol.

2.7. Ion-Exchange Chromatography on DEAE-TSK

A 40-mL DEAE-TSK column (Fractogel DEAE-650S, Pierce, The Netherlands) is packed in a glass column (internal diameter 26 mm) according to the manufacturer's instructions and equilibrated with PMD buffer containing 50 mM KCl. Cytosol is added with 50 mM KCl and applied to the column at a 300 mL/h flowrate with a peristaltic pump. The column is then washed with at least 200 mL of PMD buffer containing 90 mM KCl. Elution is performed with a 200-mL linear 90 to 400 mM KCl gradient at 80 mL/h flowrate. Fractions of 4 mL are collected. For the determination of the elution peak of estrogen receptor, 20-mL aliquots of each fraction are added to 5 nM [^{3}H]17β-estradiol, in a final vol of 500 mL. At the end of a 20-min incubation at 23°C, an equal vol of DCC suspension is added to the samples that are then incubated for 5 min and centrifuged in a bench-top microfuge. A 0.5-mL aliquot of the supernatant is counted for radioactivity. The peak fractions are pooled and aliquots are assayed for both [^{3}H]17β-estradiol binding activity and protein concentration.

3. Purification of the 67kDa Receptor Protein

3.1. Overview

The purification of the 67kDa receptor protein may be obtained by immunoaffinity chromatography, using the monoclonal antibody JS34/32 bound to Sepharose-Protein A (Medici et al., 1988), or by affinity chromatography with Superose-epoxyl-DES. By immunoaffinity chromatography, the 67kDa receptor protein can be purified either estrogen-free or estrogen-bound. By Superose-epoxyl-DES only the estrogen-bound receptor can be purified, because the elution from the resin is achieved by estradiol. Before the affinity chromatography step, either on Sepharose-Protein A-JS34/32 or on Superose-epoxyl-DES, the cytosol is preliminarily chromatographed on a column of DEAE-TSK as described in Section 2.7. For immunoaffinity chromatography purification, the DEAE-TSK eluate is further purified by chromatography on Sepharose-Heparin. The Sepharose-Heparin step is, on the other hand, the last step after Superose-epoxyl-DES purification. These procedures are described in detail next.

3.2. Purification of the Estrogen-Free Receptor Protein by Immunoaffinity Chromatography

3.2.1. Affinity Chromatography on Sepharose-Heparin of the DEAE-TSK Eluate

The pooled fraction of the estradiol binding activity of the DEAE-TSK chromatography (*see* Section 2.7) are dialysed overnight at 4°C vs 1 L of PMD buffer. The sample is diluted with an equal volume of 10 mM sodium molybdate containing the protease inhibitor cocktail, and then loaded at a flowrate of 15 mL/h on a 10-mL Sepharose-Heparin column (internal diameter 16 mm), preequilibrated with TM buffer containing 2 mM EGTA. The column is then washed for 1 h with 100 mL TMI buffer and eluted with TMI buffer containing 4 mg/mL heparin at a flowrate of 7 mL/h. The estradiol binding activity in the single fractions is assayed as described in Section 2.7. and the peak fractions are pooled.

3.2.2. Immunoaffinity Chromatography on Sepharose-Protein A-JS34/32

The pooled fractions from the affinity chromatography on Sepharose-Heparin (*see* Section 3.2.1.) are applied to a 0.5 mL Sepharose-Protein A-JS34/32 column at a flowrate of 6 mL/h. The column is washed at a flowrate of 100 mL/h with 50 mL of PM buffer containing 0.4M KCl and with

25 mL of PM buffer containing 1M guanidine. The column is eluted with 2 mL of 0.4M ammonium hydroxyde, pH 11.7. The eluate is dialyzed vs 1 L of 50 mM Tris-HCl buffer, pH 7.4 containing 10% (v/v) glycerol, 150 mM sodium chloride, and 3 mM dithiothreitol. Aliquots of the eluate are assayed for [^{3}H]17β-estradiol binding activity, as described in Section 2.2.2., protein concentration and analyzed by NaDodSO$_4$-PAGE. The results of a purification performed with this procedure are summarized in Table 2.

3.3. Purification of the Estrogen-Bound Receptor Protein by Ligand Affinity Chromatography

3.3.1. Affinity Chromatography on Superose-Epoxyl-DES

The pooled fraction of the estradiol binding activity of the DEAE-TSK chromatography (*see* Section 2.7.) are dialysed overnight at 4°C vs 1 L of PMDI buffer and then added to 0.5–1 g of Superose-epoxyl-DES equilibrated in the same buffer. The mixture is incubated for 2 h at 4°C with gentle shaking and then centrifuged at 800 rpm for 5 min. The beads are washed three times with 50 mL of PMDI buffer and five times with 25 mL of PMDI buffer containing 1M guanidine. Elution of the estrogen receptor is obtained by incubating the pelleted beads for 1 h at 4°C with 6 mL of PMDI buffer containing 0.4M sodium tiocyanate and 5 mg/mL [^{3}H]17β-estradiol (specific activity 3 Ci/mmol). After removal of the resin by centrifugation, the supernatant is dialyzed overnight vs 1 L TMDI buffer without benzamidine at 4°C.

3.3.2. Affinity Chromatography on Sepharose-Heparin

The dialysed affinity-chromatography eluate is loaded on a 1-mL Sepharose-Heparin column (flowrate 10 mL/h), preequilibrated with TMDI buffer. The column is then washed for 1 h with 100 mL TMDI buffer until no radioactivity is present in the flow through. Elution is performed with 1 mL of TMDI buffer containing 4 mg/mL heparin at a flowrate of 5 mL/h. Aliquots of the eluate are assayed for [^{3}H]17β-estradiol bound radioactivity, protein concentration and analyzed by NaDodSO$_4$-PAGE.

4. Purification of the 8–9 S Molecular Complex

4.1. Overview

The purification of the 8–9 S molecular complex is achieved by immunoaffinity chromatography, using the monoclonal antibody JS34/32 bound to Sepharose-Protein A (Medici et al., 1988). Before immunoaffinity chromatography, the cytosol is preliminarly chromatographed on a col-

Table 2
Purification of the 67kDa Estrogen Receptor Protein

Purification, step	Total volume, mL	Estrogen protein, mg	Specific receptor, pmol	Activity, pmol/mg	Yield, %	Purification, fold
Cytosol	300	1500	453	0.3	100	0
DEAE-TSK	15	52	341	6.5	74	21
Sepharose-heparin	8	1.3	174	120	35	400
Sepharose-protein A-JS34/32	2.5	0.01[a]	29	3060	7	10200

[a]Protein concentration determined by evaluation of band's intensity after NaDodSO$_4$-PAGE analysis of the sample.

umn of DEAE-TSK, as described in Section 2.7., and the TSK eluate is further submitted to ion-exchange HPLC. The same protocol can be used to purify either the estrogen-free or estrogen-bound 8–9 S complex.

4.2. Purification of the Estrogen-Free Complex

4.2.1. Ion-Exchange HPLC

The pooled fraction of the estradiol binding activity of the DEAE-TSK chromatography (*see* Section 2.7.) are dialysed overnight at 4°C vs 1L of PM buffer. The sample is the loaded on a 7.5 × 75-mm Protein Pak DEAE-5PW column (Waters Associates, USA) at a flowrate of 1 mL/min. The column is then eluted with a 40-mL linear 100 to 400 mM KCl gradient at the same flowrate. Fractions of 1 mL are collected and assayed for estradiol binding activity as described in Section 2.7. and the peak fractions are pooled.

4.2.2. Immunoaffinity Chromatography on Sepharose-Protein A-JS34/32

The pooled fractions from the ion-exchange HPLC (*see* Section 4.2.1.) are applied to a 0.5-mL Sepharose-Protein A-JS34/32 column equilibrated in PM buffer at a flowrate of 6 mL/h. The column is washed with 100 mL of PM buffer containing 0.4M KCl for 1 h and eluted with 2 mL of PM buffer containing 8M urea at 10°C. The eluate is collected in a tube containing 2 mL of PM buffer containing 3 mM dithithreitol and 2 mM β-mercaptoethanol in order to immediately dilute the high urea concentration. Aliquots of the eluate are assayed for [^{3}H]17β-estradiol binding activity (as described in Section 2.2.2.) and protein concentration. The purified receptor can be visualized by NaDodSO$_4$-PAGE analysis. The results of a purification performed with this procedure are summarized in Table 3.

4.3. Purification of the Estrogen-Bound Complex

The pooled fraction of the estradiol binding activity of the DEAE-TSK chromatography (*see* Section 2.7.) are incubated for 2 h at 0°C with 5 nM [^{3}H]17β-estradiol. After the dialysis, the sample can be applied to the ion exchange HPLC and eventually to the immunoaffinity chromatography as described in Sections 4.2.1. and 4.2.2., respectively. Aliquots of the fractions or of the purified eluate are counted for radioactivity in order to determine the estradiol binding activity.

5. Closing Remarks

One must emphasize that for achieving highest purity and yield it is of utmost importance that the uteri be put on ice or frozen immediately after removal from the animals and that storage at −80°C last no more than 3 mo. If the above mentioned specifications are not strictly observed, at the end of the purification procedure a more or less significant fraction of the receptor protein is found fragmented into polypeptides with molecular masses of 52–54kDa or smaller. These fragments still bind estrogen. Na-DodSO$_4$-PAGE analysis of the purified 8–9 S molecular complex shows one band at 67kDa, corresponding to the estrogen receptor subunit, and one band at 90kDa, corresponding to the hsp90.

When yield is a priority over purity, the ion-exchange HPLC step can be omitted from the purification of the 8–9 S molecular complex. Similarly, the Sepharose-Heparin affinity chromatography can be omitted from the purification procedure of the 67kDa receptor protein.

When an appropriate control is required for the purified estrogen receptor, the purification can be performed in parallel to a "mock" purification. Preincubation of the partially purified preparation with large excess of unlabeled steroids will prevent the biospecific adsorbtion of the estrogen receptor on the ligand affinity chromatography matrix. In this case, only contaminants will be desorbed from the resin. Alternatively, cytosol or any partially purified preparation can be depleted of its receptor content by immunoaffinity chromatography. The flow through can then be used in all the following purification steps. Also in this case, only contaminants will be present in the final eluate.

The protocols described here extensively apply immunoaffinity chromatography as a purification procedure. Covalent crosslinkage of the monoclonal antibody to Protein A on the affinity support gives resins with higher capacity than the ones prepared by direct CNBr-coupling of the immunoglobulins to Sepharose. For most purposes, immunoaffinity

Table 3
Purification of the 8–9 S Molecular Complex

Purification, step	Volume, mL	Total protein, mg	Estrogen receptor, pmol	Specific activity, pmol/mg	Yield, %	Purification, fold
Cytosol	300	1500	453	0.3	100	0
DEAE-TSK	15	57	337	6.4	73	21
DEAE-HPLC	3	5	253	49.5	56	165
Sepharose-Protein A-JS34/32	2.5	0.047[a]	27	600	16	>2000

[a]Protein concentration determined by evaluation of bands intensity after NaDodSO$_4$-PAGE analysis of the sample.

chromatography is a big advance over ligand affinity chromatography. By this technique one can, in fact, achieve higher yield and purity in shorter time. Furthermore, only by immunoaffinity chromatography can estrogen-free receptor be purified. The necessarily harsh elution condition, however, produce loss of the estrogen binding activity of the receptor protein which may be as high as 70% of total activity.

Acknowledgments

This work was supported by grants from Progetto Finalizzato Oncologia of Consiglio Nazionale delle Ricerche and from Associazione Italiana per la Ricerca sul Cancro, Italy.

References

Bradford, M. (1976) *Anal. Biochem.* **72,** 248–254, .

Bresciani, F., Sica, V., and Weisz, A. (1979) in *Biochemical Actions of Hormone,* vol. VI, (G. Litwack, ed.), Academic, New York.

Catelli, M. G., Binart, N., Jung-Testas, I., Renoir, J. M., Baulieu, E. E., Feramisco, E. E., and Welsh, W.J. (1985) *EMBO J.* **4,** 3131–3135.

Green, S., Walter, P., Kumar, V., Krust, A., Bornert, J. M., Argos, P., and Chambon, P. (1986) *Nature (London)* **320,** 134–139.

Greene, G. L, Nolan, C., Engler, J. P. and Jensen, E. (1980) *Proc. Natl. Acad. Sci. USA.***77,** 5115–5119.

King, W. J. and Greene, G. L. (1984) *Nature (London)* **307,** 745–747.

Kumar, V., Green, S., Stack. G., Berry, M., Jin, J. R., and Chambon, P. (1987) *Cell* **51,** 941–951,

Jost, J. P., Seldran, M., and Geiser, B. (1984) *Proc. Natl. Acad. Sci. USA* **81,** 429–433.

Laemmli, U. K. (1970) *Nature (London)* **227,** 680–685.

Medici, N., Moncharmont, B., Abbondanza, C., Nigro, V., Armetta, I., Minucci, N., Molinari, A. M., and Puca, G. A. (1988) *Atti Soc. Ital. Patol.* **19**, 60–63.

Molinari, A. M., Medici, N., Moncharmont, B., and Puca, G. A. (1977) *Proc. Natl. Acad. Sci. USA* **74**, 4886–4890.

Molinari, A. M., Medici, N., Armetta, I., Nigro, V., Moncharmont, B., and Puca, G. A. (1985) *Biochem. Biophys. Res. Commun.* **128**, 634–642

Moncharmont, B., Su, J. L., and Parikh, I. (1982) *Biochemistry* **21**, 6916–6921.

Moncharmont, B., Anderson, W. L., Rosenberg, B., and Parikh, I. (1984) *Biochemistry* **23**, 3907–3912.

Morrissey, J. H. (1981) *Anal. Biochem.* **117**, 307–310.

Parikh, I.,Sica, V., Nola, E., Puca, G. A., and Cuatrecasas, P. (1975) *Meth. Enzym.* **34B**, 670–688.

Puca, G. A. and Bresciani, F. (1968) *Nature (London)* **218**, 967–969.

Puca, G. A., Nola, E., and Bresciani, F. (1970a) *Res. Steroids* **4**, 319–330.

Puca, G. A., Nola, E., Sica, V., and Bresciani, F. (1970b) *Adv. Biosc.* **7**, 97–118.

Puca, G. A., Nola, E., Sica, V., and Bresciani, F. (1971) *Biochemistry* **10**, 3769–3780.

Puca, G. A., Nola, E., Sica, V., and Bresciani, F. (1975) *Meth. Enzym.* **34A**, 331–349.

Puca, G. A., Medici, N., Molinari, A. M., Moncharmont, B., Nola, E., and Sica, V. (1980) *J. Steroid Biochem.* **12**, 105–113.

Redeuilh, G., Moncharmont, B., Secco, C. M., and Baulieu, E. E. (1987) *J. Biol. Chem.* **262**, 6969–6975.

Sica, V. and Bresciani, F. (1979) *Biochemistry* **18**, 2369–2378.

Sica, V., Parikh, I., Nola, E., Puca, G. A., and Cuatrecasas, P. (1973) *J. Biol. Chem.* **248**, 6543–6558.

Sica, V., Nola, E., Puca, G. A., and Bresciani, F. (1976) *Biochemistry* **15**, 1915–1923.

Sundberg, L. and Porath, J. (1974) *J. Chromatogr.* **90**, 87–98.

Van Oosbree, T. R., Kim, U. H., and Mueller, G. C. (1984) *Anal. Biochem.* **36**, 321–327.

Yamamoto, K. R. (1985) *Ann. Rev. Genet.* **19**, 209–252.

Walker, P., Germond, J. E., Brown-Luedi, M., Givel, F., and Whali, W. (1984) *Nucleic Acid Res.* **22**, 8611–8626.

Walter, P., Green, S., Greene, G., Krust, A., Bornert, J. M., Jeltsch, J. M., Staub, A., Jensen, E., Scrace, G., Waterfield, M., and Chambon, P. (1985) *Proc. Natl. Acad. Sci. USA* **82**, 7889–7893.

Welshon, W. V., Liberman, M. E., and Gorski, J. (1984) *Nature (London)* **307**, 747–749.

Estrogen Receptor Purification

Kenneth S. Korach and Tsuneyoshi Horigome

1. Introduction

The estrogen receptor protein is a pivotal component in the mechanism of estrogen steroid hormone action. The presence of the protein within certain target tissues and cells correlates to the hormonal responsiveness. Recent studies with molecular cloning techniques have indicated a specific requirement for the estrogen receptor in gene regulation (Kumar et al., 1987). Until now, most estrogen receptor studies attempting to characterize the protein, using a variety of biochemical techniques such as sucrose density gradient centrifugation, gel filtration, isoelectric focusing, and affinity labeling, have used crude or partially purified samples. Highly purified estrogen receptor protein is an absolute requirement for gene mapping, footprinting studies, and X-ray crystallographic analysis, as well as the need to compare the properties of pure receptors with those from crude preparations.

Purification of the estrogen receptor has been difficult in most experimentally studied model systems because the estrogen receptor protein is present in low levels. These low levels are similar to other steroid receptor systems and components of membrane signal transduction mechanisms.

Receptor Purification, vol. 2   ©1990 The Humana Press

In animal tissues, moreover, the estrogen receptor appears very susceptible to proteinase. Target tissues that are a starting material for purification of the estrogen receptor contain proteinases that degrade the receptor (Horigome et al., 1987; Puca et al., 1977). The estrogen receptor, especially the unoccupied form, is susceptible to denaturation, even without degradation by proteinases. Another problem encountered during purification involves surface adsorption of the estrogen receptor. This is particularly evident with glass surfaces used in columns or containers during the procedures. Protein adsorption is generally a problem most noticeable when purifying proteins in low concentrations. The estrogen receptor also tends to aggregate, especially in the purified form. The aggregation problem needs to be eliminated for proper storage of the receptor. This tendency toward aggregation may be related to specific aspects of the receptor protein; the molecule has hydrophobic surface regions (Murayama and Fukai, 1983), and its shape is highly oblong. Reports have indicated that estrogen receptor exposed to limited proteolysis of the hydrophobic region (Murayama and Fukai, 1983) lost its tendency to aggregate (Schneider and Gschwendt, 1980).

The estrogen receptor, cDNA, has recently been cloned; enabling the expression of the receptor using a variety of expression vectors (Greene et al., 1986; Green et al., 1986; Metzger et al., 1988). Such approaches using stable gene transfections should lend themselves to large scale production of estrogen receptor protein, particularly in cases where amplification of the receptor gene under control of a highly expressed promoter will increase the absolute amount of estrogen receptor protein produced. Even with these advances, however, when increasing the amounts of estrogen receptor, purification from the cells will still require the same established procedures.

This chapter describes studies on the purification of estrogen receptor protein from the mouse uterus conducted in our laboratory, and reviews pertinent information from past studies on different approaches to estrogen receptor purification.

We have attempted to bring to the reader's attention some of the most recurrent problems involved in estrogen receptor purification, although these types of problems are not unique to the estrogen receptor but exist in the purification of other steroid hormone receptors as well. In the remaining sections, we will describe in detail procedures we are now using for estrogen receptor purification using the mouse uterus and the approaches that should more generally apply to other systems as ways to minimize the above-mentioned problems.

2. Materials

2.1. Buffers

As mentioned earlier, the estrogen receptor is easily inactivated during purification with loss of ligand binding activity. To suppress the inactivation, the buffers used for purification of the receptor were changed from the customarily used Tris buffers. Estrogen receptor exhibits multiple forms of either 4S, 5S, or 8S on sucrose gradients, depending on the cellular form and the ionic strength of the buffer. Phosphate buffer was used for purification of the 8S estrogen receptor because phosphate buffer stabilizes the 8S form (Redeuilh et al., 1985,1987), whereas Tris or HEPES buffer was generally used in isolation or preparation of the 4S receptor. Molybdate was also present in the buffers because it is essential for the purification of the 8S receptor; in its absence, the 8S receptor is not sufficiently stable to withstand purification procedures. The molybdate anion stabilizes the 8S receptor complex composed of estrogen receptor and a 90k heat shock protein. The mechanism for stabilization is not known; however, a recent study involving glucocorticoid receptor describes an endogenous molybdate factor that was purified (Bodine and Litwack, 1988a,b). In the case of the 4S estrogen receptor form, a homogenization buffer containing molybdate was also effective in suppressing the receptor inactivation during preparation from a uterine tissue cytosol fraction (Horigome et al., 1987; Lubahn et al., 1985).

EDTA was generally used in sample buffers because the receptor is actively proteolyzed by a Ca^{2+}-dependent proteinase having a neutral pH optimum that is present in many target tissues (Puca, 1977). EDTA is also effective in suppressing the oxidation of sulfhydryl groups on the receptor in combination with a low concentration of thioglycerol or dithiothreitol. Other than the reagents mentioned above, a carrier protein and proteinase inhibitors were used in the receptor preparation to suppress the nonspecific adsorption and minimize any proteolysis.

The following abbreviations are used for buffers described in this chapter. HEPES-buffer: 25 mM HEPES buffer (pH 7.5 at 25°C) containing 1 mM EDTA, 0.02% NaN_3, 10 mM Na_2MoO_4, 10 mM thioglycerol, 0.1 mM phenylmethylsulfonylfluoride (PMSF). Monothioglycerol and PMSF reagents (from a stock solution 20 mM in isopropanol) should be added to existing buffer stock and used immediately to optimize their effectiveness owing to oxidation. TEGM buffer: 10 mM Tris-HCl buffer (pH 7.6 at 20°C) containing 1.5 mM EDTA, 10% glycerol, 3.0 mM $MgCl_2$; TBS: 20 mM Tris-HCl (pH 7.5 at 25°C) containing 500 mM NaCl.

2.2. Affinity Gels

Purification of the estrogen receptor protein corresponding to greater than 10,000-fold purification has been very difficult using conventional chromatography procedures such as anion or cation exchange, gel filtration, and even high performance liquid chromatography (HPLC). Because of these apparent difficulties, affinity chromatography has been employed more successfully in such purification than the conventional procedures. This point should be emphasized, since advances and successes in steroid receptor purification were made possible following the development of suitable affinity resins.

Purification schemes using affinity chromatography with a gel resin containing estrogen as the ligand results in the greatest purification from a single chromatography step. Purification of 200– to 10,000-fold can be achieved with about 20% purity in one step, although the yields are low. A group specific chromatography step, such as heparin-Sepharose, is usually used in combination with the estrogen affinity chromatography to get highly purified preparations of receptor (Mauer and Notides, 1987).

2.2.1. Group-Specific Affinity Gels

The use of group-specific affinity chromatography usually results in a reasonable yield of receptor, but these preparations are less pure than those found by using an estrogen affinity resin. A decided advantage of using this type of resin enables the receptor to be isolated and purified in an unoccupied form, which is not possible when using estrogen affinity chromatography. The most useful group-specific affinity gel for estrogen receptor purification is heparin-Sepharose (Molinari et al., 1977; Sica and Bresciani, 1979; Van Oosbree et al., 1984; Rotondi and Auricchio, 1978; Greene, 1983; Puca et al., 1980; Ratajczak and Hähnel, 1980; Notides et al., 1985). Molinari et al. (1977) in a single step, using this resin, purified estrogen receptor 118-fold from calf uterine cytosol with a 58% yield. The estrogen receptor bound to the heparin-Sepharose can be eluted with either heparin (Molinari et al., 1977; Sica and Bresciani, 1979; Rotondi and Auricchio, 1978; Puca et al., 1980) or KCl (Van Oosbree et al., 1984). The heparin solution should be fractionated through a Sephadex G-50 column prior to use and the low mol wt fraction used to elute the receptor from the heparin-Sepharose resin. The heparin in the eluate can be removed from the receptor by passing it through a Sephadex G-50 column (Sica and Bresciani, 1979). This type of affinity column can be used prior to passage of the preparations onto an estrogen affinity column, resulting in milder elution conditions that also help maintain the binding activity of the unoccupied estrogen receptor form (Sica and Bresciani, 1979; Puca et al., 1980). A recent study, however, used the opposite sequence with success, by using estrogen affinity chromatography first followed by heparin-Sepharose (Mauer and Notides, 1987).

Binding of the receptor to heparin-Sepharose is highly dependent on the ionic strength (Molinari et al., 1977). The estrogen receptor binds to heparin but not to chondroitin sulfate (Molinari, 1977). Therefore, the binding of the receptor to heparin appears to be more than simple ionic interactions (Molinari, 1977). There are some reports using other group specific affinity resins for the purification of the estrogen receptor such as DNA-cellulose (Gschwendt, 1980; Hall et al., 1976), phenyl-Sepharose (Murayama et al., 1984), and an affinity resin with an orange triazine dye as the ligand (Bond and Notides, 1987).

Conventional column chromatography such as DEAE-cellulose (Redeuilh et al., 1985; Molinari et al., 1977; Puca et al., 1980; Murayama et al., 1984; Chong and Lippman, 1980), Sephadex G-200 (Molinari et al., 1977), Sephadex G-150 (Murayama et al., 1984), and hydroxylapatite (Murayama et al., 1984; Chong and Lippman, 1980) have been used for the purification of the estrogen receptor in combination with estrogen affinity column chromatography.

2.2.2. Specific Affinity Gels for Estrogen Receptors

Extending the earlier studies on enzyme purification using affinity resins Cuatrecasas and coworkers reported the synthesis of a series of estrogen affinity resins (Sica et al., 1973). They compared estrogen-receptor binding capacities of these resins and studied elution conditions for the estrogen receptor from the gels. Of the several different resins studied, the most effective is shown in Fig. 1A. An estrogen receptor was purified from calf uterine cytosol 1000-fold using that resin with a 14% yield (Sica et al., 1973). Estrogen receptor purification has been reported with similar affinity resins in which poly-Lys-Ala (Sica et al., 1973), ovalbumin (Sica and Bresciani, 1979), or hexamethylenediamine (Puca et al., 1980) were used as the linker arm instead of bovine serum albumin (BSA). It is known that these types of gels partially lose the steroid ligand during incubation with cytosol because the ester bond in the arm is susceptible to esterase(s) or proteinase(s) present in the cytosol (Puca et al., 1980; Chong and Lippman, 1980). For this reason, gels of this type were used as a second step in the purification procedure following chromatography that removed the enzymes. The cytosol fraction was first passed through a heparin-Sepharose column in the presence of molybdate and low ionic strength buffer to remove the enzymes, which unstabilized the linkage, and the eluate was then

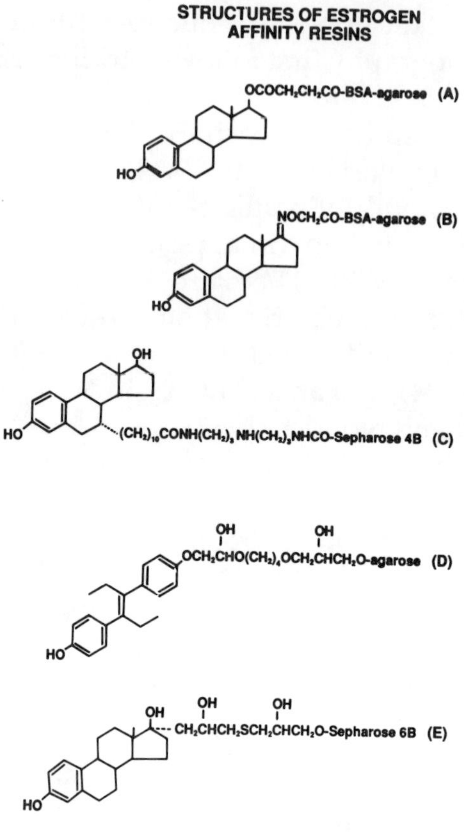

STRUCTURES OF ESTROGEN
AFFINITY RESINS

Fig. 1. Affinity gels for estrogen receptor purification.

applied onto the affinity column (Puca et al., 1980; Sica and Bresciani, 1979). A gel in which the ester bond was replaced with a more stable oxime-bond (Fig. 1B) had also been used for purification of the receptor (Chong and Lippman, 1980; Ratajczak and Hähnel, 1980).

Baulieu and colleagues also studied a group of different affinity gels for estrogen receptor purification and showed that a spacer linked to either 7α- or 17α-estradiol was most effective, and a longer spacer was better than a shorter one to get high affinity binding of the receptor to the ligand resin (Bucourt et al., 1978). They purified the estrogen receptor 868-fold directly from a trypsin-treated cytosol by using an affinity gel shown in Fig. 1C with 8.1% X yield without the use of any precolumn. The (C) gel is easier to handle than the (A) gel because a cytosol fraction can be applied directly onto the column without causing a loss of the ligand (Bucourt et al., 1978). An 8S estrogen receptor form was purified from calf uteri using a similar affinity gel and DEAE-Sephacel chromatography (Redeuilh et al., 1987,

1985,1980). A problem with these resins is that a small portion of the ligand is lost from the gels shown in (Fig. 1A–C) during chromatography when high concentrations of amine-based buffers such as TRIS are used. This is because linkage of the spacer group to the carrier matrix (Agarose or Sepharose) is through an amine bond that is susceptible to exchange with the amine groups of the buffer.

In contrast to gels A–C, gels D and E have a stable linkage between the ligand and the spacer-matrix as a result of using ether bonds. Gel D, which has diethylstilbestrol as the ligand, was first synthesized and used by Van Oosbree et al. (1984) to purify the estrogen receptor from rabbit uterus. This gel, which is very stable and easily synthesized, gave a slightly lower yield (0.5–6%) compared with other gels and required elution with *p*-sec-amylphenol and sodium thiocynate.

Greene et al. (1980) developed one of the most effective affinity gels (Fig. 1E) that had estradiol as the ligand attached to a Sepharose gel matrix at the 17α position. The estrogen receptor binds the gel extremely well and is very stable because ether and thioether bonds were used to construct the spacer group and link it to the Sepharose 6B matrix. Therefore, a cytosol fraction can be applied directly onto the gel column with no destruction of the affinity resin. This gel is now widely used to purify estrogen receptors in many laboratories, with satisfactory and reproducible results (Horigome et al., 1987; Lubahn et al., 1985; Greene, 1983; Notides et al., 1985; Greene et al., 1980). This gel is highly stable and can be washed with a sodium dodecylsulfate (SDS) solution and reused. Procedures for synthesis of this gel and purification of the estrogen receptor using this gel will be described in detail in the following sections.

There are reports of studies describing estrogen receptor purification in which affinity resins other than those shown in Fig. 1 were used. In these studies, approaches were used with dextran as a soluble carrier for the ligand (Hubert et al., 1978) and an estrogen gel resin having a linker arm with disulfide bonds (Sweet and Szabados, 1977). In another study, polyacrylamide had been used as a carrier instead of Sepharose (Bucourt et al., 1978) and a 17α-derivative of estradiol, different from that shown in Fig. 1E, was used as the ligand for an affinity resin (Moncharmont et al., 1982).

2.2.3. Preparation of Estrogen Affinity Gel

This section will describe procedures for the synthesis of the affinity gel shown in Fig. 1E. As already mentioned, this gel was first introduced by Greene et al. (1980) for purification of estrogen receptor from MCF-7 breast tumor cells. Details of the synthesis follow procedures reported by Lubahn (1983).

2.2.3.1. Preparation of 17α-Allylestradiol (I). Under a nitrogen atmosphere, M9 turnings (3.6 g: 150 mmol) were combined with allyl chloride (5.0 mL: 60 mmol) in a dry flask. After the reaction was completed, 20 mL of tetrahydrofuran (THF) were added with stirring and the mixture cooled in an ice bath. A water condensor was utilized to prevent escape of allyl chloride and THF vapor during the reaction. A solution of estrone (2.0 g: 7.4 mmol) and additional allyl chloride (7.5 mL: 90 mmol) in 75 mL of THF at 25°C was added slowly with mixing. The reaction was refluxed gently during this time. The mixture was then refluxed gently for an additional 60 min and stirred overnight at room temperature. Saturated NH_4Cl (20 mL) was added slowly, followed by an addition of enough solid NH_4Cl to remove any excess water. After filtering the mixture and washing the cake residue with ether (3×5 mL) and THF (60 mL), the solvent was removed under reduced pressure, which produced a syrup. A precipitate was obtained by dissolving the syrup product in a small vol of ethyl acetate and diluting with hexane. The product was examined on silica gel thin-layer chromatography (TLC) using two different solvents: $CHCl_3$ and 20% methanol—80% $CHCl_3$ and showed one species.

2.2.3.2. Preparation of 3-0-Acetyl-17α-Allylestradiol (II) . A solution of 17α-allylestradiol (I) (1.0 g: 3.2 mmol) and acetic anhydride (1.5 mL: 15 mmol) in 5 mL of anhydrous pyridine were incubated at 25°C overnight. The reaction mixture was then slowly added with stirring to 150 mL of ice cold 0.2M sodium acetate (pH 4.5) and produced a colorless gum. The product could be precipitated from a mixture of acetone and water. This product was analyzed to be homogeneous by silica gel TLC with 20% methanol in CH_2Cl_2.

2.2.3.3. Preparation of 3-0-Acetylestradiol-17α-Propylene Oxide (III). A solution of 3-0-acetyl-17α-allylestradiol (II) (500 mg: 1.41 mmol) and *m*-chloroperoxybenzoic acid (320 mg: 1.76 mmol) in 10 mL of CH_2Cl_2 was incubated overnight at 4°C. The reaction mixture, containing crystals of *m*-chloroperoxybenzoic acid, was then diluted with 40 mL of CH_2Cl_2 and extracted sequentially with the following aqueous solutions: 10% KI (2×25 mL), 10% sodium thiosulfate (2×25 mL), 1M $NaHCO_3$ (pH 8) (2×50 mL), and water (2×50 mL). The organic phase was dried over anhydrous Na_2SO_4, filtered, and concentrated to a syrup. The syrup was then dissolved in 40 mL of methanol and diluted with 40 mL of water yielding a white precipitate whose homogeneity was checked in three silica gel TLC systems:

1. 50% methanol—50% CHCl$_3$;
2. 20% methanol—80% CHCl$_3$; and
3. CHCl$_3$.

NMR confirmed the chemical structures.

2.2.3.4. Coupling of 3-0-Acetylestradiol-17α-Propylene Oxide (III) to Thiopropyl Sepharose. Dry Thiopropyl Sepharose 6B gel was obtained from Pharmacia and swollen to 15 mL in carbonate buffer: 0.3M NaHCO$_3$ (pH 8.4) containing 1 mM EDTA. The gel was washed in a sintered glass funnel with 1500 mL of carbonate buffer. In order to remove the protecting group, the gel was mixed with 30 mL of 1% dithiothreitol at room temperature and then washed with 1500 mL 0.3M NaHCO$_3$. Nitrogen gas was bubbled through the gel suspension for 10 min and then 27.8 mg (75 μmol) of III was dissolved in 15 mL of dimethylformamide (DMF) that was mixed with 30 mL of 0.3M NaHCO$_3$ containing 15 mL of the thiopropyl Sepharose gel. This reaction was mixed for 4 h at 34°C, then at 5°C overnight. The gel was washed with 100 mL of 33% DMF in 0.3 NaHCO$_3$ with 1 mM EDTA, and then incubated with 50 mL of 0.1M iodoacetamide in 0.3M NaHCO$_3$ for 1 h at 34°C to mask the remaining unreacted sulfhydryl groups. The gel was again washed with 100 mL of 33% DMF, 100 mL of 90% methanol, and 100 mL of water. The acetate blocking group on the 3-hydroxyl of estradiol was removed by suspending the gel in 50 mL of 0.1M Na$_2$CO$_3$, pH 12 and incubating it for 1 h at 34°C. The gel was washed with 2 L of 10% ethanol to insure the removal of all noncovalently bound estradiol derivatives. The affinity resin was packed into a 5 mL Econo-column that was stored at 4°C in 50% ethanol. Before use it was washed with 50% ethanol and then equilibrated in the HEPES affinity column sample loading buffer. After use, the column was regenerated by the following wash procedure: the resin was removed from the Econo-column, usually around a 2 mL bed vol, and washed in a 30 mL coarse sintered glass funnel sequentially with 400 mL of filtered 2M KCl, 80 mL water, 150 mL of filtered 3M NaSCN, 80 mL water, 400 mL of 50% ethanol, 80 mL water, 400 mL of NaHCO$_3$, and 80 mL water. Gels that were used several times were also treated to completely clean the resin with 5% sodium dodecylsulfate (SDS) at 45°C for 1 h and sequentially washed with 100 mL of 1% SDS, 200 mL water, 400 mL of 50% ethanol, and 200 mL water and stored in 50% ethanol.

Synthetic methods for the other estrogen affinity gels listed in Fig. 1 can be found in the following references: A (Sica et al., 1973; Puca et al.,

1980), B (Chong and Lippman, 1980), C (Bucourt et al., 1978; Redeuilh et al., 1980,1987), and D (Van Oosbree et al., 1984).

3. Special Methods

3.1. Assay of Estrogen Receptor Content

Estrogen receptor content of crude tissue preparations can be determined by using either the hydroxylapatite (HAP) (Erdos et al., 1970), dextran coated charcoal (Sanborn et al., 1971), or protamine sulfate methods (Korach and Muldoon, 1974). The HAP method, on the other hand, is recommended for analysis of estrogen receptor in purified preparations. The advantage of using HAP with purified estrogen receptor preparations is that it absorbs receptor very efficiently, incurring minimal loss. The disadvantage of charcoal with this type of sample is that charcoal can adsorb not only the steroid but also the receptor protein, particularly if the protein concentration is low, as it is in purified preparations. This nonspecific protein adsorption effect of charcoal can also occur in partially purified or dilute estrogen receptor samples. When using protamine sulfate with purified samples, the low protein concentration results in poor precipitation of the receptor protein, but this can be overcome by using a carrier protein.

The HAP assay was carried out as follows: Aliquots of 100 µl 60% (v/v) HAP slurry in TEGM buffer and 350 µL of TEGM were added to polystyrene assay tubes. A 50 µL aliquot of purified estrogen receptor preparation eluted from the steroid affinity column was put into each assay tube. Incubation was for 15 min at 4°C with vortexing every 5 min. To wash away excess estrogen, 3 mL of TEGM were added and mixed with HAP, and the tube was centrifuged at 2500g, 10 min at 4°C. The supernatant was discarded and the washing procedure repeated two more times. [^{3}H]-estradiol (S.A. 100 Ci/mmol) 10 nM in 450 µL TEGM was added to assay tubes to assess total sample binding. To determine nonspecific binding in the sample assay tubes, 450 µL TEGM containing 10 nM [^{3}H]-estradiol and 1 µM unlabeled diethylstilbestrol were added. The tubes were incubated at 30°C for 30 min while shaking to exchange label the receptor binding sites. The exchange reaction stopped by placing the tubes in ice and washing with 3 mL of cold TEGM buffer. The HAP precipitates were washed and centrifuged two additional times. After the final wash the tubes were drained and the bottom containing the HAP pellet was sliced off with a heated wire. Radioactivity was then quantified by scintillation counting in a toluene based fluid that dissolves the plastic tube and extracts the labeled steroid. Specific receptor binding in each sample was determined by subtracting the counts in the nonspecific from total binding tubes.

3.2. Suppression of Proteinase Activities

Proteolytic degradation of the estrogen receptor has been a topic of study for several years. The estrogen receptor seems very susceptible to proteinases. Some of the proteinases that appear to act on the estrogen receptor are present and stimulated by estrogen in estrogen target tissues or tumor cells (Sherman et al., 1978; Murayama et al., 1980,1984; Horigome et al., 1988).

There are a few reports describing endogenous proteinases that degrade the estrogen receptor. Early reports described a Ca^{2+}-activated proteinase, which was partially purified from calf uterus and postulated to act as an estrogen receptor-transforming factor (Puca et al., 1977). It was later shown that this factor was calpain, an enzyme inhibited by EDTA and calpastatin, and widely present in other tissues besides target tissues (Puca et al., 1977; Murayama et al., 1984; Bodwell et al., 1985). A chymotrypsin like proteinase activity was found in the cytosol fraction of an estrogen-resistant C3H mammary carcinoma. This proteinase cleaved estrogen receptor was isolated from nontumor cells and generated a 53K fragment (Baskevitch and Rochefort, 1985). The enzyme could be suppressed with tosylphenylalaninechloromethylketone and chymostatin (Baskevitch and Rochefort, 1985). Another report described a proteinase that was not Ca^{2+} requiring and that exhibited limited proteolysis of the porcine uterine receptor (Murayama et al., 1984).

Recently we reported that the highest yield of estrogen receptor purification was obtained using ovariectomized mouse uteri. The yield was decreased remarkably when purification was performed on tissue samples taken from either intact or ovariectomized animals pretreated with estrogen (Horigome et al., 1987). Immunodetection of estrogen receptor isolated from the estrogen-exposed animals indicated a large amount of low mol wt forms. A high endogenous proteolytic activity, responsible for estrogen receptor degradation, was detected in the uterine cytosol fraction of estradiol-treated ovariectomized mice compared to saline controls (Horigome et al., 1988). The proteinase(s) activity resulted in cleavage of the native estrogen receptor form; 65,000 daltons (65kDa) to a product from limited proteolysis having an apparent mol wt of 54kDa. This proteinase activity had a pH 6 optimum and was completely inhibited by 2.5 mM p-chloromercuribenzoate (PCMB) and 2.5 mM p-chloromercuriphenylsulfonate (PCMPF). It was partially inhibited by 2.5 mM iodoacetamide but not by 1 mg/mL leupeptin, 0.1 mg/mL antipain, 0.1 mg/mL chymostatin, 0.1 mg/mL pepstatin, 0.1 mg/mL E-64, 2.5 mg/mL soybean trypsin inhibitor, 2.5 mM phenylmethylsulfonylfluoride, 2.5 mM diiso-propylfluorophosphate, or 10 mM EGTA. The results suggested that the proteinase(s) had an essential thiol group for its activity and is different from estrogen-induced uterine hydrolases previously reported (Katz et al., 1976,1980; Finlay et al., 1982).

The estrogen receptor in the mouse uterine cytosol fraction appeared to be degraded by two steps in which the 65K receptor was cleaved to a 54 kDa receptor that was further degraded to a 37kDa form upon longer incubation. The second step was inhibited by leupeptin, antipain, chymostatin, E-64, and PCMB (Horigome et al., 1988). This type of proteinase(s) seems to be present in uteri of other animals because very similar fragments of the receptor are observed in receptor preparations purified from bovine, porcine, rat, and human tissue samples (Lubahn et al., 1985; Faye et al., 1986).

Considering the results from studies with these proteinases we are now using homogenization medium containing EDTA, PMSF, leupeptin, and antipain, as described in Section 4.1. These inhibitors do not completely suppress the proteinase activity responsible for the cleavage of the receptor. To suppress the thiol proteinase activity and increase the yield of receptor during purification, the addition of thiol reagents such as PCMB, cystamine, or dithiodipyridine may need to be considered. Indeed, it was reported that the addition of 1–10 mM cystamine to a cytosol sample increased the stability of the estrogen receptor (Sweet and Szabados, 1977). Some caution needs to be taken regarding thiol reagents, since their use without careful analysis can cause an alteration of the ligand binding activity, possibly because of interaction with cysteine residues in or around the ligand binding site.

It is known that molybdate stabilizes the 8S steroid hormone receptor complex and prevents the conversion of the 8S to 4S receptor forms (Bodine and Litwack, 1988b). Homogenization medium containing molybdate is also recommended for the 4S receptor purification because stabilization of the 8S form during preparation of the cytosol fraction is effective in suppressing the cleavage of the receptor by a proteinase. The proteolytic susceptibility of the estrogen receptor during purification is such a concern that tissue or cell sources containing the lowest level of proteinases should be considered if possible. Known sources with low activity are uteri from ovariectomized or immature animals or MCF-7 cells.

3.3. Affinity Labeling

Besides the use of antibodies another method of detecting the estrogen receptor protein by gel analysis is affinity labeling using a radioactive reagent with high labeling efficiency and specificity. One of the best reagents of this type is [³H]-tamoxifen aziridine developed by Katzenellenbogen et al. (1983). We have used it with a slight modification allowing the [³H]-tamoxifen aziridine to affinity label occupied mouse uterine estrogen receptor (Horigome et al., 1987).

Prior to labeling, unbound [³H]-estradiol present in the purified estrogen receptor preparation was removed by either exhaustive dialysis against 25 mM Tris-HCl (pH 7.6) containing 1 mM EDTA and 2 mM 3-[(3-cholamidopropyl) dimethylammonio]-l-propanesulfonate (CHAPS) or by desalting on a PD-10 column of Sephadex G-25 (Pharmacia). For affinity labeling of the estradiol-bound receptor eluted from an affinity column, an aliquot of the receptor preparation (2.9 nM receptor, 50 µg protein/mL) was incubated for 1 h at 30°C in 25 mM Tris-HCl (pH 7.6), 1 mM EDTA, 2 mM CHAPS, and 7% dimethyl formamide with 30 nM [³H]-tamoxifen aziridine (S.A. 20 Ci/mmol) alone or in the presence of a 100-fold molar excess of unlabeled estradiol. Conditions of this incubation allowed the exchange of bound estradiol with tamoxifen aziridine, resulting in the covalent affinity labeling of the receptor. Protein was precipitated from the reaction mixture by the addition of 4-vol acetone at –20°C. The mixture was cooled to –70°C for 1 h and the precipitated protein was isolated by centrifugation at 10,000 rpm for 10 min at 4°C. If necessary, myoglobin or insulin can also be added as a carrier protein for precipitation. The receptor precipitate was washed with 0.5 mL of 80% acetone, dissolved with 5 µL of 2% SDS, and stored at –70°C until use. The labeled receptor can be analyzed by SDS-polyacrylamide gel electrophoresis followed by fluorography (Horigome et al., 1987; Korach et al., 1988). This method is quite sensitive; approx 1 ng or less of receptor can be detected (Korach et al., 1988).

3.4. Immunochemical Detection

3.4.1. Protein-Blotting Procedure

The most common transfer method of proteins separated by SDS-polyacrylamide gel electrophoresis onto nitrocellulose sheets is by electroblotting in the presence of 20% methanol according to the method of Towbin et al. (1979). The efficiency of transfer of the estrogen receptor by this method was less than 20%, but more than 90% of the labeled BSA was transferred (Horigome et al., 1987). We have compared the efficiency of transfer of the estrogen receptor among several methods; low voltage electroblotting in the presence of 0.05% SDS, high voltage electroblotting, and diffusion blotting (Lubahn et al., 1985; Bowen et al., 1980). The diffusion blot technique gave satisfactory results and showed a 2.3-fold greater efficiency of receptor transfer compared with the electroblotting procedures (Horigome et al., 1987). The diffusion method has the additional advantage of producing two copies of the blot from one gel transfer, which creates a further opportunity for analysis. It does, however, have the apparent disadvantage with samples of low receptor concentration that the transfer bands are more diffuse and not as sharp as observed with electroblotting. Recently we have used the semidry electroblotter (Kyse-Andersen, 1984) (Nova Blot transfer unit, LKB Instruments) that produces sharp bands and better transfer efficiency for the estrogen receptor than standard electroblotting techniques.

Diffusion blotting onto nitrocellulose sheets (Schleicher and Schuell, 0.2 μm) was carried out using a threefold dilution of the HEPES buffer containing 0.15M NaCl, 0.02% NaN_3 as a transfer buffer. The gel was sandwiched between two nitrocellulose sheets and placed in a cassette used for electroblotting. The gel apparatus was submerged in 2 L of transfer buffer for 36–48 h with constant stirring. The initial transfer buffer solution was replaced with fresh buffer after 12 h. All procedures were carried out at room temperature.

3.4.2. Immunochemical Detection of Estrogen Receptor

Monoclonal antibodies have been raised against the estrogen receptor isolated from human MCF-7 tumor cells (Greene et al., 1980; Miller et al., 1982) and calf uterus (Moncharmont et al., 1982). We have used monoclonal antibody H-222 (Miller et al., 1982) against human estrogen receptor for detection of mouse estrogen receptor because it was known that this antibody had very weak species-specificity. We found it reacted well with mouse estrogen receptor. Since in certain experiments the receptor level was very low, we have optimized the sensitivity and four immunodetection methods were compared:

A. Direct staining using [^{125}I]-labeled H-222 monoclonal antibody;
B. Indirect staining using H-222 as the primary antibody and a biotinylated secondary antibody with an avidin-horseradish peroxidase color development system;
C. Indirect labeling method using H-222 as primary antibody and [^{125}I]-labeled protein A; and
D. Indirect labeling using H-222 and the [^{125}I]-labeled F(ab')$_2$ fragment of antirat-IgG as a secondary antibody.

We did not obtain satisfactory results with methods B and C because of a high amount of nonspecific staining with method B and very low sensitivity with method C. Method D appeared to provide the best results with low background and good sensitivity.

In method D, nitrocellulose blots were incubated with 300 mL TBS containing 0.05 Tween-20 for 5 min and rinsed twice with 300 mL water. After blocking nonspecific protein-binding sites by incubation with TBS containing 3% bovine serum albumin (BSA) for 1 h, the blots were incubated with 50 mL of the same solution containing 0.5 µg/mL monoclonal antibody for 4 h at room temperature. The blots were then washed for 1 h with 200 mL TBS containing 0.05% Tween-20. The wash solution was changed every 10 min with a 10-s water rinse interval. The blots were incubated with 50 mL 0.017 µg/mL antirat-IgG [^{125}I]-F(ab')$_2$ fragment of sheep IgG (0.33 µCi/mL) diluted with TBS containing 3% BSA at 4°C overnight. The blots were then washed for 2 h with 200 mL TBS containing 0.05% Tween-20, with a change every 15 min. The nitrocellulose sheet was exposed for varying periods of time to Kodak XAR-5 (Eastman Kodak, Rochester, NY) at –70°C film backed with a Cronex Quanta III intensifying screen (DuPont, Wilmington, DE). This method has a sensitivity of 3 ng estrogen receptor per gel lane (Korach et al., 1988).

3.5. Other Reagents Important for Purification Procedures

3.5.1. Adsorption

Following purification the amount of estrogen receptor usually obtained is in very low concentrations. Most proteins, however, are very sticky and unstable when they are handled at low concentrations. Therefore, it is important to keep the purified receptor at a high concentration to minimize adsorption preferably at the final step in the purification sequence. It has been difficult to effectively concentrate the receptor after purification without significant loss in yield. One report has described concentrating a purified receptor preparation using a UM-20 Diaflo membrane in the presence of heparin (Sica and Bresciani, 1979). Otherwise, to increase the concentration, the addition of a suitable carrier protein is recommended to handle the receptor protein, which has been purified more than 100-fold compared with cytosol. We are using insulin (75 µg/mL) as the carrier protein (Horigome et al., 1987) for the following reasons:

1. It moves to the dye front in SDS-PAGE procedures and does not exhibit a band in the separation area of the gel;
2. Very pure insulin is commercially available; and
3. Insulin does not bind estradiol.

Other than insulin (Horigome et al., 1987; Lubahn et al., 1985), bovine serum albumin (Schneider and Gschwendt, 1980; Shahabi et al., 1986) and ovalbumin (Redeuilh et al., 1985) have also been used as carrier proteins for receptor purifications.

It is known that estrogen receptor adsorbs to glassware. Therefore, the use of plastic ware, even columns for chromatography is recommended. If glassware, particularly columns are to be used, they should be siliconized prior to use as follows: glassware or the column is filled with a 5% solution of Surfasil (Pierce product) in CH_2Cl_2 and allowed to stand at 20°C overnight and thereafter rinsed with CH_2Cl_2, ethanol, and water. The adsorption of the purified receptor protein to dialysis membranes (Chong and Lippman, 1980) can be suppressed by the addition of a suitable carrier protein.

In cases where high performance liquid chromatography (HPLC) is used for the separation or purification of the receptor, the procedure is carried out in a stainless steel column. When purified estrogen receptor labeled with [^{3}H]-estradiol was applied onto a HPLC column of hydroxylapatite, the resolution was low, recovery of the isotope was 34%, and most estrogen receptor was adsorbed to the stainless column (unpublished result). Similar results were reported by Denner et al. (1987) for progesterone receptor. In the case of purified progesterone receptor, a glass column gave better resolution and recovery than the stainless steel column (Denner et al., 1987). Good resolution by HPLC for the analysis of the estrogen receptor has been limited to crude preparations (Shahabi et al., 1986; Hutchens et al., 1984; Brinkman et al., 1985; Pavlik et al., 1987) or an analysis of the purified estrogen receptor in the presence of SDS (unpublished results).

3.5.2. Aggregation

It is known that estrogen receptor, especially the purified preparations, have the tendency to aggregate (Sica and Bresciani, 1979; Rotondi and Auricchio, 1978; Puca et al., 1980; Madhok and Leung, 1981). Sica et al. (1980) reported that heparin prevents aggregation of purified estrogen receptor (Puca et al., 1980). Puca et al. (1980) used magnesium and Rotondi and Auricchio (1978) used NaBr, a chaotropic salt, to suppress the aggregation. The chaotropic salts are thought to alter the water solvent around the hydrophobic regions of the proteins minimizing the ability of the proteins to aggretate. We have used another chaotropic salt, NaSCN with acceptable results (Horigome et al., 1987). Storage of purified estrogen receptor preparation in buffer containing 0.5M NaSCN without desalting is better than storage after desalting for the stability of the estradiol-binding activity. A buffer containing chaotropic ion and/or heparin is, therefore, recommended for storage of purified receptor, and rapid freezing by liquid nitrogen and storage at –70°C are important to minimize the aggregation.

4. Purification of the Estrogen Receptor

4.1. Purification of the Estrogen Receptor from Mouse Uteri

Unless otherwise stated, the purification procedures were performed at 4°C. Mice were sacrificed by cervical dislocation. The uteri were rapidly removed, frozen on dry ice, and stored at –70°C until use. Under these conditions, the receptor was found to be stable for approx 6 mo. A sample of 16 g of uteri was combined with 8 vol of HEPES buffer containing 3 µg/mL leupeptin and antipain and homogenized with 15 s Polytron bursts at a setting of 6.5 with a PT-10 probe (Brinkman Instruments, Westbury, NY). The homogenate was centrifuged at 120,000g for 70 min. The supernatant was filtered through 100–125 µm Nitex to produce the cytosol fraction.

Crystalline KCl was added to the cytosol fraction to a final concentration of 0.7M. The sample was refiltered through filter paper to minimize clogging of the column. The cytosol was passed through a 1× 1.2 cm estradiol affinity column (Fig. 1E) at a 40 mL/h flowrate. The column was washed with 30 mL 0.7M KCl in HEPES buffer. After the initial 0.7M KCl wash, the column was washed sequentially with two cycles at a 120 mL/h flowrate with:

1. 20 mL 2.0M KCl in HEPES buffer;
2. 30 mL HEPES buffer; and
3. 30 mL of 10% dimethylformamide (DMF) in HEPES buffer.

The column was placed in a 30°C water bath, while the fraction collector remained at 4°C, and washing was continued using a degassed solution of 30 mL 10% DMF in HEPES buffer containing 1 µg/mL leupeptin and antipain at a 120 mL/h flowrate, followed by 15 mL of the same solution at 40 mL/h. A final wash was made with 2.5 mL of a 0.5M NaSCN buffer: HEPES buffer containing 0.5M NaSCN, 75 µg insulin/mL, 2 mM CHAPS, 10% DMF, and 1 µg/mL leupeptin and antipain. The receptor was finally eluted into polystyrene tubes with 0.5M NaSCN buffer containing 0.3 µM [^{3}H]-estradiol (S.A. 10 Ci/mmol). The fractions were monitored by the HAP adsorption assay for receptor binding activity. The fractions were pooled, quickly frozen in small aliquots in liquid nitrogen, stored at –70°C, and were stable for approx a year. Fractions were desalted on a PD-10

Table 1
Purification of Uterine Estrogen Receptor (ER)
from Mice of Different Estrogen Status by Affinity Chromatography

Fraction	Ovariectomized				Ovariectomized-E_2[a]				Intact			
	Total protein, mg	Receptor amount, μg	SA, fmol/mg	Yield, %	Total protein, mg	Receptor amount, μg	SA, fmol/mg	Yield, %	Total protein, mg	Receptor amount, μg	SA, fmol/mg	Yield, %
Cytosol[b]	239	12	720	100	1190	52	626	100	1270	35	395	100
Flow through	212	3	196	25	1140	6	79	12	1270	5	58	14
Purified ER	—	1.8 (20)[c]	—	15	—	0.8 (1.7)[c]	—	1.5	0.011d	3.4 (11.3)[c]	4.4×10^6	10

[a]Ovariectomized mice were injected with 0.6 μg estradiol (E_2) per mouse daily for 2 d, then killed on d 3.
[b]Cytosol fractions were prepared from 6 g (200 ovariectomized mice), 16 g (260 ovariectomized-E_2-treated mice) and 16 g (120 intact mice).
[c]Numbers in parentheses represent the percentage of receptor eluted from the column compared to the amount bound to the column.
[d]Receptor amounts were calculated from the amount of receptor determined by binding assay, and the purity was estimated from densitometric scan of silver stain from SDS-gel analysis. Insulin used as the carrier protein was excluded.
[e]Table from Horigome et al., 1987.

column to remove free [³H]-estradiol and NaSCN. These samples, containing the affinity purified receptor, were used to characterize the protein by a variety of techniques. Results of this simple and reproducible purification scheme are summarized in Table 1.

The strong binding and large capacity of the column for the estrogen receptor was remarkable, with over 75% of the receptor in the cytosol being removed (Table 1) even at a flowrate of 40 mL/h. The yield of total binding activity eluted in the purified pools was about 10% when intact mice uteri were used (Table 1). As mentioned above, when tissue from estrogen stimulated animals was used the yield was lower, as low as 1.5%, owing to degradation by proteinases (Horigome et al., 1988), whereas uteri of ovariectomized mice gave the highest yield, 15% (Table 1) (Horigome et al., 1987). The receptor was purified about 11,000-fold and represents about 30% purity of the noncarrier protein present in the preparation. Raising the affinity column temperature to 30°C and utilizing 0.5M NaSCN are effective in increasing the amount of receptor protein eluted from the affinity column. Unfortunately, even in the absence of estradiol, small amounts of the receptor are washed off, resulting in low yields particularly following extensive washing. The procedure described is thus a compromise between yield and purity. This method can be applied to a large scale preparation as 220 µg of estrogen receptor was purified from calf uteri with minor modifications (Lubahn et al., 1985).

4.2. Characterization of the Purified Mouse Estrogen Receptor Protein

The results of SDS-PAGE analysis of highly purified mouse estrogen receptor are shown in Fig. 2 and some properties of the receptor are summarized in Table 2. This preparation gave several protein bands by silver-staining and affinity labeling with [³H]-tamoxifen aziridine. Immunostaining with H-222 monoclonal antibody showed that the band corresponding to a 65K mol wt is the estrogen receptor (Horigome et al., 1987). Sedimentation coefficient of this native monomeric form was 4.1 S and the isoelectric point was 6.5 as judged by two-dimensional gel electrophoresis (Horigome et al., 1987). The receptor is a very hydrophobic protein; therefore, complete solubilization into a medium containing urea, for the first dimensional analysis was difficult. Saturation of the sample solution with urea is essential for getting good solubility and producing a well focused receptor spot on a two-dimensional gel (Horigome et al., 1987).

When the purified estrogen receptor was analyzed by SDS-PAGE, minor bands of 54K and 37K other than the native 65K receptor were often

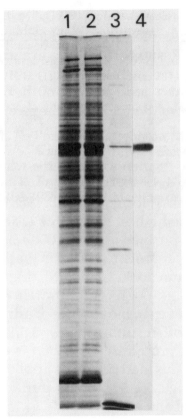

Fig. 2. Sodium dodecylsulfate-polyacrylamide gel electrophoresis (SDS-PAGE) of purified mouse uterine estrogen receptor. Samples were analyzed by SDS-PAGE and detected by silver staining (lanes 1, 2, and 3). The receptor was detected by a monoclonal antibody (H-222) using a [^{125}I]-F(ab')^2 fragment of antibody against rat IgG after blotting onto a nitrocellulose sheet (lane 4). Lane 1, mouse uterine cytosol protein (4 µg); lane 2, flow through fraction from affinity chromatography (4 µg); lanes 3 and 4, purified mouse uterine estrogen receptor (20 and 24 ng receptor protein). An arrowhead shows the 65K estrogen receptor. Double bands at the bottom of the lane 3 are carrier insulin. Figure from Horigome et al., 1987.

detected by fluorography of [^{3}H]-tamoxifen-labeled preparations and from staining with H-222 monoclonal antibody (Horigome, 1987; Horigome et al., 1988; Lubahn et al., 1985). These two minor bands were proteolytic fragments of the 65K receptor. The 54K fragment exhibited both DNA and estradiol ligand binding activity similar to the 65K receptor. However, the 37K fragment was degraded to a point where only the estradiol binding domain was present (Horigome et al., 1988). The purified receptor preparation had a K_d 1.4 nM for estradiol binding and this value is similar to that reported for crude cytosolic estrogen receptor (Horigome et al., 1987).

Table 2
Properties of Purified Mouse Estrogen Receptor

Sedimentation coefficient	4.1 s
Isoelectric point	6.5
Mol wt (SDS)	65,000
Dissociation constant (estradiol) nM	1.4

4.3. Purification of Estrogen Receptors from Other Tissue Sources

Purification of estrogen receptors from a variety of other tissue sources is summarized in Table 3. The following are three important points to consider for producing a large scale purification preparation:

1. High content of estrogen receptor;
2. Low proteolytic activity; and
3. A large amount of the starting material needs to be obtainable.

If a sufficient number of cells can be harvested, MCF-7 breast cancer cells appear to be one of the best starting materials for purification of the estrogen receptor, according to the above criteria. There are no known apparent differences between the properties of estrogen receptors of MCF-7 cells and other normal tissues.

Besides MCF-7 cells, calf uterus can also be considered as starting material. Among uteri, calf has been used in the largest number of studies since it can be easily obtained in large amounts and the receptor content of this tissue is relatively high. A proteinase activity responsible for the degradation of the receptor is lower in this tissue than in others such as porcine uteri. Other than the MCF-7 cells (Greene et al., 1980; Greene, 1983; Chong and Lippman, 1980) and calf uterus (Lubahn et al., 1985; Sica and Bresciani, 1979; Notides et al., 1985; Redeuilh et al., 1987; Bucourt et al., 1978; Moncharmont et al., 1982), estrogen receptors have also been purified from various other sources depending on the purpose needed: pig uterus (Lubahn et al., 1985; Murayama et al., 1984), mouse uterus (Horigome et al., 1987), calf mammary gland (Rotondi and Auricchio, 1978), chicken oviduct (Gschwendt, 1980), chicken liver nuclei (Schneider and Gschwendt, 1980; Gschswendt et al., 1983), and Xenopus laevis liver nuclei (Wright et al., 1983).

In some cases, estrogen receptors were purified from tissues of animals treated with estrogen (Horigome et al., 1987; Schneider and Gschwendt, 1980; Wright et al., 1983; Gschwendt et al., 1983), although, in most of those studies a truncated form of the receptor was obtained. As men-

Table 3
Purification of Estrogen Receptors (ERs) from Various Tissue Sources

	A Calf Uterus	E Calf Uterus	E Calf Uterus	E Cow Uterus	E Mouse Uterus	E Pig Uterus	D Rat Uterus	E Rabbit Uterus	E Human Uterus	E MCF-7 Cells	E MCF-7 Cells
Gel type[1]	A	E	E	E	E	E	D	E	E	E	E
Tissue origin	Calf Uterus	Calf Uterus	Calf Uterus	Cow Uterus	Mouse Uterus	Pig Uterus	Rat Uterus	Rabbit Uterus	Human Uterus	MCF-7 Cells	MCF-7 Cells
Affinity chromatography											
Cytosol (fmol ER/mg protein)	1000	520	690	270	395	133	96	490	42	1200	1410
Sample loaded with prior purification step	Yes	No	Yes	No	No	No	No	Yes	No	No	Yes
S.A. (fmol ER/mg protein)	4400	520	1800	270	395	133	96	490	42	1200	1410
Amount (mg)	35,670	16,790	550	3700	1270	1,926	321	920	–	6212	3979
Adsorption yield (%)	>80	71	–	79	86	70	79	77	–	79	>82
ER elution											
S.A. (nmol ER/mg protein)	<15.5	3.8	0.4	0.3	4.4	0.23	0.168	<0.388	–	0.897	0.152
Yield (%)	>80	9.6	54	14	10	9	7	4.5	–	40	72
Purification factor	>3542	7300	223	1100	11,100	1743	1760	<792	–	767	108
Overall											
Purification factor	15,585	7300	21,390	1100	11,100	1743	1760	792	–	767	6424
Purity[2] (%)	100	25	96	1.9	30	1.5	1.1	2.5	–	5.8	59
Yield (%)	53	9.6	22	14	10	9	7	4.5	–	40	42
Reference	14	12	20	3	2	3	3	15	12	30	17

[1] Refer to Fig. 1.
[2] Purity was calculated from a specific activity and an assumption that 65,000g ER binds 1 mole of E_2.
[3] Unpublished results.

tioned in Section 3.2., the treatment of animals with an estrogen stimulates proteinase(s) responsible for the cleavage of the receptor (Horigome et al., 1987; Horigome et al., 1988). Therefore, purification of the estrogen receptor from an estrogen-treated animal should be carried out very carefully, to minimize the effects of the proteinase activity.

The greatest increase in estrogen receptor purity is the 10,000-fold increase achieved in one step by affinity chromatography, which results in 1–30% purity. Partially purified preparations of receptor can be made using an affinity column with a single wash and elution step resulting in a 10% purity. To obtain purer preparations the combination of affinity chromatography and a second method is recommended. The most common combination is estrogen-ligand affinity chromatography followed by heparin-Sepharose chromatography (Sica and Bresciani, 1979; Van Oosbree et al., 1984; Greene, 1983; Notides et al., 1985; Mauer and Notides, 1987). Pure receptor preparations can be obtained with this combination when MCF-7 cells are used as starting material (Greene, 1983). Other than the estrogen-Sepharose and heparin-Sepharose chromatography, DEAE-cellulose (Redeuilh et al., 1987,1985; Murayama et al., 1984) and ammonium sulfate precipitation (Notides et al., 1986; Chong and Lippman, 1980; Wright et al., 1983; Gschwendt et al., 1983) are other methods that may be combined with the above two methods to get pure preparations from sources containing low levels of receptor. The use of a small bed of affinity gel and exhaustive washing before elution are key points to get higher purification of the preparations. As described in Section 4.1., in the case of mouse uterine estrogen receptor we used two washing cycles with high salt, low salt, and DMF-containing buffers in the affinity chromatography to get highly purified preparations (Table 3) (Horigome et al., 1987). Washing cycles can be changed depending on the resin and purpose of the purification and sources of the receptor, since a small but significant amount of receptor was lost during the washing cycles (Horigome et al., 1987). Washing the affinity resin with a buffer containing DMF seemed to be effective in reducing nonspecifically bound proteins that increased the purity (Horigome et al., 1987; Lubahn et al., 1985). After exhaustive washing the receptor is eluted from the resin by exchange with [^{3}H]-estradiol. A medium containing 0.5 M NaSCN and 10% DMF is usually used at 25–30°C to increase the exchange rate (Horigome et al., 1987; Lubahn et al., 1985; Wright et al., 1983; Greene et al., 1980). Similar results can be obtained with an overnight incubation at 4°C rather than exposing the receptor to 30°C and possibly resulting in some proteolysis.

In this chapter we have attempted to describe the purification of the estrogen receptor protein in sufficient detail that readers will become aware of the reagents and procedures necessary to successfully purify this or similar proteins. We have also mentioned cautions, problems, solutions, and hints from our own experiences as additional guidance for others attempting this type of endeavor. Good luck.

References

Baskevitch, P. P. and Rochefort, H. (1985) *Eur. J. Biochem.* **146**, 671–678.
Bodine, P. V. and Litwack, G. (1988a) *J. Biol. Chem.* **263**, 3501–3512.
Bodine, P. V. and Litwack, G. (1988b) *Proc. Natl. Acad. Sci. USA* **85**, 1462–1466.
Bodwell, J. E., Holbrook, N. J., and Munck, A. (1985) *J. Biol. Chem.* **260**, 2601–2604.
Bond, J. P. and Notides, A. C. (1987) *Anal. Biochem.* **163**, 385–390.
Bowen, B., Steinberg, J., Laemmli, U. K., and Weintraub, H. (1980) *Nucl. Acid Res.* **8**, 1–20.
Brinkman, A. O., Vries, J. B.-de, Boer, W. D., Lindh, L. M., Mulder, E., and Molen, H. J. V. D. (1985) *J. Steroid Biochem.* **22**, 85–90.
Bucourt, R., Vignau, M., Torellil, V., Richard-Foy, H., Geynet, C., Secco-Millet, C., Redeuilh, G., and Baulieu, E.-E. (1978) *J. Biol. Chem.* **253**, 8221–8228.
Chong, M. T. and Lippman, M. (1980) *Cancer Res.* **40**, 3172–3176.
Denner, L. A., Weigel, N. L., Schrader, W. T., and O'Malley, B. W. (1987) *Anal. Biochem.* **161**, 291–299.
Erdos, T., Best-Belpomme, M., andBessada, A. R. (1970) *Anal. Biochem.* **37**, 244–252.
Faye, J. C., Fargin, A., and Bayard, F. (1986) *Endocrinology* **118**, 2276–2283.
Finlay, T. H., Katz, J., and Levitz, M. (1982) *J. Biol. Chem.* **257**, 10914–10919.
Green, S., Walter, P., Kumar, V., Krust, A., Bornert, J.-M., Argos, P., and Chambon, P. (1986) *Nature* **320**, 134–139.
Greene, G. L. (1983) in *Gene Regulation by Steroid Hormones II* (Roy, A. K. and Clark, J. H., eds.), Springer-Verlag, New York, pp. 191–200.
Greene, G. L., Gilna, P., Waterfield, M., Baker, A., Hort, Y., and Shine, J. (1986) *Science* **231**, 1150–1154.
Greene, G. L., Nolan, C., Engler, J. P., and Jensen, E. V. (1980) *Proc. Natl. Acad. Sci. USA* **77**, 5115–5119.
Gschwendt, M. (1980) *Biochim. Biophys. Acta* **627**, 281–289.
Gschwendt, M., Hähnel, R., and Ratajczak, T. (1983) *Biochim. Biophys. Acta* **760**, 238–245.
Hall, C., Thrower, S., Lim, L., and Davison, A. N. (1976) *Biochem. Soc. Trans.* **4**, 766–769.
Horigome, T., Golding, T. S., Quarmby, V. E., Lubahn, D. B., McCarty, K., Sr., and Korach, K. S. (1987) *Endocrinology* **121**, 2099–2111.
Horigome, T., Ogata, F., Golding, T. S., and Korach, K. S. (1988) *Endocrinology* **123**, 2257–2266.
Hubert, P., Mester, J., Dellacherie, E., Neel, J., and Baulieu, E. E. (1978) *Proc. Natl. Acad. Sci. USA* **75**, 3143–3147.
Hutchens, T. W., Gibbons, W. E., and Besch, P. K. (1984) *J. Chromatog.* **297**, 283–299.
Katz, J., Finlay, T. H., Tom, C., and Levitz, M. (1980) *Endocrinology* **107**, 1725–1730.
Katz, J., Troll, W., Levy, M., Filkins, K., Russo, J., and Levitz, M. (1976) *Arch. Biochem. Biophys.* **173**, 347–354.
Katzenellenebogen, J. A., Carlson, K. E., Heiman, D. F., Robertson, D. W., Wie, L. L., and Katzenellenbogen, B. S. (1983) *J. Biol. Chem.* **258**, 3487–3495.

Korach, K. S., Horigome, T., Tomooka, Y., Yamashita, S., Newbold, R. R., and McLachlan, J. A. (1988) *Proc. Natl. Acad. Sci. USA* **85,** 3334–3337.

Korach, K. S. and Muldoon, T. G. (1974) *Endocrinology* **94,** 785– 793.

Kumar, V., Green, S., Stack, G., Berry, M., Jin, J-R., and Chambon, P. (1987) *Cell* **51,** 941–951.

Kyse-Andersen, J. (1984) *J. Biochem. Biophys. Methods* **10,** 203–209.

Lubahn, D. B. (1983) *Dissertation,* Duke University, Durham, North Carolina.

Lubahn, D. B., McCarty, K. S. Jr., and McCarty, K. S. Sr. (1985) *J. Biol. Chem.* **260,** 2515–2526.

Madhok, T. C. and Leung, B. S. (1981) *J. Steroid Biochem.* **15,** 299–305.

Mauer, R. A. and Notides, A. C. (1987) *Mol. Cell Biol.* **7,** 4147–4254.

Metzger, D. White, J. H., and Chambon, P. (1988) *Nature* **334,** 31–35.

Miller, L. S., Tribby, T. I. E., Miles, M. R., Tomita, J. T., and Nolan, C. (1982) *Fed. Proc.* **41,** 520.

Molinari, A. M., Medici, N., Moncharmont, B., and Puca, G. A. (1977) *Proc. Natl. Acad. Sci. USA* **74,** 4886–4890.

Moncharmont, B., Su, J.-L. and Parikh, I. (1982) *Biochemistry* **21,** 6916–6921.

Murayama, A. and Fukai, F. (1983) *FEBS Lett.* **158,** 255–258.

Murayama, A., Fukai, F., Hazato, T., Yamamoto, T. (1980) *J. Biochem.* **88,** 955–961.

Murayama, A., Fukai, F., and Murachi, T. (1984) *J. Biochem.* **95,** 1697–1704.

Notides, A. C., Sasson, S., and Callison, S. (1985) in *Molecular Mechanism of Steroid Hormone Action* (Moudgil, V.K., ed.), Walter De Gruyter, Berlin, New York, pp. 173–197.

Pavlik, E. J., Nelson, K., Nagell, J. R. V. Jr., Gallion, H. H., and Baranczuk, R. J. (1987) in *Recent Advances in Steroid Hormone Action* (Moudgil, V. K., ed.), Walter de Gruyter, Berlin, New York, pp. 477–498.

Puca, G. A., Medici, N., Molinari, A. M., Moncharmont, B., Nola, E., and Sica, V. (1980) *J. Steroid Biochem.* **12,** 105–113.

Puca, G. A., Nola, E., Sica, V., and Bresciani, F. (1971) *Biochemistry* **10,** 3769–3780.

Puca, G. A., Nola, E., Sica, V., and Bresciani, F. (1977) *J. Biol. Chem.* **252,** 1358–1366.

Ratajczak, T. and Hähnel, R. (1980) *J. Steroid Biochem.* **13,** 439–444.

Redeuilh, G., Moncharmont, B., Secco, C. and Baulieu, E. E. (1987) *J. Biol. Chem.* **262,** 6969–6975.

Redeuilh, G., Richard-Foy,. R. Secco, C., Torelli, V., Bucourt, R., Baulieu, E. E., and Richard-Foy, H. (1980) *Eur. J. Biochem.* **106,** 481–493.

Redeuilh, G., Secco, C., and Baulieu, E. E. (1985) *J. Biol. Chem.* **260,** 3996–4002.

Rotondi, A. and Auricchio, F. (1978) *Biochem. J.* **178,** 581–587.

Sanborn, B. M., Rao, B. R., Korenman, S. G. (1971) *Biochemistry* **10,** 4955–4960.

Schneider, W. and Gschwendt, M. (1980) *Biochim. Biophys. Acta* **633,** 105–113.

Shahabi, N. A., Hyder, S. M., Wiehle, R. D., and Wittliff, J. L. (1986) *J. Steroid Biochem.* **24,** 1151–1157.

Sherman, M. R., Pickering, L. A., Rollwagen, F. M., and Miller, L. K. (1978) *Fed. Proc.* **37,** 167–173.

Sica, V. and Bresciani, F. (1979) *Biochemistry* **18,** 2369–2378.

Sica, V., Parikh, I., Nola, E., Puca, G. A., and Cuatrecasas, P. (1973) *J. Biol. Chem.* **248,** 6543–6558.

Sica, V., Puca, G. A., Molinari, A. M., Buonaguro, F. M., and Bresciani, F. (1980) *Biochemistry* **19,** 83–88.

Sweet, F. and Szabados, L. (1977) *Steroid* **29,** 127–144.

Towbin, H., Staehelin, T., and Gordon, J. (1979) *Proc. Natl. Acad. Sci. USA* **76,** 4350–4354.

Van Oosbree, T. R., Kim, U. H., and Mueller, G. C. (1984) *Anal. Biochem.* **136,** 321–327.

Wright, C. V. E., Wright, S. C., and Knowland, J. (1983) *EMBO J.* **2,** 973–977.

Progesterone Receptor Purification

J. M. Renoir and E. E. Baulieu

1. Introduction

At the end of 1950s, the estradiol receptor was first discovered in the rat uterus (Jensen and Jacobson, 1962). However, because of its low abundance, efforts to purify it were unsuccessful. Purification of the estradiol receptor from calf uterus was limited by the vagaries of slaughterhouse material. Other steroid hormone receptors, believed (rightly) to share many physicochemical properties with the estrogen, were also described (review in Baulieu et al., 1971). The progesterone receptor (PR) of the chick oviduct gained considerable interest, because of biological responses amenable to biochemical analysis (reviewed in Schrader et al., 1981). Indeed, PR rapidly became one of the most suitable receptors to purify: Its synthesis is increased under estrogen stimulation (Milgrom and Baulieu, 1970; O'Malley et al., 1970), thus providing relatively large amount of receptor free of its own ligand (estrogen stimulus given to immature or castrated animals); regulatory aspects involve not only estrogen simulation but also down regulation related to progesterone action (Milgrom et al., 1973); specific proteins are under progesterone control, as those of the egg white in birds (Shimke et al., 1975; Schrader et al., 1981), or uteroglobin in the rabbit endometrium (Beier, 1968). However, several difficulties were encountered. The first technical problem was to separate PR from other progesterone binding proteins as transcortin (Milgrom and Baulieu, 1969) present in uterus and oviduct extracts, and synthetic progestins, i.e., R5020 and ORG 2058, became operationally very useful. Then, two progesterone binding subunits, A and B (Sherman et al., 1970; Schrader et al., 1972), were described differently from other steroid hormone receptors, leading to hypothetical specific and distinct role for each subunit in the transcription initiation response (Schrader et al., 1981). In addition, the purification procedure that was employed did not permit antibodies to be obtained after injection in different animals species (Shrader et al., 1977); hormone affinity chromatography was not used, and it is probable that, instead of receptor, a receptor-associated protein was purified (Sargan et al., 1986). Differently, with appropriate selective technique, the 110kDa progesterone binding protein B of the receptor (sedimentation coefficient ~4S) could be obtained in both the chick oviduct (Renoir et al., 1984b) and the rabbit uterus (Logeat, 1985). A controversial issue was the size and the composition of the large, nontransformed form of PR, referred to as "aggregate" (Shrader et al., 1977). We had initially described the nontransformed large formed the guinea pig PR (Milgrom and Baulieu, 1970), and believed in its oligomeric structure, as for other steroid receptors (Baulieu et al., 1971), including the estrogen receptor (Toft et al., 1967). Our interest in the "8S"- form of receptors (the sedimentation coefficient actually varies between 7–10S according to the case) was owing to its larger size, which would render the receptor separable from other proteins sedimenting in majority at ~4S (3–5S). When the stabilizing effect of molybdate on the 8S form of the glucocorticosteroid receptor (Leach et al., 1979) and the PR (Nishigori and Toft 1980; Wolfson et al., 1980) was discovered, purification became easier. This large oligomeric protein could be separated by affinity chromatography, and significant amounts were easily obtainable, leading to the description of the hsp 90-containing structure of nontransformed 8S-PR (Baulieu et al., 1983; Catelli et al., 1985a).

It was also known that high salt (>0.3M KCl) and heating lead to formation of a 4S species from 8S-receptors, parallel to the acquisition of DNA binding (for review, *see* Milgrom, 1981). Therefore, it was also possible to purify the 8S-form of the chick oviduct and of the rabbit uterus, in order to secondarily obtain the 4S receptor from the purified 8S-PR (Fig. 1). The basis of the procedure for receptor purification described in this article is steroid affinity chromatography; buffers and other chromatographic techniques used throughout the purification depend on the desired form of PR.

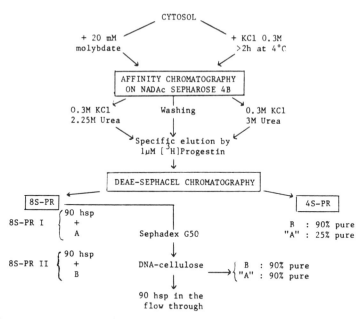

Fig. 1. Purification flowchart of the chick oviduct molybdate-stabilized 8S- non-transformed and 4S-transformed PR.

2. Steroid Affinity Chromatography

2.1. NADAc Sepharose

Sepharose CL 4B is activated by cyanogen bromide in 5M phosphate buffer, pH 11.9 rather than 2M sodium carbonate buffer. The spacer arm (1, 2-diaminododecane) is solubilized in 40 mM absolute ethanol, and diluted with an equal volume of 0.2 M bicarbonate buffer, just prior addition to the gel (final diamine concentration: 0.02M). The entire BrCN activation is performed as quickly as possible, and the washing procedure has to take less than 90 s. The resuspended gel (in 100 mL of precooled diaminododecane solution) is rotated overnight at 4°C. The gel is then washed on a sintered glass funnel with the following sequence of solutions: sodium acetate buffer 0.1M pH 4.0, 2M urea in 0.5M NaCl, NaCl 0.5M in sodium bicarbonate pH 10.0, deionized water, sodium azide 0.1%. We then wait at least 1 wk before steroid coupling (storage at 20°C). Deoxycorticosterone (DOC) is then coupled to diaminododecane after periodic oxydation. ^{3}H-DOC 30 nCi/μmole is used to check the efficiency of coupling. The 17β-carboxylic derivative was then coupled to amino-Sepharose after activation by N-hydroxybenzotriazole-dicyclohexylcarbodiimide, according to the protocol of Formstecher et al. (1980).

After coupling and washing with ethanol, aliquots of the grafted gel are tested for the efficiency of coupling. A radioactive assay indicate the total amount of [³H] DOC bound/mL of gel. The differential alkylamino-sidechain assay (Lustenberger et al. 1980), before and after steroid coupling, reveals the amount of alkylaminogroup coupled at the correct positions. The difference between the total DOC bound and the unreacted alkylaminosidechains yields the amount of [³H] DOC incorrectly bound on the gel on hydroxyl groups by ester linkage. Such unwanted and unavoidable coupling is generally moderate and without troublesome consequences. The gel and its supernatant solution are regularly checked during storage, by radioactivity counting. The NADAc gel (*N*-12-amino-dodecyl-3oxo-4-androstene-17β carboxamide-diaminododecane-sepharose CL 4B, Renoir et al., 1982a) generally is stable for years.

2.2. Sterogel A

Another gel very similar to NADAc gel has been obtained in Toft's laboratory (Grandics et al., 1982), and also is used to purify the chick oviduct nontransformed PR. The ligand is DOC, modified at the C-21 level, and linked to Agarose (Sepharose 2B) via a stable spacer arm by an epoxide technique. This affinity gel is commercialized by GK-Biochemicals under the name of "Sterogel A affinity resin" for purification of progesterone and glucocorticoid receptors. It has been used with success to purify the chick oviduct, nontransformed 8S-PR (Puri et al., 1982; Dougherty et al., 1984).

3. The Chick Oviduct Receptor

3.1. The Nontransformed, Molybdate-Stabilized 8S-PR

3.1.1. Hormone Affinity Chromatography

The nonactivated, nontransformed 8S-PR form is stabilized by addition of 20mM Na_2MoO_4 in all buffers. Homogenates of estrogen-stimulated 5-wk-old chicken is prepared from oviducts in cooled (0–2°C) Tris or phosphate buffer 10 mM, glycerol 10%, EDTA 1.5 mM, 1α-thioglycerol 12 mM, 20 mM Na_2MoO_4, HCl pH 7.4–7.8, as described in Renoir et al. (1984a). All operations are performed at 0–2°C. Cytosol was obtained after centrifugation for 1 h at 105,000 g. Protease inhibitors (PMSF, leupeptin, pepstatin A, aprotinin) and cortisol (1 μM) are added to cytosol, to block proteo-lysis and the binding of either transcortin or glucocorticosteroid receptor to the affinity gel.

Cytosol (100 mL) is loaded overnight on 20 mL NADAc gel equilibrated with the appropriate buffer (5 mL/h flowrate). Gels are washed with the following buffers: cytosol buffer (3 column volumes—same flowrate), cytosol buffer containing 0.3M KCl (1 column vol; rate; 300 mL/h), cytosol buffer alone (6 column volumes; 300 mL/h), cytosol buffer containing 2.25M urea (1 column vol; 300 mL/h), and finally the original buffer (6 column volumes) to reequilibrate the column. The concentration of urea has to be kept <2.25M, since above this concentration a significant proportion of the PR is transformed (Buchou et al., 1983). The above washing procedure remove most all of the contaminating proteins.

The PR bound to the affinity gel is eluted by exchange (Fig. 1) with 1 column vol of buffer containing Na_2MoO_4 and 2 μM of either [^{3}H] progesterone, [^{3}H]-Org 2058 (5–10 Ci/mmol), or [^{3}H] R5020 (87 Ci/mmol). Steroid containing elution buffer is added to the sedimented gel and shaked for 18–20 h at 4°C. After separation of the eluate, addition of one column vol of molybdate containing buffer allows a second extraction of the adsorbed PR. After the initial affinity chromatography, the yield of the 8S-PR is 30–35%, and the purity 5–15%, calculated on the basis of one steroid binding site per 100,000 Da protein (Renoir et al., 1984a).

3.1.2. Ion-Exchange Chromatography

The second step of purification is DEAE-Sephacel chromatography. Eluates from affinity chromatography are loaded on 1mL of preequilibrated DEAE (at 30mL/h). After washing with 10 mL of molybdate containing buffer, to remove most of free steroid, the PR is eluted with a 50 mL 0–0.5M KCl linear gradient, still in the presence of molybdate. Fractions (1 mL) are collected and the PR located by measurement of radioactivity. The receptor is eluted in two peaks at 0.1 and 0.16M KCl (Fig. 2, Renoir et al., 1984a), each containing the receptor in 8S-form, whose mol wt, calculated on the basis of hydrodynamic parameters, is ~245kDa. The two peaks correspond to the forms I and II of the nontransformed 8S-PR described by Dougherty et al. (1982, 1984).

Following purification, purity of the receptor never exceeds 30%, calculated according to one steroid hormone binding site per 100,000 Da protein. Purified 8S chick oviduct PR is routinely obtained in a 25 ± 5% yield (16 separate purifications with the same batch of NADAc gel). We have also purified the crosslinked 8S chick oviduct PR. After irreversible crosslinking of the cytosol receptor with dimethylpimelimidate in low salt buffer containing no molybdate (Aranyi et al., 1988), and purification on NADAc affinity gel, the purified PR had the same Rs (8.0 nm), sedimentation coefficient (~8.5 S) and thus M_r (~290,000) as the original cytosol 8S-PR.

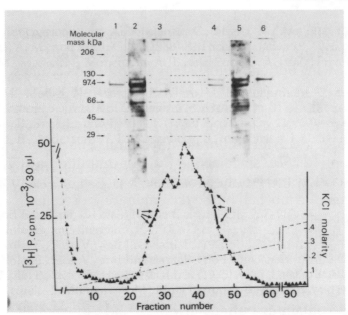

Fig. 2. DEAE-Sephacel chromatography of affinity purified chick oviduct 8S-PR. 8S-PR isolated from the NADAc affinity gel was adsorbed on DEAE-Sephacel in the presence of molybdate and eluted with a 0-0.5M KCl. Ascending fractions (27–29) and descending fractions of the radioactive peak (44–46) were studied by SDS-PAGE and immunoblotting. Lanes 1 and 4, silver stained gel after transfer of proteins into two nitrocellulose papers; Lanes 2 and 5, immunoblots of the transferred proteins with IgG–G3; Lanes 3 and 6 immunoblot with IggG-RB. The presence of the hsp 90 in 8S- PI and 8S-PR II is revealed by silver staining and by IgG-G3 in lanes 1 and 2 and lanes 4 and 5, respectively. The A protein in 8S-PR I and B protein 8S-PR II are detected by the three techniques.

3.1.3. Subunit Composition

Silver staining of SDS-PAGE and immunoblots with a rabbit polyclonal antiboby IgG-RlB (Tuohimaa et al., 1984), also demonstrated the presence of subunit A in 8S-PR I and subunit B in 8S-PRII (Fig. 2), confirming the results of Dougherty et al. (1982). In addition to A and B, a strong signal at 90kDa, was observed in both 8S-PR I and II and it was originally (and erroneously) believed to be the PR itself (Renoir et al., 1982a; Puri et al., 1982, 1984a). Our polyclonal IgG-G3 antibody (Renoir et al., 1982b; Renoir and Mester, 1984) reacted with hsp90 and A in 8S-PR I and hsp90 and B in 8S-PRII. SDS-PAGE fluorography of 8S-PR covalently linked to [^{3}H] R5020 by UV irradiation, demonstrated that the 90kDa protein does not bind progestin (Renoir et al., 1984a). In fact, we found this 90kDa protein in all steroid receptors tested in the chick oviduct cytosol (Baulieu et al., 1983; Joab et al., 1984). The 90kDa protein was identified as a heat shock protein by biochemical, biological, and cloning experiments (Catelli et al., 1985a,b;

Sanchez et al., 1985). Hsp 90 was found to be a ubiquitous component of steroid hormone receptors (Housley et al., 1985; Okret et al., 1985; Mendel et al., 1986; Renoir et al., 1986; Redeuilh et al., 1987) and even the aquatic fungus *Achlya ambisexualis* (Riehl et al., 1985). Originally, a very important tool in the recognition of a nonsteroid binding, 90kDa protein in the 8S-PR was a monoclonal antibody obtained after immunization with 8S-PR (Radanyi et al., 1983), and that appears specific of the chick hsp 90. The purified nontransformed 8S-PR (Rs ~7.1 nm; S_{20},w ~7.9) reacts with only one molecule of the BF_4 antibody, even if biochemical analysis suggests that the 8S-PR includes two molecules of hsp 90 and one of either B or A subunit (Radanyi et al., 1989). After crosslinking of the purified 8S-PR with dimethylpimelimidate (Aranyi et al., 1988) the BF_4-positive, progesterone binding complexes have S_{20},w ~7.5 S, with Rs of 6.3 and 5.8 nm on AcA 34 and HPLC, giving M_r ~200,000 and 184,000, respectively, and suggesting a PR form containing one molecule of hsp 90 and one of progesterone binding unit.

3.2. Transformed or Activated Chick Oviduct 4S-PR

3.2.1. Hormone Affinity Chromatography

Purification is initiated by 0.3M KCl treatment (≥2 h at 0°C) of cytosol (prepared without molybdate), followed by NADAc affinity chromatography (Fig. 1). Adsorbed receptor is washed as described above, with buffers devoid of molybdate, and including 3M instead of 2.25M urea.

Under these conditions, both the hormone binding activity and the sedimentation coefficient are stable (Buchou et al., 1983). After exchange elution with progestin, we recover ~30% of the initial receptor (Renoir et al., 1984b).

3.2.2. Ion Exchange Chromatography

The final step of the purification is ion exchange chromatography on 1 mL DEAE- Sephacel column. Eluates from the NADAc are loaded overnight, similarly to the procedure described for isolation of the 8S-PR. Two radioactive peaks are eluted at 0.08 (A) and 0.2M (B) KCl (Fig. 3). Both migrate at ~4S in sucrose gradients (Renoir et al., 1984b). Quantities of B are always larger than those of A, presumably because of loss of A during affinity chromatography. Washing with 0.08–0.01M KCl prior to the 0.1– 0.5M KCl gradient, yields greater quantities of A. The B subunit, can be purified 5–6,000-fold with ≥95% purity, calculated on the basis of one molecule of progestin bound per 100,000 Da protein (Renoir et al., 1984b). Homogeneity of the purified B subunit was assessed by SDS-PAGE and

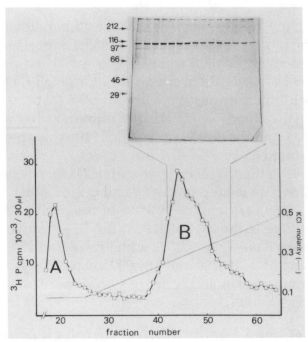

Fig. 3. Homogeneity of B PR purified by sequential NADAc affinity and DEAE-Sephacel chromatographies. A and B chick oviduct PR were biospecifically eluted by [³H] P (SA = 8.9 Ci/mmol), from NADAc gel according to the protocol of Fig. 1. After chromatography on 1ml DEAE-Sephacel, the PR A and B proteins eluted at 0.09M KCl (peak A) and 0.25M KCl (peak B). Aliquots (100 mL) of fractions of peak B (42–55) were boiled in 1% SDS and run on a 7.5-15% SDS-PAGE. Silver staining revealed only the 110 kDa B PR. Protein content at the top of B peak was ~ 17 mg.

silver staining: a single protein band migrates at ~110kDa. In some preparation we noted minor impurities at ~200 and ~160kDa (Renoir and Baulieu, 1987). We never detected hsp90 in our preparations of A or B receptors.

We routinely obtain 100–200 µg of pure B protein from 50 estrogenized stimulated chickens in 4 d-work. Several batches of NADAc gel have been employed over the 4 y of these studies without loss of capacity (in 55 purifications, the purity of the B PR was calculated to be 89 ± 6% with a 17 ± 5% yield). The hydrodynamic parameters of the pure B-PR are identical to those of the crude B -PR (S_{20},w = 4.2; Rs = 6.1 nm). Photoaffinity-labeling of B-PR labeled with [³H] R5020, revealed similar mol wt for both the purified receptor and the starting cytosol, suggesting a lack of degradation during purification (Renoir and Baulieu, 1987). In such purified B-PR preparations, no hsp108 contaminant (Peleg et al., 1985) was observed. SDS-PAGE analysis of chick oviduct PR forms at each purification step is shown in Fig. 4.

Using purified preparations of B-PR, we developed the polyclonal antibody IgGRlB (Tuohimaa et al., 1984). This antibody was employed to indicate immunological similarities between B and A proteins (Renoir et al., 1984a,b), and to immunohistochemically localize the PR in the nuclei of target cells in the absence of hormone (Gasc et al., 1983,1984). This last observation, together with others (Welshons et al., 1984; King and Greene, 1984; Perrot-Applanat et al., 1985), was important to reconsider the notion of nuclear translocation of steroid hormone receptors (Gasc et al., 1989), and to screen a chicken cDNA bank from which the chicken PR was cloned (Gronemeyer et al., 1987). Furthermore, purified chick oviduct B-PR was used to demonstrate binding to DNA-cellulose of this PR subunit (Renoir and Baulieu, 1985), a controversial issue (Schrader et al., 1981), and to study two PR binding sites to DNA of the 5' upstream end of the lyzozyme gene, in comparison to the glucocorticosteroid receptor (Von der Ahe et al., 1986). It is known now that the entire sequence of A subunit (659aa) is contained in the sequence of B (786aa), and thus that two different sites of transcription may operate (Gronemeyer et al., 1987; Conneely et al., 1987).

3.3. Purification of Activated 4S-PR from Purified 8S-PR

Key to the following procedure is to rid the receptor of molybdate ions prior to chromatography.

3.3.1. DEAE-Sephacel Chromatography

When using a steep 0–0.5M KCl gradient, 8S-PR I and 8S-PR II elute as a single radioactive peak. This mixture of two receptors forms can be transformed to a mixture of 4S species, namely, the B and A subunits by the following procedure. The pool of radioactive 8S-PR is diluted five-fold in KCl free buffer containing 20 mM molybdate. After loading on a 1 mL DEAE-Sephacel column, the adsorbed receptor is washed with 10 mL molybdate-free buffer, provoking a dissociation of hsp90 from A and B receptors, which remain bound to the column. A linear 0–0.5M KCl gradient (50 mL) in molybdate-free buffer elutes two separate radioactive peaks at 0.1 (A subunit) and 0.25M KCl (B subunit). Hsp90 is eluted as a large peak, partially overlapping the B receptor. Stepwise elution allows better yield of A (Renoir et al., 1984a).

3.3.2. DNA-Cellulose Chromatography

The purified molybdate-stabilized 8S-PR and the crude molybdate stabilized 8S-PR (Puri and Toft, 1984b; Renoir and Baulieu, 1985) do not bind to DNA-cellulose. If prior to DNA-cellulose, molybdate is removed

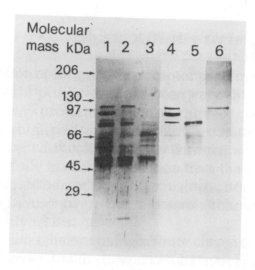

Fig. 4. SDS-PAGE of purified PR forms. Lanes 1, 3 and 4: chick oviduct molybdate-stabilized 8S "non-transformed" PR; lanes 2, 5 and 6: KCl-"activated" PR. Samples in lanes 1-3 were eluted from affinity columns; in lane 1; 8S-PR affinity purified from molybdate containing cytosol; lane 2; 4S-PR affinity purified from 0.3*M* KC) treated cytosol: lane 3: "mock purification" of 8S-PR-(id. 1, but 2m*M* unlabeled P were added to cytosol); note that neither the non-hormone binding hsp 90 not the B and A proteins are present); lanes 4-6 show the protein composition of the 8S-PR, A and B receptor containing fractions of the DEAE-Sephacel purified affinity column eluates, respectively.

from 8S-PR (purified by DEAE-Sephacel) by rapid Sephadex G50 filtration in the presence of 0.3*M* KCl (no molybdate) then PR is eluted in the Vo of the Sephadex G50 column. After dilution with molybdate-free, KCl-free buffer to bring the ionic strength ≤0.05*M*KCl, this PR is loaded onto a 15 mL DNA-cellulose column equilibrated in molybdate-free phosphate buffer. The column is washed with 50 mL of the same buffer, and a 0–0.5*M* linear KCl gradient is applied. PR is eluted in two radioactive peaks at 0.1 (B) and 0.3*M* KCl (A) (Fig. 5). Autoradiography after covalent labeling of [^{3}H] R5020 to both peaks, revealed bands at 110 (B) and 75 kDa (A) (Renoir and Baulieu, 1985), whereas hsp90 is recovered in the flowthrough fraction.

3.4. Purification of B and A Subunits by Preparative Electrophoresis

This purification procedure has been used for PR from laying hens. In this instance, the hormone binding site is occupied by endogenous progesterone, precluding the use of affinity chromatography. Preparative gel electrophoresis also was used by Gronemeyer et al. (1985) as a final purification step following differential DNA-cellulose chromatography of both form B and A receptors that had been previously UV crosslinked to

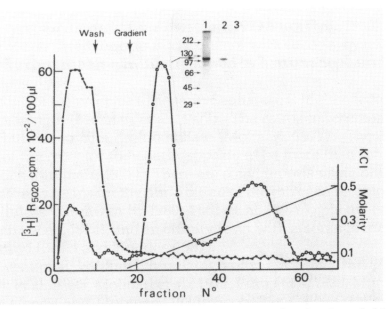

Fig. 5. DNA-cellulose chromatography of non activated and activated chick oviduct PR. After purification of 8S PR in the presence of molybdate, nonactivated PR (SA = 3.02 pmol [^{3}H] R5020/mg protein) was chromatographed on a 15mL DNA-cellulose column (Renoir and Baulieu 1985). A similar sample of 8S-PR was filtered on Sephadex G50 without molybdate and transformed to 4S-PR. After dilution to decrease the KCl to < to 0.05M, 8S-PR (l – l) and 4S-PR (m – m) were eluted with KCl gradients.

the synthetic tritiated R5020. Photolabeled B and A receptors were concentrated by ammonium sulfate, resuspended in SDS sample buffer, and dialyzed against 250 mL of sample buffer, prior to heating 2 min at 95°C.

Bromophenol blue is added to the dialyzed and heated cytosol sample (<5mg protein) and loaded on to a precast preparative gel (1100 PG, Bethesda Research Laboratories). The SDS-PAGE system uses 4 mL of 7.5% separating and 2 mL of 4% stacking gels. The upper and lower chambers contain 0.025M Tris, 0.192M glycine and 0.1% SDS. Electrophoresis is performed under constant current (3.0 mA, 76V initially) at a flowrate of 1 mL/5 min (Gronemeyer et al., 1985). Buffer eluting after Bromophenol Blue is collected in 1 mL fractions from which aliquot (5–20 μL) were assayed for radioactivity. The purity of the peak fractions was analyzed by SDS- PAGE. Progesterone receptor B (M_r ~110 kDa) is routinely observed as a pure protein (one silver stained band), whereas it is necessary to reapply receptor A (M_r ~79) onto a second preparative SDS gel to reach a purity of ~95% after the first preparative gel.

Time consuming and cumbersome, this technique allows the production of high amount of B and A receptors permitting peptide sequencing (Simpson et al., 1987).

4. The Rabbit Uterus Receptor

4.1. The Nontransformed Molybdate-Stabilized 8S-PR

4.1.1. Hormone Affinity Chromatography

A procedure very similar to that described earlier for the purification of the 8S chick oviduct PR can be applied to the purification of nontransformed rabbit uterus receptor (Fig. 6).

Nontransformed rabbit uterus PR form is stabilized by inclusion of 20 mM molybdate in the buffer (10 mM phosphate, 10% glycerol, 1.5 mM EDTA, 12 mM 1α thioglycerol, pH 7, 8). However, even in presence of molybdate, a small portion of the 8S-PR is transformed to 4S-PR during the purification (Renoir et al., 1986). Receptors were eluted in similar quantity whether agonist (R5020, Org 2058) or antagonist (RU486) are used for elution from affinity gel. The use of progesterone is not recommended, since P-8S-PR complexes are unstable and P dissociates at a high rate (Moguilewski and Philibert, 1985).

4.1.2. Ion-Exchange Chromatography

Subsequent purification of the receptor by DEAE-Sephacel chromatography (Fig. 7) is performed as described for the chick oviduct PR (*see* Section 3.1.2.). KCl elution yields a single radioactive peak at 0.15M KCl. This 8S-PR sediments at ~8.5S in sucrose gradient, does not bind to DNA-Cellulose (Renoir et al., 1986), and its purity based on one binding site per ~100,000 Da was 19%, the yield ~20%. Silver staining of SDS-PAGE, reveals, in addition to two proteins of ~120 and 85kDa, an abundant 9kDa protein (Fig. 8). Photoaffinity labeling of purified [^{3}H] R5020-8S PR complexes identified that the ~120 and 85kDa proteins represent the B and A receptors, and that the 90kDa protein does not bind progesterone. No p59 was observed in this purified rabbit uterus PR preparation, although this protein binds to the nontransformed form of the receptor (Tai and Faber, 1985). This is because of urea and KCl washing of the NADAc affinity gel (unpublished data). The 90kDa protein is most likely hsp90, as in the case of the chick oviduct 8S-PR. Use of protease inhibitors in all buffers, suppresses largely the occurence of the 85kDa ("A" like) hormone binding unit. Therefore, this form of the rabbit uterus PR, absent in the initial cytosol, according to Logeat et al. (1985), may be, at least partly, a proteolytic product of the larger 120kDa unit. Using our protocol (Fig. 6), starting with 20–30 estrogen-stimulated immature rabbits, we can routinely purify 1–2 nmoles of 8S-PR (1,000-fold purification).

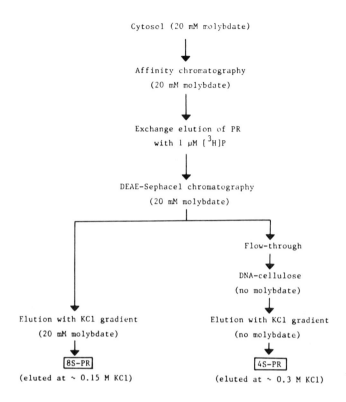

Fig. 6. Purification flowchart of rabbit uterus 8S- and 4S-PR forms.

4.1.3. Immunoaffinity Chromatography

A monoclonal antibody (KN 382/EC$_1$) directed against a 59kDa protein, was developed after injection of nontransformed PR of rabbit uterus (Nakao et al., 1984). This antibody does not crossreact with rabbit hsp90 or B and A receptors (Tai et al., 1986). An Affigel 10-immunoadsorbant has been obtained with 1 mg KN 382/EC$_1$ coupled per mL of gel. Rabbit uterus cytosol labeled with [^{3}H] R5020 was immunoadsorbed and eluted under different conditions. Immunological and autoradiographic analysis of the immunoadsorbed material revealed B and A PR subunits, hsp90 and p59 as components of the 8S-PR (Tai and Faber, 1984). Furthermore, the 59kDa protein was immunologically detected in other nontransformed steroid receptors (estrogen, androgen and glucocorticosteroid), thus leading to the notion that the 59kDa protein is commonly associated with many nontransformed steroid hormone receptors (Tai et al., 1986).

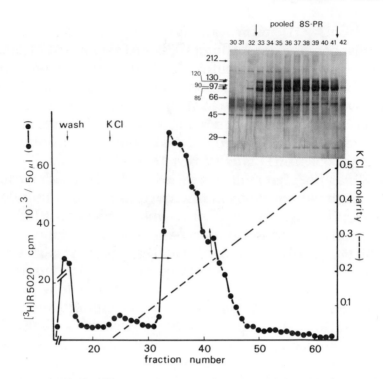

Fig. 7. DEAE-Sephacel chromatography of rabbit uterus affinity-purified 8S-PR. Elution profile of the 8S-PR. Uterine cytosol was prepared and purified by NADAc chromatography as mentioned in Renoir et al., 1986. After elution with [³H] R5020, the 8S-PR (in a vol of 120 mL) was purified on a 1-mL ion-exchange gel. Nine milliliter fractions from fraction number 1 to fraction number 15, and then 1.1-mL fractions, were collected, respectively. A 50 mL portion of each fraction was counted for radioactivity (l). KCl concentration was determined from the conductivity of each fraction (—). Fractions between the double-headed arrows were pooled. (Inset) SDS gel electrophoresis of individual fractions 30– 42 containing 8S-PR. Portions (100mL) of fractions of the DEAE-Sephacel column effluent were boiled in SDS and subjected to SDS-PAGE according to Laemmli (1970). Arrows indicate the mol wt of the markers (× 10-3).

4.2. The Transformed or Activated 4S-PR

4.2.1. DNA-Cellulose Chromatography

As previously stated, despite the stabilization of 8S-PR by molybdate ions, some 4S-PR is always found in NADAc affinity gel eluates.

This 4S-PR does not bind anionic ion-exchange columns, regardless of the buffer system, and is always recovered in the flowthrough fraction (Renoir et al., 1986). Its hormone binding is unstable, except when labeled with [³H] RU 486. In that case, it is a 4–5S complex devoid of hsp90. Then, it does bind well to DNA-cellulose, irrespective of the ligand (progestin or even antiprogesterone RU486). Taking advantage of this property, we em-

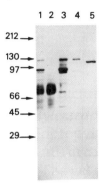

Fig. 8. SDS-polyacrylamide gel electrophoresis of rabbit uterus purified PR at each purification step. Uterine cytosol was prepared in the presence of molybdate and of protease inhibitors; after affinity chromatography and biospecific elution, the eluted 8S-PR was rechromatographed on DEAE-Sephacel and the flowthrough was loaded on DNA-cellulose. An aliquot of each PR fraction was boiled in SDS and loaded on top of the gel. Lane 1, 8S-PR (0.9 µg of protein) eluted from the affinity gel; lane 2, eluate from a parallel mock purification (0.9 µg); lane 3, pooled 8S-PR eluted from the DEAE-Sephacel column (1µg); lane 4, pooled 4S-PR eluted from the DNA-cellulose column (0.4 µg); lane 5, B subunit of the chick oviduct PR (1 µg) (Renoir et al., 1984a) as a comparison of different M_r.

ploy DNA-cellulose chromatography to further purify the 4S complex. Material isolated from DEAE-Sephadex is loaded into a 20 mL DNA-cellulose column, preequilibrated in phosphate buffer containing 20 mM molybdate. The adsorbed material is washed with 50 mL of the same buffer lacking molybdate. A 0–0.5M KCl linear gradient results in elution of a radioactive peak at 0.3M KCl (Fig. 9). Receptor contained in this peak migrated as a ~4–5S moiety in density gradient. Silver staining of SDS-PAGE of the peak material revealed only the 120 and 85kDa PR units.

Starting with 20–30 estradiol stimulated immature rabbits, this procedure yielded routinely 30–50 µg (~0.4nmole) of [³H] R5020-4S PR at 60% purity, calculated similarly as for the chick oviduct purified PR (Renoir et al., 1986).

DNA-cellulose chromatography can also be performed in order to obtain the 4S-PR from purified 8S-PR by successive steroid affinity and ion exchange chromatographies (Fig. 10). Radioactively labeled 8.5S PR is treated as indicated in the legend of Fig. 10, and the proteins eluted by the KCl gradient contain both B and A subunits. In addition, hsp 90 is found in the flowthrough of the column, since it does not bind to DNA cellulose. These observations argue in favor of in vitro activation of the nontransformed PR, prevention of DNA binding of the receptor by hsp 90 contained in the 8S-PR, and inability of hsp 90 to bind DNA-cellulose.

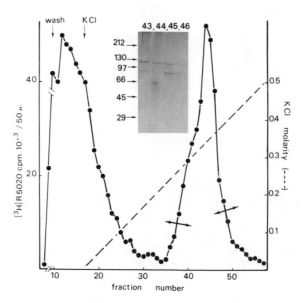

Fig. 9. DNA-cellulose chromatography of the flowthrough of the DEAE-Sephacel column. Elution profile of the purified 4S-PR. The DEAE flow-through was loaded of the DNA-cellulose column, and chromatographed as described in Renoir et al., (1986). Nine-millliliter fractions from fraction number 1 to fraction number 19, and then 1.1-mL fractions, were collected, respectively. A 50 mL portion of each fraction was counted for radioactivity (1) and the KCl concentration (—) was measured as in Fig. 7. The fractions between the double-headed arrows were pooled and referred to as pooled 4S-PR. (Inset) SDS electrophoretis analysis of 4S-PR. Aliquots (130 mL) of fractions 43–46 (top of the radioactive peak) were electrophoresed in SDS buffer as in Fig. 8.

4.2.2. Immunoaffinity Chromatography

Monoclonal antibodies have been developed against rabbit (Feil, 1983; Logeat et al., 1983) and chicken (Sullivan et al., 1985) PR. Logeat et al. (1985) synthesized a very high capacity immunoaffinity matrix, by crosslinking anti-rabbit PR monoclonals antibody to protein A-Sepharose through the Fc fragment of the immunoglobulin. Using this matrix, the rabbit uterus PR can be purified to apparent homogeneity. A 1 mL column (containing 7 mg of monoclonal antibody) can bind 1600 pmol of steroid receptor complexes, of which 79% were eluted by 50 mM Tris HCl pH 10.5 buffer, containing 20% glycerol, with an overall yield of ~49%. The purified receptor consists of a mixture of B and A forms. The latter appeared to be a product of proteolysis of the larger form B, occurring during purification (Loosfelt et al., 1984). Immunopurification is of great interest since it allows receptor to be obtained free of ligand. Such preparations could be helpful to study the effect of agonists or antagonists on receptor structure, and in experiments devised to unravel the mechanism of action of hormones and antihormones.

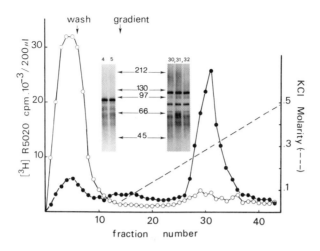

Fig. 10. Activation of purified 8S-PR. An aliquot (~ 0.4 nmol) of [³H] R5020-labeled purified by DEAE-Seephacel 8S-PR was loaded on a 10 mL DNA-cellulose column pre-equilibrated in phosphate containing molybdate buffer (O). In a parallel experiment, 5mL of the same 8S-PR preparation was incubated for 1h at 0° C with 0.3*M* KCl, diluted 7 times in buffer A, and chromatographed on a identical column (●). After washing with 50 mL of buffer, a 0-0.5*M* KCl gradient (50mL) was applied. Fractions of 2 and 5 mL were collected between fractions 1 and 6 (O and ●, respectively), 10 mL between fractions 7 and 15 for both columns and 1 mL during the gradient. Aliquots of 200 μL were assayed for radioactivity. (Inset) SDS-PAGE analysis (7.5% acrylamide) of proteins eluted or not from DNA-cellulose chromatographies. Aliquots (150 μL) of fractions 1–9 and 28–36 of the DNA-cellulose columns loaded with the KCl-treated purified 8S-PR (O) were boiled in SDS, electrophoresed, and silver stained. Only representative protein profiles of fractions 4, 5 and 30–32 are shown. Molecular weight markers (× 10–3) are indicated between the two parts of the gel.

5. Pending Problems

If cloning techniques allow us to know the primary structure and some aspects of receptor domain functions (reviewed in Evans, 1988), the secondary and tertiary structure of steroid hormone receptors is not yet described. It is probable that the DNA binding domain (DBD) includes two zinc fingers, and that, besides DNA, it probably interacts with hsp90 (Sabbah et al., 1987; Baulieu and Catelli, 1988). The hydrophobic ligand binding domain (LBD) is likely involved in the binding of hsp90 of steroid receptors (Pratt et al., 1988; Chambraud et al., unpublished results). However, examination of the primary structure of LBD of various receptors and parent proteins has not yet indicated that amino acids are involved in hormone binding specificity of different receptors, including differences between PR of different species: chick and mammalian PRs do show differences in the binding of progestins (Smith et al., 1974), and RU486, a high affinity ligand for human, rat, and rabbit PR, does not appear to bind chick, hamster, and marsupial PR (Groyer et al., 1985; Gray and Leavitt, 1987; Short, personal communication) (in spite, for instance, of 87% aa identity in LBD of rabbit and chicken PR) (Gronemeyer et al., 1987).

It is not yet known whether the B and A subunits of PR, particularly well studied in chicken and human cells (Horwitz and Alexander, 1983; Gronemeyer et al., 1987; Conneely et al., 1987) are produced by one or several mechanisms (differential biosynthesis, proteolysis). Even more important, it is not known if the two proteins B and A have or not the same function, or if there is regulation of their respective concentration in different tissues, and so on.

The association of PR to hsp90 has been seminal to open the chapter of receptor-associated proteins (Baulieu, 1987; Baulieu and Catelli, 1989; Baulieu et al., 1989). Since DNA binding, which is a prerequisite for hormone action, is dependent on the release of hsp 90 from 8S-steroid receptors, it follows that detailed study of the interactions between steroid receptor and hsp90 is of primary importance. In addition, since steroid hormone agonists appears to favor, and steroid hormone antagonist RU 486 to slow down, the release of hsp 90 from the corresponding 8S-receptor, the LBD-hsp90 interaction has to be better defined. However, even the stochiometry of the nontransformed, heterooligomeric, 8S-receptor forms is still not clear. More has to be learned, since our and other preliminary evidence play in favor of two hsp 90 molecules and one hormone binding unit in progesterone and glucocorticosteroid 8S-receptors (reviewed in Baulieu, 1987), and that two hsp 90 molecules seem to interact with two estradiol receptor molecules in the 8S-estradiol receptor (Redeuilh et al., 1987). In addition, the significance of the nonhormone binding protein of 59,000kDa (p59), initially found as associated to the 8S-PR of rabbit uterus (Tai and Faber, 1984), is also not understood. Its selective nuclear localization has been observed (Gasc et al., in press).

Most of these questions await the obtention of a large amount of steroid hormone receptors, probably thanks to molecular genetics techniques, and X-ray crystallography. More many biochemical, molecular genetics, and biological experiments are needed to understand both steroid hormone and antihormone action.

References

Aranyi, P., Radanyi, C., Renoir, J. M., Devin, J., and Baulieu, E. E. (1988) *Biochemistry* **27**, 1330–1336.

Baulieu, E. E., Alberga, A., Jung-Testas, I., Lebeau, M. C., Mercier-Bodard, C., Milgrom, E., Raynaud, J. P., Raynaud-Jammet, C., Rochefort, H., Truoung, H., and Robel, P. (1971) *Rec. Progr. Horm. Res.* **27,** 351–419.

Baulieu, E. E., Binart, N., Buchou, T., Catelli, M. G., Garcia, T., Gasc, J. M., Groyer, A., Joab, I., Moncharmont, B., Radanyi, C., Renoir, J. M., Tuohimaa, P., and Mester, J. (1983) in *Steroid Hormone Receptors: Structure and Function* (H. Erikson and J. A. Gustafsson, eds.), Nobel Symposium no. 57. Elsevier, Amsterdam, New York, Oxford, pp. 45–72.

Baulieu, E. E. (1987) *J. Cell. Biol.* **35,** 161–174.

Baulieu, E. E. and Catelli, M. G. (1989) in *"Stress-Induced Proteins"*. Proceedings of a Hoffman-La Roche-Director's Sponsors-UCLA Symposium, Keystone, Colorado, April 10–16, 1989, (M. L. Pardue, J. R. Feramisco, and S. Lindquist, eds.), Liss, New York, pp. 203–219.

Baulieu, E. E., Binart, N., Cadepond, E., Catelli, M. G., Chambraud, B., Garnier, J., Gasc, J. M., Groyer-Schweizer, G., Oblin, M. E., Radanyi, C., Redeuilh, G., Renoir, J. M., and Sabbah, M. (1989) in *The Steroid/Thyroid Hormone Receptor Family and Gene Regulation* (J. A. Gustafsson and H. Eriksson, eds.), Birkhauser Verlag (Basel), pp. 301–318.

Beier, H. M. (1968 *Biochem. Biophys. Acta* **160,** 289–291.

Buchou, T., Mester, J., Renoir, J. M., and Baulieu, E. E. (1983) *Biochem. Biophys. Res. Commun.* **114,** 479–487.

Catelli, M. G., Binart, N., Jung-Testas, I., Renoir, J. M., Baulieu, E. E., Feramisco J. R., and Welch, W. J. (1985a) *EMBO* **4,** 3131–3135

Catelli, M. G., Binart, N., Feramisco, J. R., and Helfman, D. (1985b) *Nucl. Acid. Res.* **13,** 6035–6047.

Conneely, O. M., Maxwell, B. L., Toft, D. O., Schrader, W. T., and O'Malley, B. W. (1987) *Biochem. Biophys. Res. Commun.* **149,** 493–501.

Dougherty, J. J. and Toft, D. O. (1982) *J. Biol. Chem.* **257,** 3113–3120.

Dougherty, J. J., Puri, R. K., Toft, D. O. (1984) *J. Biol. Chem.* **259,** 8004–8009.

Evans, R. M. (1988) *Science* **240,** 889–895.

Feil, P. D. (1983) *Endocrinology* **112,** 396–398.

Formstecher, P., Lustenberger, P., and Dautrevaux, M. (1980) *Steroids* **35,** 265–272.

Gasc, J. M., Ennis, B. W., Baulieu, E. E., and Stumpf, W. E. (1983) *C.R. Acad. Sci. Paris* **297,** 477–482.

Gasc, J. M., Renoir, J. M., Radanyi, C., Joab, I., Tuohimaa, P., and Baulieu, E. E. (1984) *J. Cell. Biol.* **99,** 1193–1201.

Gasc, J. M., Delahaye, F., and Baulieu, E. E. (1989) *Exp. Cell Res.* **181,** 492–504).

Gasc, J. M., Renoir, J. M., Faber, L. E., Delahaye, F., and Baulieu, E. E. (1989) *Exp. Cell. Res.,* in press.

Grandics, P., Puri, R. K., and Toft, D. O. (1982) *Endocrinology* **110,** 1055–1057.

Gray, G. O. and Leavitt, W. W. (1987) *J. Steroid Biochem.* **28,** 493–497.

Gronemeyer, H., Govindan, M. V., Chambon, P. (1985) *J. Biol. Chem.* **260,** 6916–6925.

Gronemeyer, H., Turcotte, B., Quirin-Stricker, G., Bocquel, M. T., Meyer, M. E., Krozowski, Z., Jeltsch, J. M., Lerouge, T., Garnier, J. M., and Chambon, P. (1987) *EMBO J.* **6,** 3985–3994.

Groyer, A., Le Bouc, Y., Joab, I., Radanyi, C., Renoir, J. M., Robel, P., and Baulieu, E. E. (1985) *Eur. J. Biochem.* **149,** 445–451.

Horwitz., K. B. and Alexander, P. (1983) *Endocrinology* **113,** 2195–2201.

Housley, P. R., Sanchez, E. R., Westphal, H. M., Beato, M., and Pratt, W. B. (1985) *J. Biol. Chem.* **260,** 13810–13817.

Jensen, E. V. and Jacobson, H. I. (1962) *Rec. Progr. Hormone Res.* **18**, 387–414.

Joab, I., Radanyi, C., Renoir, J. M., Buchou, T., Catelli, M. G., Binart, N., Mester, J., and Baulieu, E. E. (1984) *Nature* **308**, 850–853.

King, W. J. and Greene, G. L. (1984) *Nature* **207**, 745–747.

Leach, K. L., Dahmer, M. K., Hammond, N. D., Sando, J. J., and Pratt, W. B. (1979) *J. Biol. Chem.* **254**, 11884–11890.

Logeat, F., Vu Hai, M. T., Fournier, A., Legrain, P, Buttin, G., and Milgrom, E. (1983) *Proc. Natl. Acad. Sci. USA* **78**, 1426–1430.

Logeat, F., Pamphile, R., Loosfelt, M., Jolivet, A., Fournier, A., and Milgrom, E. (1985) *Biochemistry* **24**, 1029–1035.

Loosfelt, H., Logeat, F., Vu Hai, M. T., and Milgrom, E. (1984) *J. Biol. Chem.* **259**, 14196–14202.

Lustenberger, P., Formstecher, P., and Dautrevaux, M. (1980) *J. Chromotogr.* **193**, 451–457.

Mendel, D. B., Bodwell, J. E., Gametchu, B., Harrisson, R. W., and Munck, A. (1986) *J. Biol. Chem.* **261**, 3758–3763.

Milgrom, E. and Baulieu, E. E. (1969) *Biochim. Biophys. Acta* **194**, 602–605.

Milgrom, E. and Baulieu, E. E. (1970) *Biochem. Biophys. Res. Commun.* **40**, 723–730.

Milgrom, E., Atger, M., and Baulieu, E. E. (1973) *Biochemistry* **12**, 5198–5205.

Milgrom, E. (1981) in *Biochemical Actions of Hormones* (G. Litwack , ed.), Academic, New York, pp. 466–491.

Moguilewski, M. and Philibert, D. (1985) (E. E. Baulieu and S. J. Segal, eds.), Plenum, New York and London, pp. 87–101.

Nakao, K., Myers, J. E., and Faber, L. E. (1984) *Can. J. Biochm. Cell. Biol.* **63**, 33–40.

Nishigori, H. and Toft, D. O. (1980) *Biochemistry* **19**, 77–83.

Okret, S., Wikström, A. C., and Gustafsson, J. A. (1985) *Biochemistry* **24**, 6581–6586.

O'Malley, B. W., Sherman, M. R., and Toft, D. O. (1970) *Proc. Natl. Acad. Sci. USA* **67**, 501–508.

Peleg, S., Schrader, W. T., Edwards, D. P., Mc.Guire, W. L., and O'Malley, B. W. (1985) *J. Biol. Chem.* **260**, 8492–8501.

Perrot-Applanat, M., Logeat, F., Groyer-Picard, M. T., and Milgrom, E. (1985) *Endocrinology* **116**, 1473–1484.

Pratt, W. B., Jolly, D. J., Pratt., D. V., Hollenberg, S. M., Giguere, V., Capedond, F., Schweizer-Groyer, G., Catelli, M. G., Evans, R. M., and Baulieu, E. E. (1988) *J. Biol. Chem.* **263**, 267–273.

Puri, R. K., Grandics, P., Dougherty, J. J., and Toft, D. O. (1982) *J. Biol. Chem.* **257**, 10831–10837.

Puri, R. K., Dougherty, J. J., and Toft, D. O. (1984a) *J. Steroid Biochem.* **20**, 23–29.

Puri, R. K. and Toft, D. O. (1984b) *Endocrinolgy* **115**, 2453–2463.

Radanyi, C., Joab, I., Renoir, J. M., Richard-Foy, H., and Baulieu, E. E. (1983) *Proc. Natl. Acad. Sci. USA* **80**, 2854–2858.

Radanyi, C., Renoir, J. M., Sabbah, M., and Baulieu, E. E. (1989) *J. Biol. Chem.*, **264**, 2568–2573.

Redeuilh, G., Moncharmont, B., Secco, C., and Baulieu, E. E. (1987) *J. Biol. Chem.* **262**, 5530–5535.

Renoir, J. M., Yang, C. R., Formstecher, P., Lustenberger, P., Wolfson, A., Redeuilh, G., Mester, J., Richard-Foy, H., and Baulieu, E. E. (1982a) *Eur. J. B.* **127**, 71–79.

Renoir, J. M., Radanyi, C., Yang, C. R., and Baulieu, E. E. (1982b) *Eur. J. B.* **127**, 81–86.

Renoir, J. M., Buchou, T., Mester, J., Radanyi, C., and Baulieu, E. E. (1984a) *Biochemistry* **23**, 6016–6023.

Renoir, J. M., Mester, J., Buchou, T., Catelli, M. G., Tuohimaa, P., Binart, N., Joab, I., Randanyi., C., and Baulieu, E. E. (1984b) *Biochem. J.* **217**, 685–692.
Renoir, J. M. and Mester (1984) *Mol. Cell. Endocr.* **37**, 1–13.
Renoir, J. M. and Baulieu, E. E. (1985) *C R Acad. Sci. Paris* **301**, 859–864.
Renoir, J. M., Buchou, T., and Baulieu, E. E. (1986) *Biochemistry* **25**, 6405–6413.
Renoir, J. M. and Baulieu, E. E. (1987) in *Steroid and Sterol Hormone Action* (T. C. Spelsberg and R. Kumar, eds.), Martinus Nijhoff, Boston, pp. 41-59.
Rielh, R. M., Sullivan, W. P., Vroman, B. T., Bauer, V. J., Pearson, G. R., and Toft, D. O. (1985) *Biochemistry* **24**, 6586–6591.
Sabbah, M., Redeuilh, G., Secco, C., and Baulieu, E. E. (1987) *J. Biol. Chem.* **262**, 8631–8635.
Sanchez, E. R., Toft, D. O., Schlesinger, M. J., and Pratt, W. B. (1985) *J. Biol. Chem.* **260**, 12398–12401.
Sargan, D. R., Tsai, M. J., and O'Malley, B. W. (1986) *Biochemistry* **25**, 6252–6258.
Schrader, W. T. and O'Malley, B. W. (1972) *J. Biol. Chem.* **247**, 2401–2407.
Schrader, W. T., Heuer, S. S., and O'Malley, B. W. (1975) *Biol. Reprod.* **12**, 134–142.
Schrader, W. T., Coty, W. A., Smith, R. G., and O'Malley, B. W. (1977) *Ann. N.Y. Acad. Sci.* **286**, 64–80.
Schrader, W. T., Birnbaumer, M. E., Hughes, M. R., Weigel, N. L., Grody, W. W., and O'Malley, B. W. (1981) *Rec. Progr. Hormone Res.* **37**, 583–633.
Sherman, M. R., Corvol, P. L., and O'Malley, B. W. (1970) *J. Biol. Chem.* **245**, 6085–6096.
Shimke, R. T., Mc Knight, G. S., Shapira, D. J., Sullivan, D., and Palacios, R. (1975) *Rec. Progr. Hormone R es.* **31**, 175–211.
Simpson., R. J., Grego, B., Govindan, M., S., and Gronemeyer, H. (1987) *Mol. Cell. Endocr.* **52**, 177–184.
Smith, M. E., Smith, R. G., Toft, D. O., Neergaard, J. R., Burrows, E. P., and O'Malley, B. W. (1974) *J. Biol. Chem.* **249**, 5924–5932.
Sullivan, W. T., Vroman, B. T., Bauer, V., Puri, R. K., Riehl, R. M., Pearson, G. R., and Toft, D. O. (1985) *Biochemistry* **24**, 4214–4222.
Tai, P. K. K. and Faber, L. E. (1984) *Can. J. Biochem. Cell. Biol.* **63**, 41–49.
Tai, P. K. K., Maeda, Y., Nakao, K., Wakim, N. G., Duhring, J. L., and Eaber, L. E. (1986) *Biochemistry* **25**, 5269–5275.
Toft, D. O., Shyamala, G., and Gorski, J. (1967) *Proc. Natl. Acad. Sci. USA* **57**, 1740–1743.
Tuohimaa, P., Renoir, J. M., Radanyi, C., Mester, J., Joab, I., Buchou, T., and Baulieu, E. E. (1984) *Biochem. Biophys. Res. Commun.* **119**, 433–439.
Von der Ahe, D., Renoir, J. M., Buchou, T, Baulieu, E. E., and Beato, M. (1986) *Proc. Natl. Acad. Sci.* **83**, 2817–2821.
Welshons, W. V., Lieberman, M. E., and Gorski, J. (1984) *Nature* **207**, 747–749.
Wolfson, A., Mester, J., Yang, C. R., and Baulieu, E. E. (1980) *Biochem. Biophys. Res. Commun.* **95**, 1577–1584.

Purification of Nontransformed Avian Progesterone Receptor

David F. Smith and David O. Toft

1. Introduction

As with other steroid receptors, the chick progesterone receptor in vitro can exist in a variety of forms dependent on status of the receptor in vivo prior to tissue disruption, and physicochemical conditions to which the receptor is exposed during and after preparation of a tissue extract. Some of these receptor forms are artifactual aggregates or breakdown products that are not biologically significant, existing only in the unnatural milieu of the test tube. Other forms are significant in that functional states of receptor are reflected, at least partially, in heterogeneous associations and modifications of receptor components that can be maintained during purification.

1.1. Monomeric Forms
of Chick Progesterone Receptor

In vivo and in the absence of bound progesterone, the chick progesterone receptor is an inactive transcriptional regulator localized to the nucleus (Ennis et al., 1986; Isola et al., 1987), but is readily extracted into the cytosol fraction by oviduct homogenization. If bound to hormone in vivo, the receptor is "transformed" to a functional state that is active in regulating gene transcription and remains in the nuclear fraction when oviduct is

Receptor Purification, vol. 2   ©1990 The Humana Press

homogenized in low ionic strength buffer. The "non-transformed" and "transformed" functional states of progesterone receptor correspond to structural differences in the receptor forms in vitro, as detailed below.

There are two monomeric forms, A and B, of the avian progesterone receptor. The apparent sizes of these forms determined by SDS gel electrophoresis are 110kDa for form B and 79kDa for form A. The gene for the avian progesterone receptor has been cloned and sequenced (Gronemeyer et al., 1987; Conneely et al., 1987a). From molecular studies (Gronemeyer et al., 1987; Conneely et al., 1987b), it appears that the two forms are products of alternate translation start sites from a single mRNA, and thus differ in their primary structure by a deletion from the amino terminus of form B as compared to the smaller form A. By sequence analysis, the sizes of the receptors are: B—86kDa and A—72kDa. The functional significance of there being two receptor forms is unresolved, although there is evidence that specificity exists between the two forms for the regulation of certain target genes (Tora et al., 1988).

1.2. Composition of Nontransformed Progesterone Receptor

It is unlikely that receptor ever exists in vivo in monomeric form, unassociated with other cellular components. Isolating receptor while maintaining native associations has been one of our objectives toward understanding the regulation of progesterone receptor function. Before considering specific protocols for the purification of receptor complexes, we present here a brief discussion on the composition of nontransformed receptor complexes and experimental conditions that affect the stability of these receptor complexes in vitro.

Nontransformed receptors in cytosol extracts exist in heteromeric complexes with the 90kDa heat shock protein (hsp90) (Sullivan et al., 1985; Joab et al., 1984). There appear to be two hsp90 subunits for each receptor subunit, with forms A and B of receptor residing in distinct complexes (Renoir et al., 1984; Mendel and Orti, 1988). It is also possible that other polypeptides and nonprotein factors are part of this complex. As shown in data presented later, we find a protein (p70) related to the 70kDa heat shock protein as a component of purified receptor complexes. Faber and colleagues (Tai et al., 1986) have demonstrated the presence of a 59kDa protein with rabbit progesterone receptor complexes, but we have not detected a similar protein in chicken receptors. There is also evidence for the association of RNA with a number of steroid receptors (Ali and Vedeckis, 1987; Economidis and Rousseau, 1985; Tymoczko et al., 1987).

The role of any of these components in receptor function remains to be clarified, but there is a factor whose role has become clear. From molecular genetic studies (Weinberger et al., 1985; Evans, 1988, for review) it has been shown that steroid receptors contain two potential "zinc-fingers" in their DNA-binding domains. The necessity of Zn^{2+}-stabilized fingers for proper function of the glucocorticoid receptor has been demonstrated recently (Freedman et al., 1988). It is highly likely that progesterone receptor and other steroid receptors have a similar requirement for Zn^{2+} to establish the proper conformation of the DNA-binding domain.

1.3. Stabilization of Progesterone Receptor Complexes

The dissociation of hsp90 from progesterone receptor in vitro has been correlated with transformation of the receptor to a form that binds DNA. The mechanism of receptor transformation in vivo is not understood, but the dissociation of hsp90 from receptor appears to be one step in this transformation although there are clearly additional modifications to the transformed receptor, such as increased phosphorylation (Logeat et al., 1985; Sullivan et al., 1988; Wei et al., 1987), which have not been mimicked in vitro. A number of conditions promote the dissociation of hsp90 from receptor complexes (Groyer et al., 1987; Mendel et al., 1986; Sanchez et al., 1987); among these conditions are heat and high ionic strength. For reasons still unknown, molybdate has been found to stabilize the binding of hsp90 to receptor (Nishigori and Toft, 1980; Dahmer et al., 1981), and is commonly included in buffers during and after homogenization to maintain nontransformed hsp90:receptor complexes.

The functional significance of progesterone receptor binding to p70 is unknown; we find p70 associated with nontransformed receptor as well as with receptor transformed either in vitro or in vivo. This association with p70 is not disrupted by high ionic strength or heat alone, unlike hsp90's association, but p70 can be dissociated in the presence of ATP and divalent cations, particularly Mn^{2+} (Kost et al., 1989).

1.4. Proteolysis of Progesterone Receptor

Besides considerations for maintaining receptor complexes in vitro, precautions should also be taken to prevent proteolytic degradation of receptor. We have found (Sullivan et al., 1988) that proteolytic activity in oviduct cytosol varies greatly with homogenization conditions and whether oviduct had been frozen prior to homogenization. Such proteolytic degradation is illustrated by the Western blot data shown in Fig. 1. The details of our homogenization procedures are given in an earlier

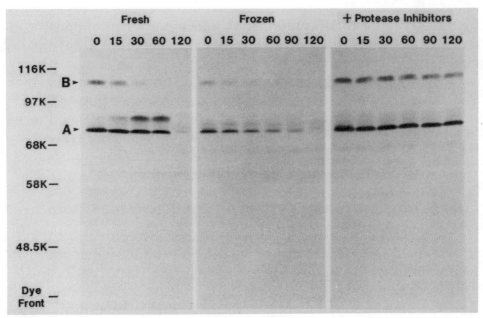

Fig. 1. Conditions enhancing in vitro proteolysis of receptor. Cytosol, plus or minus protease inhibitors, was prepared from oviduct that was excessively homogenized (Fresh) or homogenized normally but from oviduct tissue frozen at –70°C (Frozen). Cytosol was also prepared from frozen oviduct in the presence of protease inhibitors (+Protease Inhibitors). The cytosols were then incubated at 37°C for the number of minutes indicated. Progesterone receptor was examined in Western blotted samples (2 µL cytosol per lane) by immunostaining with PR13. Taken from Sullivan et al., 1988.

publication (Sullivan et al., 1988); briefly, though, we normally homogenize fresh oviduct in 4 volumes of homogenization buffer using a Polytron tissue homogenizer for 2 three-second bursts. Progesterone receptor in carefully prepared cytosol is stable to proteolytic breakdown after 3 h at 37°C. However, adding additional homogenizer bursts or homogenizing small volumes promotes the proteolytic degradation of receptor as shown in the time-course study (Fig. 1, left series). The degradation is characterized by the rapid depletion of receptor form B concomitant with the appearance of a receptor fragment at 85kDa. Smaller fragments are not seen, probably because of the loss of epitope for PR13, the antibody used for immunostaining. The inclusion of protease inhibitors in homogenization buffer completely blocks this breakdown (not shown). The use of frozen oviduct tissue for cytosol preparation also promotes proteolytic degradation of receptor as shown in the next time-course study (Fig. 1, center series). The pattern of receptor breakdown is somewhat different than that caused by excessive homogenization, but the breakdown can be retarded by protease inhibitors (Fig. 1, right series).

1.5. Phosphorylation
of Progesterone Receptor

Progesterone receptor is a phosphoprotein that exhibits increased phosphorylation after receptor transformation in vivo (Logeat et al., 1985; Sullivan et al., 1988; Wei et al., 1987), though the role(s) of phosphorylation in regulating receptor activities has not been demonstrated. Studies from our laboratory (Sullivan et al., 1988) have shown that progesterone receptor form A transformed in vivo has a slower electrophoretic mobility than nontransformed receptor. The low mobility structure is readily reversed in the absence of phosphatase inhibitors, apparently through the activity of endogenous phosphatases. Therefore, the use of phosphatase inhibitors such as molybdate, fluoride, and EDTA may be essential for the purification of native receptor.

2. Steroid Affinity Purification
of Nontransformed Progesterone Receptor

Our laboratory has developed purification procedures that rely primarily on the use of a steroid affinity resin to isolate nontransformed progesterone receptor (Grandics et al., 1982; Puri et al., 1982; Dougherty et al., 1984). By combining the use of steroid affinity resin with other physicochemical steps, we have purified progesterone receptor complexes from oviduct cytosol 6000-fold with a yield of almost 20%, while also separating receptor complexes containing either receptor form A or form B (Puri et al., 1982; Dougherty et al., 1984). Table 1 summarizes the purification data from one such protocol.

Each step in a commonly used purification protocol, somewhat modified from that summarized in Table 1, is discussed below. These steps are described in detail in earlier reports (Grandics et al., 1982; Puri et al., 1982; Dougherty et al., 1984). Cytosol is prepared from diethylstilbestrol-stimulated chick oviducts in the presence of 10 mM sodium molybdate in order to stabilize nontransformed receptor complexes. Molybdate is present in all buffers throughout the protocol for this reason. Also included during tissue homogenization is a mixture of protease inhibitors [0.1 mM leupeptin, 0.1 mg/mL bacitracin, 77 µg/mL aprotinin, 0.5 mM phenylmethylsulfonyl fluoride, 1.5 µM pepstatin, and 5 mM EDTA] that have been shown to minimize proteolytic cleavage of receptor (Sullivan et al., 1988). The following protocols assume 30 mL of oviduct cytosol is the starting vol.

Table 1
Purification of Avian Progesterone Receptor

Step	Volume, mL	Total protein, mg	Total bound hormone, pmol	Specific activity, pmol/mg	Purification factor, yield
Cytosol	36	728	454	0.62	
35% ammonium sulfate	9	70	222	3.2	5.2 (49%)
Steroid affinity chromatogaphy	10	0.2	130	650	1,050 (29%)
Gel filtration	16	0.078	128	1,640	2,650 (28%)
DEAE-Sephadex[a]					
Peak I	7	0.017	60	3,530	5,690
Peak II	5	0.0098	23	2,350	3,790

[a]*See* Puri et al., 1982, for experimental details.

2.1. *Phosphocellulose Chromatography*

It has been shown (Birnbaumer et al., 1983) that phosphocellulose chromatography can effectively remove from crude cytosol much of the proteolytic activity to which receptor is sensitive. Though transformed receptor will bind to phosphocellulose and to other polyanionic media such as DNA and RNA, the nontransformed, molybdate-stabilized receptor complexes are not retained on phosphocellulose, although important proteases do bind. The use of phosphocellulose as an initial step in the protocol yields very little receptor purification, but does minimize the possibility of proteolytic cleavage of receptor. Typically, 5 mL bed vol of phosphocellulose is used for 30 mL cytosol.

2.2. *Steroid Affinity Chromatography*

The steroid affinity resin contains modified deoxycorticosterone (DOC) linked through a spacer arm to agarose beads, as illustrated in Fig. 2. The synthesis of this resin and its characteristics have been described in detail (Grandics et al., 1982); it is commercially available as Sterogel™ (Sterogene, San Gabriel, CA). This resin has also been used for the purification of glucocorticoid receptors (Grandics et al., 1984).

To prevent the retention of glucocorticoid receptor and plasma transcortin on DOC resin, cytosol is adjusted to 3 μM cortisol prior to phosphocellulose treatment. To distinguish proteins that bind nonspecifically to DOC resin, a control preparation can be used where progesterone is in-

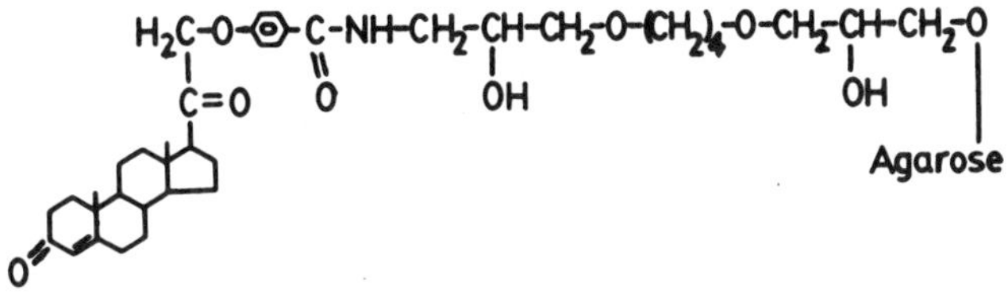

Fig. 2. The structure of deoxycorticosterone affinity resin.

cluded in the sample prior to chromatography, thereby precluding the binding of receptor to DOC resin.

The phosphocellulose flowthrough fraction is added to DOC resin (3 mL bed vol) and rocked at 4°C for at least 2 h. The resin is then washed 3 times each with homogenization buffer [typically, 50 mM potassium phosphate, pH 7.0, 10 mM monothioglycerol, 10 mM sodium molybdate, plus a mix of protease inhibitors] supplemented with 250 mM KCl, DOC wash buffer [10 mM potassium phosphate, pH 7, 10 mM monothioglycerol, 10 mM sodium molybdate, 1 mM EDTA] plus 60% glycerol, and finally DOC wash buffer plus 10% glycerol.

Receptor is eluted from the DOC resin using DOC wash buffer plus 10% glycerol and 10 µM progesterone (tritiated progesterone when label is required) as the elution buffer. A looped elution system is set up in which a heparin agarose column (1–2 mL bed vol) is attached in tandem to the DOC resin column with a peristaltic pump recirculating elution buffer through the column system. Five to 10 mL of elution buffer are sufficient to establish the loop that is then circulated overnight at 4°C. As receptor is released from the DOC resin by exchange of DOC for progesterone, the receptor complexes are flushed through and bind to heparin (Puri and Toft, 1986). Free steroid does not bind heparin.

The heparin-agarose column is washed with 50 mL TMME buffer [10 mM Tris, pH 7.4, 10 mM monothioglycerol, 10 mM molybdate, 1 mM EDTA] containing 100 mM KCl. Receptor complexes are eluted from the column with 5 mL TMME buffer containing 300 mM KCl. Purified receptor complexes may be TCA precipitated at this point for concentration and analysis by gel electrophoresis. For the separation of nontransformed receptor form A from form B, an additional DEAE chromatography step is used.

2.3. DEAE-Sephadex Chromatography

The eluate from heparin-agarose is diluted or dialyzed against TMME buffer to lower the KCl concentration and then loaded onto DEAE-Sephadex (1 mL bed vol). We should note that DEAE-Sephadex gives a better separation of receptor forms than does DEAE-cellulose (Puri et al., 1982;

Dougherty et al., 1984). It is also important to thoroughly equilibrate the resin with running buffer as described (Dougherty et al., 1984). The column is washed with 25 mL TMME and then eluted with a 40-mL gradient of 0–400 mM KCl in TMME. Figure 3A is an elution profile of receptor forms from DEAE-Sephadex. Peak fractions are precipitated with TCA or chloroform/methanol for concentration prior to gel analysis.

2.4. Electrophoretic Analysis of Nontransformed Progesterone Receptors

Shown in Fig. 3B is an SDS gel of DEAE-fractionated nontransformed progesterone receptor. The lane numbers correspond to the fraction numbers shown in Fig. 3A. The identity of receptor bands has been verified by photoaffinity labeling with ^{3}H-R5020 (Dougherty et al., 1984). The presence of form A of receptor (79kDa) corresponds to peak I from DEAE chromatography (fraction 3), and form B (110kDa) corresponds to peak II (fraction 8). The major protein band is hsp90, whose abundance peaks with both form I and form II of receptor. Two other prominent bands are seen. One is a band at 67kDa (fraction 3), which likely is the p70 protein related to hsp70, as discussed below. The other is a band at 104kDa (fraction 7); this protein is not associated with receptor complexes since it is present in mock-purified samples in which progesterone receptor binding to steroid affinity matrix was blocked by the addition of progesterone prior to steroid affinity chromatography (Dougherty et al., 1984).

In a later experiment, receptor complexes purified by steroid affinity chromatography were analyzed by SDS gel electrophoresis and by Western blot analysis using antibodies to progesterone receptor, hsp90, and hsp70. This analysis is shown in Fig. 4. The samples analyzed were similar to those shown above except that the final DEAE chromatography step kDa was omitted. The first lane, from a Coomassie blue-stained gel, shows the total proteins in a steroid affinity-purified sample (T). Lane 2 is a sample purified in parallel, but pretreated with progesterone to block binding of receptor complexes and distinguish nonspecifically bound proteins (NS). Duplicate lanes from this gel were electroblotted to nitrocellulose and immunostained with the following antibodies: lanes 3 and 4—PR22, a monoclonal antibody to progesterone receptor; lanes 5 and 6—7D11, a monoclonal antibody to chicken hsp90; and lanes 7 and 8—rabbit antiserum to chicken hsp70 (Kelley and Schlesinger, 1982). Comparing lanes 1 and 2, receptor forms A and B and hsp90 are the only protein bands decreased by pretreatment with progesterone. As seen in immunostained strips, pretreatment did not completely block binding of receptor to DOC

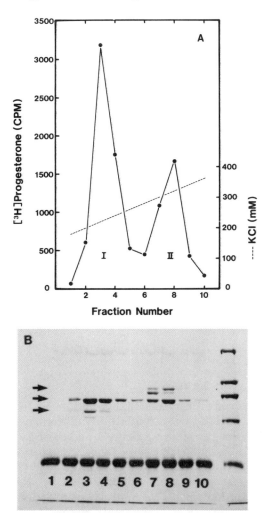

Fig. 3. Comparison of [³H]progesterone binding activity and protein profiles follow-ing DEAE-Sephadex chromatography of steroid affinity-purified receptor. After pur-ification using DOC-agarose and heparin-agarose, progesterone receptor was chroma-tographed on DEAE-Sephadex. Aliquots of each fraction were counted for radioac-tivity (A), and the remainder was analyzed by SDS gel electrophoresis and Coomassie blue stain (B). Arrows at the left point out the 110, 90, and 75kDa receptor components. Ovalbumin was added to all column fractions as carrier; mol wt standards (rightmost lane) are myosin (212kDa), β-galactosidase (116kDa), phosphorylase (97kDa), bovine serum albumin (68kDa), and ovalbumin (kDa). Taken from Puri et al., 1982.

resin, probably because of incomplete binding of hormone to receptor or exchange to the DOC resin. PR22 immunostains receptor very strongly (lanes 3 and 4) and clearly detects minor proteolytic fragments of receptor migrating intermediate to forms A and B (Sullivan et al., 1988). Monoclonal antibody 7D11 does not immunostain hsp90 as strongly, thus the relative

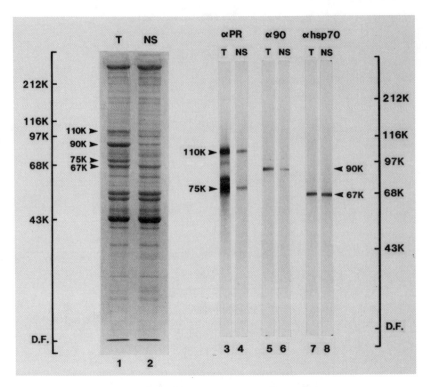

Fig. 4. Analysis of steroid affinity-purified progesterone receptor components. Receptor complexes were purified by steroid affinity chromatography and analyzed by electrophoresis and Coomassie staining (lanes 1 and 2) or by Western blot immunostaining (lanes 3 through 8). In each pair of lanes, receptor components were determined by comparing total purified protein (T) with protein which bound nonspecifically (NS) after preabsorbing receptor with progesterone. Receptor components are identified by arrows to the left of lane 1. Progesterone receptor forms B and A are identified by immunostaining with PR22 (lanes 3 and 4); hsp90 is identified with 7D11 (lanes 5 and 6); and p70 is identified with antiserum against hsp70 (lanes 7 and 8).

abundance of hsp90 to receptor forms is more accurately reflected by Coomassie Blue staining. By both Coomassie Blue staining and immunostaining (lanes 7 and 8), the recovery of p70 is undiminished by progesterone pretreatment, suggesting that p70 is unassociated with receptor complexes. However, we present additional data below supporting an association of p70 and receptor. Forms of hsp70 and its cognate proteins tend to bind nonspecifically to various resins (Welch, personal communication); since p70 is present in oviduct cytosol in great abundance over receptor (Smith, Kost, and Toft, personal observation), nonspecific binding of p70 possibly obscures the specific association of p70 with receptor.

3. Immune Purification
of Nontransformed Progesterone Receptor

We have prepared mouse monoclonal antibodies to chick progesterone receptor (Sullivan et al., 1986). The antigen used in the production of these antibodies was oviduct progesterone receptor purified by steroid affinity chromatography. Five monoclonal antibodies were produced, four of which recognize both forms A and B of receptor (named PR11, PR13, PR16, and PR22) and one of which recognizes only the larger progesterone receptor form B (PR6). Each of these antibodies binds nontransformed receptor complexes as well as monomeric receptor and has therefore proven very useful in purifying progesterone receptor.

Nontransformed receptor complexes are obtained to a high degree of purity from crude oviduct cytosol using a single-step immunoaffinity purification. We have indirectly precipitated receptor complexes from cytosol by incubating with one of our monoclonal antibodies followed by an incubation with either protein A-agarose or goat anti-mouse IgG-agarose. Also, we have covalently coupled antireceptor antibody directly to Actigel™ (Sterogene, San Gabriel, CA) or to immobilized protein A (Schneider et al., 1982; Logeat et al., 1985) to avoid contaminating the resin eluate with antibody.

One potential problem with the use of immunoaffinity purification of progesterone receptor is the difficulty in dissociating native receptor from antibody while avoiding denaturing receptor. For elution of receptor from antibody resin, we have tried high pH buffers or $2M$ sodium thiocyanate; both methods elute receptor from our antibodies, but in our experience the recovery of native receptor is low. Antibody PR6 has been used to isolate native progesterone receptor from human T47D cells (Estes et al., 1987); in this case, elutions either at pH 11.5 or with $1M$ thiocyanate were successful. For electrophoretic and Western analyses of receptor and associated polypeptides, we normally add SDS sample buffer to receptor immobilized on antibody resin and boil the mixture to completely dissociate receptor–antibody complexes.

A typical immune purification protocol for the isolation of nontransformed receptor complexes begins with the preparation of cytosol in PMM buffer (50 mM potassium phosphate, pH 7.0, 10 mM monothioglycerol, and 10 mM sodium molybdate). Sufficient antibody is added to cytosol to bind all receptor and incubated on ice for at least 2 h. Sufficient protein A or goat anti-mouse IgG resin to absorb all antibody is then added and the mixture is gently rocked for an additional 2 h at 4°C and then pelleted by low-speed centrifugation. The usual proportion of ingredients is 1 mL oviduct cytosol, 20 µg monoclonal antibody, and 50 µL protein A resin.

The resin is washed, by resuspension and centrifugation, in buffer that may contain salt or detergent to reduce nonspecific adsorption. In the following experiments protein A-Sepharose was used as the resin. After incubation with primary antibody-receptor complexes, the resin was washed three times with 50 mM Tris, pH 7.5, 10 mM thioglycerol, and 300 mM KCl followed by 2 washes in this buffer without KCl.

For electrophoretic analysis, the washed, pelleted resin is resuspended in an equal vol of 2X SDS sample buffer and diluted as needed with 1X SDS sample buffer. This mixture is placed in boiling water for 2 min and then loaded onto gels or stored frozen at –25°C.

Figure 5 is a Coomassie Blue-stained gel of immune-purified, nontransformed progesterone receptor; a different antibody was used to purify proteins as depicted in each lane. With thorough washing, the non-specific binding of proteins to resin alone (lane 1) is quite low, although silver staining does reveal a number of protein bands. Some of this background is caused by adsorption of proteins to the tube walls. This varies with the type of tube and can be minimized by transferring the preparation to a fresh tube between the washing steps. Monoclonal antibody PR22 (lane 2) specifically binds nontransformed complexes of the two receptor forms—110 and 79kDa— and the associated proteins hsp90 and p70. Also stained is the heavy chain (55kDa) of PR22, which dissociates from the secondary antibody resin in SDS buffer. The 52kDa band migrating just below PR22, heavy chain is seen consistently in immune-purified preparations of nontransformed receptor. The status of this protein as a component of nontransformed receptors is under further study. Immune-purification using PR13 covalently coupled to agarose (lane 3) isolates the same proteins, as seen with PR22, except that no antibody band is present. PR6 (lane 4), which binds only the 110kDa B form of receptor, also isolates the receptor-associated proteins hsp90 and p70. PR22 was used to subsequently purify receptor complexes from the PR6 unbound fraction (lane 5). Immune-isolation of hsp90 (lane 6) was performed on cytosol using AC88, a monoclonal antibody with a broad species specificity for hsp90 (Riehl et al., 1985). Similarly, p70 was immune-purified (lane 7) using monoclonal antibody N27 directed against hsp70 (Kost et al., 1989).

Western blot analysis of purified, nontransformed receptor complexes was used to verify the identity of protein components. Figure 6 presents this data using material immune-purified from oviduct cytosol

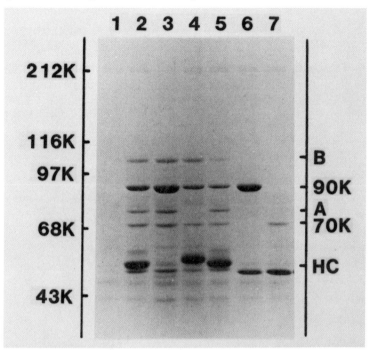

Fig. 5. Immune-purification of progesterone receptor complexes and components. Protein was isolated from oviduct cytosol by immune-precipitation followed by SDS gel elctrophoresis and Coomassie blue staining. One mL aliquots of cytosol were treated with 20 μg antibody plus 50 μL protein A-Sepharose, except for lane 3 where the antibody was covalently linked to resin. The antibodies used were: lane 1, resin control without antibody; lane 2, antireceptor PR22; lane 3, antireceptor PR13; lane 4, antireceptor PR6 (specific for receptor form B); lane 5, PR22 adsorption after extraction with PR6; lane 6, antibody 7D11 against hsp90; lane 7, antibody N27 against hsp70. The positions of receptor components and antibody heavy chain are indicated on the right.

with PR22 (lanes 1–3) or with a control antibody (EC1) having no specificity for progesterone receptor (lanes 4–6). The immunostained bands seen in each of the lanes at 50–55kDa and at the bottom of each lane are the heavy and light chains, respectively, of either PR22 or EC1. PR13 was used to identify progesterone receptor forms (lane 1), including B and A, major proteolytic fragments of receptor that migrate intermediate to B and A, and minor fragments migrating around 40kDa. D7a, a monoclonal against hsp90 (Brugge et al., 1983; Schuh et al., 1985), was used to identify the 90 kDa receptor component (lane 2), and N27 was used to identify p70 (lane 3). None of these proteins was detected in the control sample (lanes 4–6).

The ability to immobilize purified receptor complexes on a resin matrix can be exploited to readily test the effect of various treatments on the composition and stability of the complexes (Kost et al., 1989). As demonstrated in Fig. 7, hsp90 and p70 can be dissociated independently from

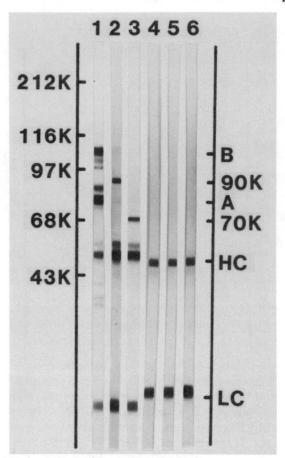

Fig. 6. Western blot analysis of immune-purified nontransformed progesterone receptor complexes. Oviduct cytosol was fractionated by immune-precipitation either with PR22 (lanes 1 through 3) or with EC1,a control antibody that has no chicke antigen (lanes 4 through 6). The immune isolated fractions were examined by Western blot analysis using PR13 (lanes 1 and 4), D7a against hsp90 (lanes 2 and 5), and N27 against hsp70 (lanes 3 and 6).

immobilized, nontransformed receptor complexes. Nontransformed receptor complexes were first purified from crude oviduct cytosol by incubation with PR22-protein A resin; resin was then washed thoroughly. Separate aliquots of immobilized receptor were warmed at 22°C for 30 min in the presence of resin buffer plus an added reagent. Following additional washes, the polypeptide components in receptor complexes were analyzed by SDS gel electrophoresis. The gel profile of immobilized receptor incubated with resin buffer alone (lane 1) is identical to the profiles of complexes incubated with molybdate (lane 2) or progesterone (lane 3). Incubating receptor complexes at elevated ionic strength caused a marked depletion of hsp90 (lane 4), whereas p70 was depleted by incubation with ATP (lane 5).

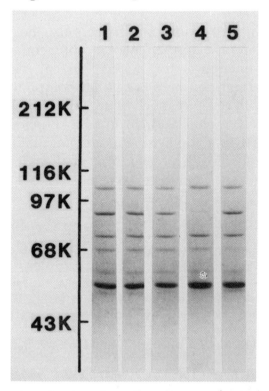

Fig. 7. Effects of various treatments on the recovery of proteins from immune-isolated receptor complexes. Following immobilization of nontransformed receptor complexes onto PR22:protein A resin, 50 μL of resin was rocked for 30 min at 22°C in 200 μL of 50 mM Tris, pH 7.5, 10 mM thioglycerol, and 5 mM MnCMgCl$_2$ plus one of the following additions: none (lane 1), 10 mM sodium molybdate (lane 2), $10^{-7}M$ progesterone (lane 3), 300 mM KCl (lane 4), or $10^{-4}M$ ATP (lane 5). After two washes in low salt buffer at 4°C, the components of receptor complexes were analyzed by SDS gel electrophoresis and Coomassie stain.

4. Conclusions

Avian progesterone receptor complexes have been highly purified by either of two protocols, the first relying primarily on the use of steroid affinity chromatography and the second using immunoprecipitation of receptor. With the availability of high affinity and high specificity monoclonal antibody, immunoprecipitation provides a much more rapid means of isolating receptor complexes than the multiple steps required for comparable purification by steroid affinity chromatography.

The composition of nontransformed chick progesterone receptor complexes is, at a qualitative minimum, either of the two steroid-binding receptor subunits, A or B, and hsp90. Also, a protein related to hsp70 is associated with purified receptor complexes. The association of hsp90 and p70 is sensitive to in vitro conditions. There is also the good possibility that RNA, found by others as a component of receptor complexes in vitro, plays a role in receptor function.

Certainly, there are numerous cellular components that are transiently associated with receptor, such as enzymes involved in phosphorylation, cytoskeletal elements, and chromatin. Elucidating these associations and reconstructing the synthetic and regulatory maze of steroid receptor mechanics will be a longstanding challenge.

References

Ali, M. and Vedeckis, W. V. (1987) *J. Biol. Chem.* **262**, 6771–6777.
Birnbaumer, M., Schrader, W. T., and O'Malley, B. W. (1983) *J. Biol. Chem.* **255**, 1637–1644.
Brugge, J., Yonemoto, W., and Darrow, D. (1983) *Mol. Cell. Biol.* **3**, 9–19.
Conneely, O. M., Dobson, A. D. W., Tsai, M.-J., Beattie, W. G., Toft, D. O., Huckaby, C. S., Zarucki, T., Schrader, W. T., and O'Malley, B. W. (1987a) *Mol. Endo.* **1**, 517–525.
Conneely, O. M., Maxwell, B. L., Toft, D. O., Schrader, W. T., and O'Malley, B. W. (1987b) *Biochem. Biophys. Res. Commun.* **149**, 493–501.
Dahmer, M. K, Quasney, M. W., Bissen, S. T., and Pratt, W. B. (1981) *J. Biol. Chem.* **256**, 9401–9405.
Dougherty, J. J., Puri, R. K, and Toft, D. O. (1984) *J. Biol. Chem.* **259**, 8004–8009.
Economidis, I. V. and Rousseau, G. G. (1985) *FEBS Lett.* **181**, 47–52.
Ennis, B. W., Stumpf, W. E., Gasc, J.-M., and Baulieu, E.-E. (1986) *Endocrinology* **119**, 2066–2075.
Estes, P. A., Suba, E. J., Lawler-Heavner, J., Elashry-Stowers, D., Wei, L. L., Toft, D. O., Sullivan, W. P., Horwitz, K. B., and Edwards, D. P. (1987) *Biochemistry* **26**, 6250–6262.
Evans, R. M. (1988) *Science* **240**, 879–895.
Freedman, L. P., Luisi, B. F., Korszun, Z. R., Basavappa, R., Sigler, P. B., and Yamamoto, K. R. (1988) *Nature* **334**, 543–546.
Grandics, P., Puri, R. K., and Toft, D. O. (1982) *Endocrinology* **110**, 1055–1057.
Grandics, P., Miller, A., Schmidt, T. J., Mittman, D., and Litwack, G. (1984) *J. Biol. Chem.* **259**, 3173–3180.
Gronemeyer, H., Turcotte, B., Quirin-Stricker, C., Bocquel, M. T., Meyer, M. E., Krozowski, Z., Jeltsch, J. M., Lerouge, T., Garnier, J. M., and Chambon, P. (1987) *EMBO J.* **6**, 3985–3994.
Groyer, A., Schweizer-Groyer, G., Cadepond, F., Mariller, M., and Baulieu, E.-E. (1987) *Nature* **328**, 624–626.
Isola, J., Pelto-Huikko, M., Ylikomi, T., and Tuohimaa, P. (1987) *J. Steroid Biochem.* **26**, 19–23.
Joab, I., Radanyi, C., Renoir, M., Buchou, T., Catelli, M.-G., Binart, N., Mester, J., and Baulieu, E.-E. (1984) *Nature* **308**, 850–853.
Kelley, P. M. and Schlesinger, M. J. (1982) *Mol. Cell. Biol.* **2**, 267–274.
Kost, S. L., Smith, D., Sullivan, W., Welch, W. J., and Toft, D. O. (1989) *Mol. Cell. Biol.* **9**, 3829–3838.
Logeat, F., Le Cunff, M., Pamphile, R, and Milgrom E. (1985) *Biochem. Biophys. Res. Commun.* **131**, 421–427.

Logeat, F., Pamphile, R., Loosfelt, H., Jolivet, A., Fournier, A., and Milgrom, E. (1985) *Biochemistry* **24**, 1029–1036.

Mendel, D. B. and Orti, E. (1988) *J. Biol. Chem.* **263**, 6695–6702.

Mendel, D. B., Bodwell, J. E., Gametchu, B., Harrison, R. W., and Munck, A. (1986) *J. Biol. Chem.* **261**, 3758–3763.

Nishigori, H. and Toft, D. O. (1980) *Biochemistry* **19**, 77–83.

Puri, R. K. and Toft, D. O. (1986) *J. Biol. Chem.* **261**, 5651–5657.

Puri, R. K., Grandics, P., Dougherty, J. J., and Toft, D. O. (1982) *J. Biol. Chem.* **257**, 10831–10837.

Renoir, J.-M., Buchou, T., Mester, J., Radanyi, C., and Baulieu, E.-E. (1984) *Biochemistry* **23**, 6016–6023.

Riehl, R. M., Sullivan, W. P., Vroman, B. T., Bauer, V. J., Pearson, G. R, and Toft, D. O. (1985) *Biochemistry* **24**, 6586–6591.

Sanchez, E. R, Meshinchi, S., Tienrungroj, W., Schlesinger, M. J., Toft, D. O., and Pratt, W. B. (1987) *J. Biol. Chem.* **262**, 6986–6991.

Schneider, C., Newman, R. A., Sutherland, D. R., Asser, U., and Greaves, M. F. (1982) *J. Biol. Chem.* **257**, 10766–10769.

Schuh, S., Yonemoto, W., Brugge, J., Bauer, V. J., Riehl, R. M., Sullivan, W. P., and Toft, D. O. (1985) *J. Biol. Chem.* **260**, 14292–14296.

Sullivan, W. P., Beito, T. G., Proper, J., Krco, C. J., and Toft, D. O. (1986) *Endocrinology* **119**, 1549–1557.

Sullivan, W. P., Vroman, B. T., Bauer, V. J., Puri, R. K, Riehl, R. M., Pearson, G. R., and Toft, D. O. (1985) *Biochemistry* **24**, 4214–4222.

Sullivan, W. P., Smith, D. F., Beito, T. G., Krco, C. J., and Toft, D. O. (1988) *J. Cell. Biochem.* **36**, 103–119.

Tai, P.-K. K, Maeda, Y., Nakao, K., Wakim, N. G., Duhring, J. L., and Faber, L. E. (1986) *Biochemistry* **25**, 5269–5275.

Tora, L., Gronemeyer, H., Turcotte, B., Gaub, M.-P., and Chambon, P. (1988) *Nature* **333**, 185–188.

Tymoczko, J. L., Anderson, E. E., Lee, K. A., and Unger, A. L. (1987) *Biochim. Biophys. Acta* **930**, 114–121.

Wei, L. L., Sheridan, P. L., Krett, N. L., Francis, M. D., Toft, D. O., Edwards, D. P., and Horwitz, K. B. (1987) *Biochemistry* **26**, 6262–6272.

Weinberger, C., Hollenberg, S. M., Rosenfeld, M. G., and Evans, R. M. (1985) *Nature* **318**, 670–672.

Purification
and Characterization
of Chicken Progesterone
Receptor

Nancy L. Weigel

1. Introduction

The chicken progesterone receptor contains two hormone binding components, PRa (M_r = 72,000) and PRb (M_r = 86,000) (Schrader and O'Malley, 1972), which are produced from the same transcript by initiation of translation at alternate sites (Conneely et al., 1987). PRa is, therefore, a truncated version of PRb. Although both proteins can induce cotransfected reporter genes (Carson et al., 1987; Tora et al., 1988), it is not yet known whether these proteins have different functions in vivo.

In the absence of progesterone, PRa and PRb are found in cytosol prepared from chicken oviduct and are each associated with a heat shock protein, hsp90 (Dougherty et al., 1984) and possibly other nonhormone binding components. These complexes sediment as 8S complexes on sucrose gradients and are referred to as 8S complexes. Treatment of cytosol with salt, heat, or ammonium sulfate dissociates the complexes; the monomer hormone binding components PRa and PRb sediment as 4S components on sucrose gradients (Schrader and O'Malley, 1972). Injection of chickens with progesterone results in a redistribution of receptor with a portion of the PRa and PRb binding tightly to nuclear components. However, immunocytochemical studies suggest that the receptor is always localized in the nucleus (Isola et al., 1986; Ennis et al., 1986). Thus, it is not clear whether the 8S receptor complexes exist in vivo and play a role in receptor function. Whether or not these complexes exist in vivo, their existence in cytosol can be utilized to purify the individual progesterone binding components.

Oviducts from both immature and mature or diethylstilbestrol (DES) treated animals contain progesterone receptor. However, the oviducts in immature untreated animals are too small (about 10 mg) to be a practical source of receptor for purification. Moreover, the concentration of progesterone receptor is substantially lower. The oviducts of mature hens and DES-stimulated chicks contain approx equal amounts of PRa and PRb when analyzed by immunoblotting, but the oviducts of immature untreated animals appear to contain more PRb than PRa (Bingman et al., 1988). This chapter describes some of the methods used in our laboratory for the purification and analysis of the chicken progesterone receptor PRa and PRb proteins.

2. Purification Procedures

2.1. General Methods

2.1.1. Tissue Sources

Oviducts from 5–7-wk-old female White Leghorn chicks that have been injected daily (beginning at 3 wk of age) with 5 mg of diethylstilbestrol in sesame oil, as described (Schrader and O'Malley, 1972), or injected weekly with 20 mg of DES in polyethylene glycol (prepared in cartridges by Mattox and Moore, Inc., Indianapolis, IN) (Coty et al., 1979) are used as a source of receptor that does not contain endogenous hormone. Oviducts from these chicks typically weigh between 1 and 2 g.

Oviducts from laying hens (typically about 20 g each) can also be used as a source of receptor for some purposes. However, receptor concentrations tend to be lower and some of the receptor contains endogenous progesterone.

2.1.2. Buffers and Resins

Buffer A (10 mM Tris pH 7.5, 1 mM EDTA, 12 mM thioglycerol) with the indicated concentration of NaCl is used in most purification steps. The original purification procedures utilized KCl. However, NaCl is more convenient for direct analysis of samples by SDS gel electrophoresis since

NaSDS is soluble and there is no detectable difference in the purification protocol. Buffer B (10 mM N-2 Hydroxyethylpiperazine-N'-2-ethanesulfonic acid (HEPES) pH 7.9, 0.1 mM EDTA, 10 mM DTT, and 20% glycerol) is used for final purification steps if the receptor is to be stored frozen.

Whatman P-11 (phosphocellulose) is precycled using NaOH and HCl, as recommended by the manufacturer, and equilibrated with Buffer A. Whatman DE52 (DEAE cellulose) is prepared as recommended and equilibrated with Buffer A.

DNA cellulose is prepared by a modification of the procedure of Alberts and Herrick (1971). Calf thymus DNA from Worthington Diagnostic Systems Inc., Freehold, NJ is dissolved in 15 mM NaCl, 1.5 mM Na citrate pH 7.0 at 0.75 mg/mL and mixed with washed Cellex N-1 from Bio-Rad (about 100 g dry resin/g of DNA). The mixture is allowed to dry and lumps of resin are broken up using a mortar and pestle. The resin is suspended in 95% ethanol and stirred gently during the irradiation step. The resin is irrradiated for 1 h using a General Electric 85 W mercury vapor arc lamp placed 2–3" above the surface of the resin slurry. The resin is then washed with 1M KCl followed by Buffer A and stored either at 4°C in Buffer A containing 0.04% Na azide or kept as a frozen suspension. Bound DNA concentrations of greater than 2 mg DNA/g resin can be obtained. The resin can be reused several times if it is washed immediately after each use with buffer containing 1M NaCl to remove contaminating proteins.

2.1.3. SDS Gel Analysis and Immunoblotting

Slab gel electrophoresis is performed as described by Laemmli (1970) using gels containing 7.5% acrylamide, 0.2% bisacrylamide, and 0.1% SDS. Gels are either stained with a silver stain as described by Wray et al. (1981) or the protein is transferred to nitrocellulose using an LKB Novablot apparatus following the manufacturer's directions. Receptor is detected as described (Conneely et al., 1986) using primary antibody, rabbit anti-mouse IgG, and [125I] Protein A.

2.2. Purification of Receptor Using Conventional Chromatography

This protocol takes advantage of the differences in chromatographic properties of the 8S receptor complexes and the separate hormone binding components. The procedure takes 3 d and yields highly purified PRa and somewhat less purified PRb. The receptors retain both hormone binding and DNA binding activity and are not subjected to denaturing conditions. These preparations are used as a source of receptor for functional assays including in vitro transcription studies.

2.2.1. Preparation of Oviduct Cytosol

DES treated chicks are sacrificed by cervical dislocation and the oviducts removed and immediately placed in ice cold 0.9% NaCl. Typically, 50 chicks are used, resulting in 70–80 g of tissue. All other steps unless indicated are performed at 4°C. The oviducts are blotted with paper towels, and any mesentary and shell gland is removed and discarded. The oviducts are always used immediately. Although frozen tissue retains its hormone binding, the receptor cannot, in our hands, be successfully purified from frozen tissue. The oviducts are minced with scissors or passed through a meat grinder and suspended in 4 mL of Buffer A / g of tissue. The tissue is homogenized using a Polytron (Brinkmann Instruments, PT-10). Five or six 5-s bursts are used with 30 s of cooling in an ice water bath between each burst. Bits of tissue caught in the probe should be removed during the 30 s cooling periods to ensure good homogenization. The homogenate is centrifuged at 10,000g (usually in a Beckman JA14 rotor at 10,000 rpm) for 15 min. The floating fat layer is carefully removed by aspiration and the remaining supernatant is centrifuged in a Beckman Ti45 rotor at 105,000 g for 1 h. Floating fat is again removed by aspiration.

2.2.2. Phosphocellulose and DNA Cellulose Chromatography

The 8S receptor complexes found in oviduct cytosol do not interact either with phosphocellulose or with DNA cellulose, whereas the dissociated hormone binding components PRa and PRb do. Therefore, the cytosol is quickly passed through these columns in order to remove the majority of the phosphocellulose and DNA binders found in cytosol. Although only a small purification results at these steps, these are the proteins that otherwise would copurify with the receptor in later steps. Although molybdate has been found to stabilize the 8S cytosolic complexes, these first steps are carried out quickly and molybdate is not required. The cytosol is applied to a phosphocellulose column in Buffer A (1 mL resin/1 mL cytosol). The flow through and one column of wash is collected. These are then applied to a DNA cellulose column equilibrated in Buffer A (1 mL resin/1 mL original cytosol) and the flow through and 1 column vol of wash is collected.

The flow through is labeled overnight with 30 nM [^{3}H] progesterone. Since the [^{3}H] progesterone is sold dissolved in organic solvent, it must first be dried under a stream of nitrogen and redissolved in ethanol prior to addition. The final concentration of ethanol in the receptor preparation must be kept at a minimum (0.1–0.2%) to avoid denaturing the hormone binding site. The receptor preparation must be carried through this step on the first day, since storage in a more dilute form in the absence of hormone results in a reduced yield. Although we typically label overnight for convenience, 2 h at 4°C is sufficient to bind the hormone, and the preparation can be continued after 2 h. However, we have found that the labeled cytosol is the most stable form of the receptor, and we find that we obtain higher yields if the preparation is stopped at the labeling step rather than after the ammonium sulfate fractionation.

The phosphocellulose and DNA cellulose columns are washed immediately in Buffer A containing $1M$ NaCl and can then be reused after reequilibration with Buffer A.

2.2.3. Ammonium Sulfate Fraction

Saturated ammonium sulfate in Buffer A adjusted to pH 7.5 is added to give a final concentration of 40%. The solution is stirred for 15 min and centrifuged for 1 h at 3400g. Because of the large volumes at this step (as much as 2 L), the sample is usually centrifuged in a Beckman J6B centrifuge at 4000 rpm. The bottles are inverted and wiped with kimwipes to remove excess solution. The pellets are resuspended in Buffer A (typically 1 mL/g of original tissue) and centrifuged at 8000g (a Beckman JA20 rotor at 10,000 rpm) for 15 min to remove any particulate material. The clarified supernatant is diluted with Buffer A to a conductivity equivalent to $0.1M$ NaCl in Buffer A. This step concentrates the receptors, dissociates the 8S complexes so that the hormone binders will bind to DNA cellulose and phosphocellulose, and provides some purification.

2.2.4. DNA Cellulose Chromatography

The diluted supernatant is applied at 1 mL/min to a DNA cellulose column equilibrated in Buffer A containing $0.1M$ NaCl (1 mL resin/8 g tissue). The column is washed with 5 column volumes of the same buffer. The flow through and wash contain most of the PRb and are purified separately as described below. The DNA cellulose column is eluted with Buffer A containing $0.3M$ NaCl. Typically, 10 fractions containing a total of four column volumes of eluate are collected and 10 µL of each fraction is counted for [³H] progesterone. This eluate contains only a small amount of protein (*see* Table 1), since most of the DNA binders were removed by the previous DNA cellulose column. The PRa is usually kept at this stage until the next day and then the preparation is completed.

2.2.5. Purification of DNA Binding Receptor

The $0.3M$ NaCl eluate contains predominantly PRa but some of the PRb also binds to the column. Although both PRa and PRb bind well to specific DNA (*see* DNA binding assay below), much of the PRb does not bind to DNA at this step (Peleg et al., 1988). The PRa and PRb are separated by chromatography using DEAE cellulose, as described below.

Table 1
Purification of PRa from Chicken Oviducts[a]

Step	Total protein,[b] mg	Total receptor,[c] μg	μg Receptor/ mg protein	Yield, %	Purification, fold
Cytosol	7250	467	0.065	100	1.0
PCDT	5500	288	0.052	62	0.8
DNADT	4681	264	0.056	56	0.9
Ammonium sulfate	310	300*	0.97	64	15
DNA pool	0.288	102*	354	22	5440
DNA final	0.068	40*	588	8.5	9050

[a]Samples were assayed using a charcoal assay. Samples with* were counted directly for [³H]progesterone. Table is from Weigel et al., 1989; courtesy of Houston Biological Associates.
[b]Determined using Bradford Assay.
[c]Based on one hormone binding site/molecule.

2.2.5.1. DEAE Cellulose Chromatography. Since highly purified PRa appears to aggregate or to denature in low ionic strength, it is not possible to dilute the DNA eluate sufficiently to allow direct binding and elution of PRa from DEAE in good yield. Therefore, the DNA eluate is diluted to a conductivity equivalent to $0.15M$ NaCl in Buffer A and applied to a DEAE column equilibrated in the same buffer (1 mL resin/80 g tissue). Under these conditions, the PRa will not bind but PRb binds well. The column is washed with 2–5 column volumes of Buffer B and eluted with Buffer B containing $0.3M$ NaCl. The eluate contains any PRb that bound to the DNA column.

2.2.5.2. Concentration of PRa Using DNA Cellulose Chromatography. The $0.15M$ NaCl flowthrough and wash of the previous DEAE column contains most of the PRa. Typically, this pool is diluted to a conductivity equivalent to Buffer A containing $0.1M$ NaCl and applied to a small (0.1–0.2 mL/80 g tissue) DNA cellulose column equilibrated in the same buffer. The column is washed with Buffer B and eluted with Buffer B containing $0.3M$ NaCl. This provides a concentrated PRa preparation (typically, 10–50 μg/mL) suitable for DNA binding and other experiments. Somewhat better yields may be obtained with a larger column, but we have found it to be more convenient to have concentrated preparations that can simply be diluted to appropriate salt concentrations as needed.

The purification of PRa is summarized in Table 1. The hormone binding assays for the unlabeled samples—cytosol, phosphocellulose dropthrough (PCDT), and DNA dropthrough (DNADT) were performed by incubating aliquots of the fraction in 200 μL of Buffer A containing 10 nM

[³H]progesterone with or without 100-fold excess cold progesterone. The samples were then assayed using a charcoal assay, as described (Schrader and O'Malley, 1972). The purification protocol is a modification of a protocol previously described by Coty et al. (1979). Our protocol results in a somewhat higher yield of purified receptor. Moreover, since theoretical purity is 1000 µg/mg protein, this preparation at 588 µg/mg is about 60% receptor based on hormone binding. If any of the receptor has lost its hormone, then the preparation is more highly purified. In contrast, the protocol of Coty et al. (1979) resulted in a preparation containing 16% receptor by hormone binding. Thus, this protocol produces receptor that retains its hormone binding as well as its DNA binding activity.

2.2.6. Purification of PRb

2.2.6.1. Phosphocellulose Chromatography. The DNA flowthrough (*see* Section 2.2.5.) containing PRb and usually a small amount of PRa is applied directly to a phosphocellulose column equilibrated with Buffer A containing 0.1*M* NaCl (1 mL resin/8 g tissue). The column is washed with 3–5 column volumes of the same buffer and eluted with a linear gradient (16 column volumes total from 0.1*M* NaCl in Buffer A to 0.5*M* NaCl in Buffer A). PRb elutes at about 0.27*M* NaCl. Aliquots (10 µL) of each sample are counted and the peak fractions pooled. The PRb is concentrated and any remaining PRa removed using a small DEAE Cellulose column.

2.2.6.2. Concentration of PRb by DEAE-Cellulose Chromatography. The phosphocellulose pool is diluted to a conductivity equivalent to Buffer A containing 0.1*M* NaCl and applied to a small DEAE column (0.1 mL resin/80 g tissue) equilibrated in the same buffer. The column is washed with Buffer B and eluted with Buffer B containing 0.15*M* NaCl to elute any remaining PRa. PRb is then eluted using Buffer B containing 0.3*M* NaCl. PRb is typically about 5–10% of the total protein in this type of preparation (*see* Section 2.2.7).

2.2.7. Characterization of Purified Receptor Fractions

Table 1 shows a typical purification of receptor. Because the protein concentrations of the final receptor fractions are quite low, a more accurate assessment of the purity of the receptor preparations can be obtained by examining SDS gels that have been stained with silver. Figure 1 shows typical PRa and PRb preparations. The PRa fraction contains a major band at the mol wt of PRa as well as a few minor bands. In contrast, the PRb routinely contains a number of contaminants in addition to the PRb. These preparations have also been characterized by immunoblotting. Two proteins, hsp70, and a protein termed A antigen that was originally thought to be PRa (Compton et al., 1983) have mol wts similar to the mol wt of PRa

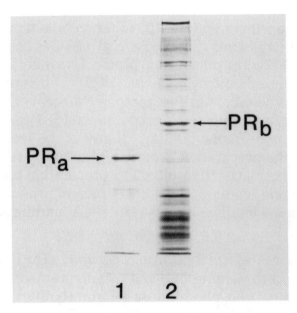

Fig. 1. SDS gel electrophoresis of purified PRa and PRb. PRa (0.22 µg) and PRb (0.27 µg) were analyzed by SDS gel electrophoresis and the proteins detected by silver staining. From Weigel et al., 1989; courtesy of Houston Biological Associates.

and also bind DNA. Therefore, we have analyzed the receptor preparations for these proteins as well. Figure 2 shows immunoblot analysis of receptor preparations with monoclonal antibodies to receptor. The left panel shows the reaction of the receptor preparations with PR22, a monoclonal antibody that recognizes both PRa and PRb (Sullivan et al., 1986) that has been generously provided by David Toft. The PRa preparation does not contain PRb. In contrast, a small amount of PRa appears to be present in the PRb preparation. The right panel shows the reaction of the same PRa and PRb preparations with PR6, a monoclonal antibody that recognizes only PRb (Sullivan et al., 1986). PRa contains a small amount of a proteolytic fragment of PRb that still contains the PRb epitope. Mapping studies have shown that this epitope resides just to the amino-terminal side of the initiation site for PRa in PRb. In contrast, the PRb does not contain any of this fragment. Therefore, the lower mol wt immunoreactivity seen in the PRb preparation probed with PR22 is apparently authentic PRa. Anti-hsp 70 (Riabowol et al., 1988) (N27F3-4, graciously provided by William J. Welch, Departments of Medicine and Physiology, University of California, San Francisco) reacts with chicken hsp70 found both in chicken cytosol and in salt extracts of nuclei (Fig. 3), but is not detectable in either the purified PRa or PRb. Thus hsp70 is either absent or very scarce in these purified

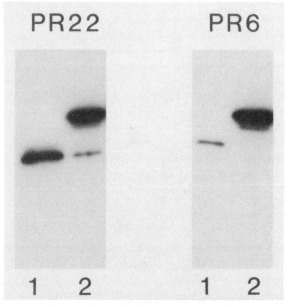

Fig. 2. Immunoblot analysis of PRa and PRb. PRa and PRb were separated by SDS gel electrophoresis, transferred to nitrocellulose, and probed with PR22 (left panel) or PR6 (right panel). Lane 1, 0.11 µg PRa; Lane 2, 0.14 µg PRb. From Weigel et al., 1989; courtesy of Houston Biological Associates.

preparations. Figure 4 shows the reaction of similar amounts (by silver staining) of A Ag and PRa with anti-A-Antigen polyclonal antibody 6261 prepared by Orla Conneely and specific for this protein (personal communication) (Panel A). There is very little antigen detected in the PRa fraction and none detectable in the PRb fraction (not shown). In contrast, purified A antigen produces a strong signal. The minor lower mol wt bands are a result of proteolysis of A antigen and increase in intensity with the age of the preparation. The A antigen itself is free of PRa (Panel B).

The receptor is quick frozen in a dry ice-ethanol bath and thawed as needed. The properties of this receptor are indistinguishable from fresh receptor although repeated freezing and thawing does cause gradual inactivation. In order to test the stability of the hormone binding and the effect of freezing on receptor aggregation, the receptor was frozen and then thawed. Figure 5 shows that PRa that is frozen and then thawed immediately retains its hormone binding when analyzed by size exclusion chromatography and does not appear to aggregate. It is likely that the combination of relatively high ionic strength (typically 0.2–0.3M NaCl) and glycerol is stabilizing these activities. This receptor is suitable for DNA binding studies, as described below.

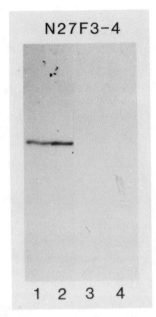

Fig. 3. Analysis of oviduct samples for hsp70. The indicated samples were subjected to SDS gel electrophoresis, transferred to nitrocellulose, and hsp70 was detected with N27F3-4. Lane 1, 10 μL of chicken oviduct cytosol; Lane 2, 10 μL chicken oviduct salt extract; Lane 3, 0.94 μg PRa; Lane 4, 0.33 μg PRb.

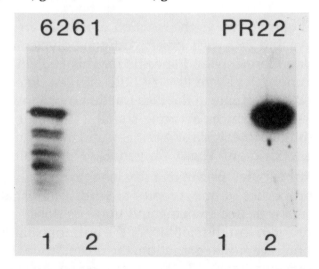

Fig. 4. Analysis of PRa for A Antigen. Equal amounts of A antigen and PRa (about 0.2 μg) were subjected to SDS gel electrophoresis, transferred to nitrocellulose, and probed with antibodies. Left Panel, polyclonal antibody 6261, which is specific for A antigen. The film was exposed for 2 h. Lane 1, A antigen; Lane 2, PRa. Right Panel, PR22. Lane 1, A antigen; Lane 2, PRa. The right panel was exposed overnight in order to detect any trace contamination of A Ag by PRa.

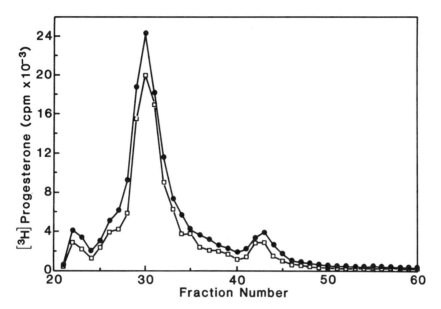

Fig. 5. TSK 3000 Size Exclusion Analysis of PRa. An aliquot of PRa (50 μL) was diluted with 150 μL of Buffer A containing 0.3*M* NaCl and applied to a GlasPac TSK 3000 column equilibrated in Buffer A containing 0.2*M* NaCl. A second aliquot was frozen in a dry ice ethanol bath, thawed, and diluted. The column was run at 0.5 mL/min and 0.5 min fractions were collected. The amount 0.2 mL of each fraction was counted for [³H]progesterone; (●) control receptor; (□) frozen receptor.

2.3. Purification of PRa and PRb Using Immunoaffinity Columns

The protocol described above is useful for preparing active receptor. However, the yields are low and the procedure takes 3 d. Moreover, the procedure is only useful for receptor that is in 8S complexes in the initial purification steps. Immunoaffinity purification can be used to rapidly and efficiently obtain purified protein for protein sequencing, peptide mapping, and even for some functional studies, such as DNA binding.

2.3.1. Preparation of Immunoaffinity Columns

Resin for immunoaffinity purification is prepared by the method of Schneider (1982). Antibody (7 mg/mL of resin) is incubated with Protein A Sepharose and crosslinked using dimethylpimelimidate. If the primary antibody binds poorly to Protein A Sepharose, a bridging antibody such as rabbit anti-mouse or rabbit anti-rat IgG can be used. This protocol offers several advantages over the use of CNBr activated sepharose. First, the Protein A Sepharose can be used as a purification step. The IgG binds tightly but not covalently to the Protein A Sepharose and can be washed extensively with solutions containing salt or small amounts of detergent to remove contaminating enzymes in the IgG preparation prior to covalent crosslinking. Second, the IgG is oriented by the specific binding of the Protein A Sepharose, which leaves the antigen binding sites free. A limitation of this resin as well as other immunoaffinity resins is its capacity for antigen. In our experience, only a portion of the IgG (about 20%) purified from ascites preparations of PR22 reacts with receptor, yet all of the IgG interacts with the Protein A Sepharose. Since the excess IgG does not recognize other antigens even in crude preparations, it does not reduce the specificity of the column, but does reduce the potential capacity. Moreover, the crosslinking procedure appears to inactivate a portion of the IgG, presumably by reacting with active regions of the IgG. Using PR22 as the primary antibody, we have been able to obtain immunoaffinity resins with capacities for receptor up to 100 µg/mL of resin. The resin can be reused numerous times if appropriate precautions are taken.

2.3.2. Purification of PRa and PRb

Cytosol is prepared as described for the conventional purification of receptor except that protease inhibitors are added prior to loading the cytosol on the immunoaffinity column.

Mix 1: 1 mg/mL of each of the following components in dimethylsulfoxide—leupeptin, antipain, aprotinin, and benzamidine HCL.

Mix 2: 1 mg/mL chymostatin and pepstatin in dimethylsulfoxide.

One µL of each mix and 2.5 µL of 0.2M PMSF in dioxane is added to each mL of cytosol. The receptors themselves appear to be reasonably stable in the absence of the protease inhibitors, but the inhibitors appear to prolong the lifetime of the immunoaffinity columns. We have also found that omitting the ultracentrifugation step in the preparation of the cytosol greatly decreases the usable lifetime of the columns.

The immunoaffinity column should be run with a Protein A Sepharose precolumn that can be linked in tandem with the immunoaffinity column. This column will bind nonspecific proteins and remove particulate matter that was not removed by centrifugation. Although a number of investigators also use a second precolumn containing an irrelevant IgG, we have not found this to be necessary to prepare highly purified receptor.

The receptors are typically applied in Buffer A at a rate of 5–10 column volumes/h and washed with 10 column volumes of Buffer A containing 0.5M NaCl. The columns are then disconnected and the immunoaffinity column is washed with the following solutions:

1. 10 column volumes of Buffer A containing 0.5M NaCl;
2. 10 column volumes of Buffer A;
3. 20 column volumes of IPB (0.5M NaCl, 50 mM Tris pH 7.5, 0.02% sodium azide, and 0.2% Triton X 100);
4. 10 column volumes of Buffer A;
5. 10 column volumes of H$_2$O.

The washes can all be performed at the maximum flowrate of the columns. The column is then eluted with 10 column volumes of 0.1M glycine-HCl pH 2.5 and the eluate collected as 1 mL fractions in silanized Eppendorf tubes. Sixty μL of 1M Tris base is immediately added to each fraction to neutralize the sample. Alternatively, the column can be eluted with 1M acetic acid and the eluates dried. We typically use the acetic acid elution procedure for preparative protein chemistry procedures and the glycine-HCl procedure for direct gel analysis and for functional studies.

The Protein A Sepharose column should also be eluted with acid and both columns are washed with 20 column volumes of Buffer A containing 0.5M NaCl. If the columns are used infrequently, 0.02% sodium azide should be included in the final wash. The purification procedure should be completed in one day or at least completed through the IPB wash step. Prolonged contact of the resin with cytosol results in loss of column capacity.

The eluted protein contains predominantly PRa and PRb (Fig. 6). The protein does not bind hormone because an acid elution step was used. However, the protein does bind DNA specifically and can be used in the gel retardation assays described below. Other elution procedures can be used (NaSCN or alkaline pH) that will preserve hormone binding, but we have not found a procedure that gives a good yield of receptor. Elution procedures are antibody specific, and it is likely that a lower affinity antibody would prove to be more useful for immunopurification.

2.4. Other Purification Methods

Neither of the procedures described above yield highly purified native PRb. We routinely use the affinity chromatography method described by Renoir et al. (1984), which is discussed elsewhere in this book, to prepare more highly purified PRb. We find, however, that an additional phosphocellulose column chromatography step performed as described above for PRb purification results in a substantially more purified preparation of PRb.

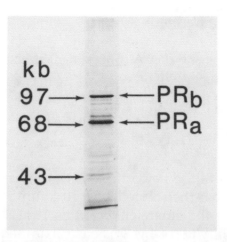

Fig. 6. SDS gel analysis of immunopurified receptor. A portion of the PR22 column eluate was subjected to SDS gel electrophoresis and the proteins were detected using a silver stain. From Weigel et al., 1989; courtesy of Houston Biological Associates.

3. HPLC Analyses of Progesterone Receptor

High-performance liquid chromatography is a very powerful technique for the purification of proteins. However, although a number of laboratories have been successful in analyzing crude receptor preparations (Pavlik et al., 1982; Hutchens et al., 1984; Wiehle et al., 1984), attempts to chromatograph partially purified receptor were unsuccessful because the yields of receptor were extremely low. We have examined a number of HPLC systems as well as numerous columns and have found that the stainless steel columns appear to be responsible for the low yields of receptor (Denner et al., 1987).

3.1. Preparation of Partially Purified Receptor

Hen oviduct progesterone receptor was purified, as described previously (Weigel et al., 1981), except that the first DNA cellulose column was omitted. Fractions from the DNA column were used for analyses and contained PRa that was about 2% of the total protein. Chicken oviduct receptor was purified by a modification of the procedure described previously (Coty et al., 1979). PRa prior to DEAE cellulose chromatography was used and was about 200-fold purified. PRb from the phosphocellulose column was about 1000-fold purified and about 10% of the total protein.

3.2. HPLC Systems and Columns

Three HPLC systems were compared in this study. The LKB Ultro-Chrome glass/titanium system (GTi) with a 2152 controller and 2150 pump with dual titanium heads and the LKB stainless steel system with a

2152 controller and a 2150 pump with dual stainless steel heads were virtually identical except that the metals were different. A Beckman HPLC with a model 421 controller and dual 110A stainless steel pumps was also used. All three HPLC systems were used in a cold room maintained at 4°C.

The 3000 SW size exclusion columns all contained the same resin, a proprietary silica based resin prepared by Toya Soda. This resin will dissolve in alkaline conditions and should be exposed only to a pH lower than pH 8. The GlasPac column is made of glass with inert frits and is connected with inert tubing. The stainless steel columns have stainless steel frits and stainless steel connecting tubing. The PW series of columns are polymer based and are in stainless steel columns. All columns were run in Buffer A containing 0.2M NaCl at 0.5 mL/min. Table 2 shows that good yields of PRa can be obtained from the GlasPac column, but not from any of the stainless steel columns. Although the receptor was detected by hormone binding in this experiment, similar experiments using immunoassays show that the receptor is not simply denatured but appears to bind to the column. The receptor can be eluted as denatured protein by washing the column with 0.1% SDS. Although the recoveries from the stainless steel HPLC systems were somewhat lower than the recovery from the titanium system, it is clear that the choice of column support is crucial. Similar comparisons of DEAE columns in glass and stainless steel indicate that the recovery from glass columns is again better (95%) compared to the stainless steel column (45%).

3.3. Analysis of Receptor Forms

3.3.1. Preparation of Receptor Samples

HPLC columns are easily clogged by small particles and frequently cannot be repaired. Many HPLC columns are used with special precolumns to minimize this problem. Since GlasPac precolumns are not available, great care must be taken to remove insoluble material from receptor samples. Although this is not a problem with more purified samples, cytosols and ammonium sulfate fractions frequently contain insoluble material. Samples are routinely centrifuged for 15 min at 8000g prior to injection. Samples containing high protein concentrations can be chromatographed successfully, but these columns are limited to 0.5 mL of sample/analysis to obtain good resolution.

3.3.2. Resolution of PRa and PRb

Chick receptor was prepared as described above. Samples of PRa (closed circles) and PRb (open circles) were chromatographed on the GlasPac TSK3000 column (Fig. 7 panel A) and the receptor was detected by

Table 2
Recovery of Hen Progesterone Receptor A
from Size-Exclusion HPLC on Different Columns and Systems[a]

Column type	Distributor	Resin	HPLC system		
			LKB glass/ titanium[b]	LKB stainless steel[c]	LKB stainless steel[d]
GlasPac Stainless steel	LKB	TSK 3000 SW	80	64	56
	LKB	TSK 3000 SW	<5	<5	<5
	Altex	TSK 3000 SW	<5	<5	<5
		TSK 4000 SW	<5		<5
		TSK 4000 PW	<5		
		TSK 5000 PW	<10	<5	<5

[a]Values represent percentage recovery of cpm in receptor peak relative to cpm injected. Table is from Denner et al., 1987; courtesy of Academic Press.
[b]LKB UltroChrome glass/titanium system with 2152 controller and 2150 pump with dual titanium heads.
[c]LKB stainless-steel system with 2152 controller and 2150 pump with dual stainless-steel heads.
[d]Beckman stainless-steel system with 421 controller and dual 110A stainless-steel pumps.

counting aliquots of the fractions for [³H]progesterone. This column binds free progesterone, and, consequently, a peak of free progesterone is not seen in these profiles. Panel B shows the chromatography of a mixture of PRa and PRb. Panel C shows the Stokes' radii of PRb and PRa determined by comparison with the indicated standards. This column provides better resolution of PRb and PRa than can be obtained with conventional size exclusion chromatography. Moreover, the procedure is rapid since each analysis takes approx 30 min.

3.3.3. Effect of Ionic Strength on Size Exclusion Chromatography

Resins for HPLC are subjected to substantially higher pressures than are resins used for conventional chromatography. This limits the choice of matrices used and the resins used for HPLC tend to be less inert with respect to interaction with proteins. Panel A of Fig. 8 shows that the elution position of PRa from the GlasPac 3000 column is very dependent on the ionic strength of the buffer used. In Buffer A alone, the receptor elutes at the void vol suggesting that the receptor is aggregated. However, increasing the salt concentration to 0.1M resulted in a very different elution pattern. The receptor appears to interact with the matrix and elutes near the

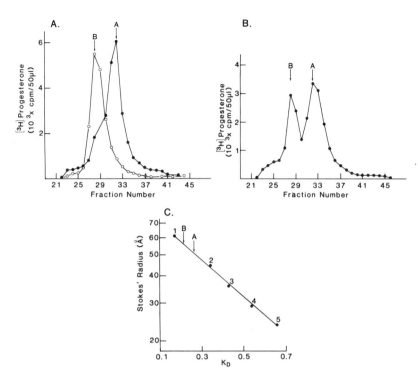

Fig. 7. Resolution and Stokes' radii of receptors PRa (A) and PRb (B) by glass column size-exclusion HPLC. Chicken oviduct receptors labeled with [³H] progesterone were applied to a GlasPac TSK 3000 SW column and run as described in Fig. 5. **A.** Preparation of PRa (●–●) and PR b (O–O) were analyzed in separate runs. **B.** PRa and PRb were mixed and chromatographed. **C.** The log of the apparent Stokes' radius for several standard proteins was plotted against K_d #1, ferritin; 2, aldolase; 3, bovine serum albumin; 4, ovalbumin; 5, soybean trypsin inhibitor. Apparent Stokes' radii for PRa and PRb are indicated by arrows. From Denner et al., 1987; courtesy of Academic Press.

total vol. At 0.2*M* NaCl and above the receptor elutes somewhat after PRb, as would be expected. The change in elution appears to be caused by interaction with the column rather than irreversible changes in receptor structure. Receptor that has been chromatographed at 0.1*M* NaCl will elute at the position expected for receptor chromatographed at higher ionic strength if it is rechromatographed at higher ionic strength. This change in elution position can be used to further purify proteins. However, the effect of ionic strength on elution position should be analyzed carefully before attempting to determine the Stokes radius of an unknown protein. Panel B of Fig. 8 shows similar studies with PRb. It also appears to aggregate in the absence of added salt but does not show the unusual interaction with the column at 0.1*M* NaCl. Analyses of the standard proteins indicated

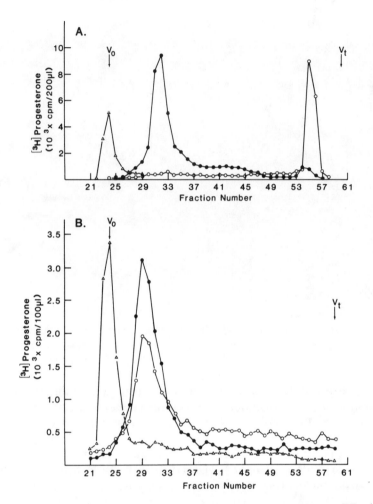

Fig. 8. Ionic strength-dependent mobility of progesterone receptors PRa (A) and PRb (B) on glass column size-exclusion HPLC. Extensively purified chicken oviduct PRa (A) or PRb (B) labeled with [³H]progesterone was applied to a GlasPac TSK 3000 SW column as described in Fig. 5, except that elution was performed with Buffer A containing 0 (Δ–Δ), 0.1M (O–O), or 0.2M NaCl (●–●). From Denner et al., 1987; courtesy of Academic Press.

that all of the standards tested eluted in the void vol in the absence of added salt. This suggests that there may be some residual charge on the resin that must be neutralized by the addition of salt in order to allow the column to function as a size exclusion column.

3.3.4. Isolation of Receptor Fragments

Proteolysis has been a major problem in receptor purification and analysis. The TSK 3000 column can be used to analyze preparations for proteolytic fragments and to separate them from intact receptor.

Chicken oviduct cytosol contains a Ca^{2+} activated protease that cleaves the receptor to produce Form IV ($M_r = 45,000$), which contains the hormone and DNA binding domains and meroreceptor ($M_r = 23,000$), which contains only the hormone binding domain (Vedeckis et al., 1980). Chick oviduct cytosol, labeled with [^{3}H]progesterone was incubated for 10 min at 4°C with 10 mM $CaCl_2$ to produce Form IV and for 1 h with 0.1M $CaCl_2$ to produce meroreceptor. Digestion was terminated by removing the Ca^{2+} by chromatography on Sephadex G-25. Form IV was concentrated on a DEAE column, whereas meroreceptor was found in the flow through of the DEAE column. Figure 9 shows the chromatography of these samples and of a mixture of the two. The fragments are separated from each other and elute much later than the intact receptor. Thus, this column can be used both to analyze preparations for proteolytic fragments and to separate them from intact receptor.

4. Functional Assays for Receptor

Unfortunately, good assays for all of the predicted functions of receptor have not been developed. However, hormone binding and a number of DNA binding assays can be used to examine some of the properties of purified receptor.

4.1. Hormone Binding

Hormone binding of purified preparations can be examined using hydroxylapatite assays (Birnbaumer et al., 1983) or by gel filtration using either small conventional columns or HPLC columns as described above.

4.2. DNA Binding

General DNA binding activity can be examined by testing the ability of receptor to bind to DNA cellulose as described above in the purification procedure. However, of much greater interest, is the ability of the receptor to bind specifically to *cis*-acting enhancer sequences called steroid response elements. These elements have been shown to be sufficient to cause genes to be induced by steroid receptors (Strahle et al., 1987). The sequence of the steroid response elements for glucocorticoid receptor (GRE) and for progesterone receptor (PRE) may be identical (Strahle et al., 1987) or are at least very similar. In contrast, the estrogen receptor recognizes a different sequence (Klock et al., 1987).

We have developed a gel retardation assay to examine the DNA binding properties of the various receptor forms (Tsai et al., 1988). For these

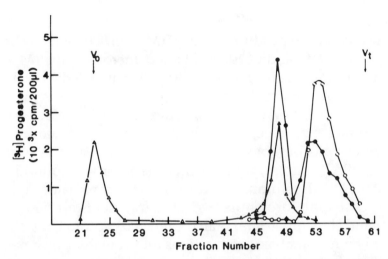

Fig. 9. Size-exclusion HPLC of progesterone receptor proteolytic fragments. Chick oviduct cytosol was labeled with [³H]progesterone and treated with calcium to cause proteolysis of the receptors, which produces Form IV and meroreceptor. Meroreceptor (O–O) was analyzed by HPLC on the GlasPac TSK 3000 SW column. One preparation of Form IV (Δ–Δ) contained some undigested aggregated receptors in the void volume (V₀). A preparation containing only Form IV was mixed with meroreceptor and then chromatographed (●–●). From Denner et al., 1987; courtesy of Academic Press.

studies a 27 mer containing the GRE/PRE sequence of the tyrosine amino transferase gene (Jantzen et al., 1987) was used (Tsai et al., 1988) as the specific DNA.

The GRE/PRE is end-labeled using Klenow enzyme and [a-^{32}P] dGTP. The specific activity is $1–5 \times 10^8$ cpm/µg. pBR322 is digested with Hinf I and is used as a source of nonspecific DNA. Incubations generally contain 0.1–0.2 ng of PRE/GRE, 50 ng of pBR322/Hinf I in 10 mM HEPES pH 7.9, 5–10 mM dithiothreitol, 20% glycerol, 0.1 mM EDTA, 5 mM MgCl$_2$, 60–80 mM NaCl, 0.1 mg/mL nuclease free bovine serum albumin, and 2% Ficoll 400 in a final volume of 10 µL. The receptor fraction is added and incubated at room temperature for 10 min. The reactions are analyzed as described (Tsai et al., 1988) using 5% acrylamide gels prepared and run in 50 mM Tris base, 1 mM EDTA, 50 mM boric acid. The gel is run at constant voltage, first at 200V for 15 min and then at 160 V for about 90 min. Bromphenol Blue and xylene cyanol are used as marker dyes. The free probe moves somewhat ahead of the Bromphenol Blue. The gels are dried and typically exposed to XAR-5 film overnight at -70°C. Figure 10 shows the DNA binding properties of PRa. The free DNA runs near the bottom of the gel and the protein DNA complexes near the top. The assays are always adjusted so that there will be excess free DNA. Lane 1 shows the complex made by PRa and the PRE. The complex clearly contains receptor, since it is retarded

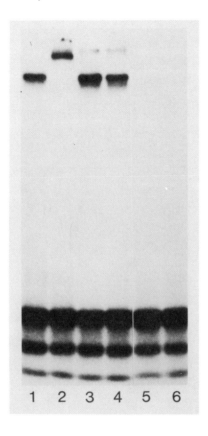

Fig. 10. Characterization of PRa preparation binding to GRE/PRE. Each incubation contained 0.2 ng of [^{32}P] endlabeled GRE/PRE, 50 ng of pBR322/HinfI in transcription mix and the following additions. Lane 1, 9 ng PRa; Lane 2, 9 ng PRa + 3 µg PR22; Lane 3, 9 ng PRa + 2 µg PR6; Lane 4, 9 ng PRa + 4 µg PR6; Lane 5, 2 µg PR22; Lane 6, 4 µg PR6.

to a greater extent when PR22, a receptor specific antibody is added (Lane 2). Very little of the complex reacts with PR6 (Lanes 3 and 4), the B specific antibody. Figure 2 showed that the PRa contains no intact PRb and only a small amount of a PRb fragment. It is presumably this fragment that is interacting both with the DNA and with PR6. Our studies (Tsai et al., 1988) have shown that this reaction is sequence specific and that the receptor apparently binds as a dimer to this small piece of DNA. Figure 11 shows the binding of PRb and immunopurified receptor to the GRE/PRE. The PRb preparation appears to produce multiple bands (Lane 1), but all of these have been shown to contain receptor since all of the bands are shifted with specific monoclonal antibodies (Tsai et al., 1988). The PRb contains predominantly PRb (Lane 2), as expected from the analyses in Fig. 2. The immunopurified receptor shows good reactivity (Lane 3) and interaction

Fig. 11. Characterization of PRb preparation and immunopurified receptor binding to GRE/PRE. The incubations were prepared as described in Fig. 10 with the following additions: Lane 1, 28 ng PRb; Lane 2, 28 ng PRb + 2 µg PR6; Lane 3, 10 ng immunopurified receptor; Lane 4, 10 ng immunopurified receptor + 2 µg PR22; Lane 5, 10 ng immunopurified receptor + 2 µg PR6.

with antibodies demonstrates that both PRa and PRb are active in this preparation (Lanes 4 and 5).

One puzzling aspect of this work is the observation that the PRb used for these studies was originally isolated from a nonDNA binding fraction (*see* purification procedure). It still does not bind to DNA cellulose when further purified, but binds quite well in the gel retardation assay. This suggests that the modes of interaction with nonspecific and specific DNA may differ. Frequently, the signal obtained with PRb is weaker than that obtained with comparable amounts of PRa, suggesting there may be some differences in their interactions with DNA. However, methylation interference studies suggest that PRa and PRb interact with the same nucleotides (Tsai et. al., 1988).

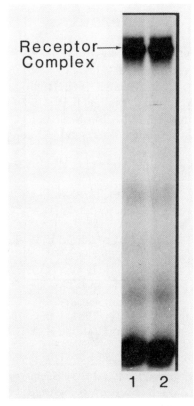

Fig. 12. Effect of freezing and thawing on PRa. Receptor was frozen in a dry ice ethanol bath and then thawed. Equal aliquots of receptor were incubated as described in Fig. 10 and analyzed for DNA binding. Lane 1, control; Lane 2, frozen receptor.

We frequently also observe an enhancement in the intensity of the protein DNA complex signal upon addition of specific antibodies (*see* Fig. 11), suggesting that the antibodies stabilize the complexes.

Finally, we examined the effect of freezing and thawing on the DNA binding properties of PRa. Figure 12 shows that there is no change in signal as a result of a single freeze/thaw cycle.

Although we used highly purified receptor preparations for these studies, relatively impure preparations can be used as well. Even if a few additional complexes are observed, the specific receptor complexes can be identified by addition of specific antibodies. Thus, this assay should prove useful for a variety of studies of receptor–DNA interaction.

In summary, we have developed a number of purification procedures that produce highly purified receptor suitable for studies of both the structure and function of chicken progesterone receptor.

References

Alberts, B. and Herrick, G. (1971) in *Meth. Enzymol.* vol. 21 (L. Grossman and K. Moldave, eds.), Academic, New York, pp. 198–217.

Bingman, W. E. III, Ayres, M. L., and Weigel, N. L. (1988) *Endocrinology Suppl.* **122,** 246.

Birnbaumer, M., Schrader, W. T. and O'Malley, B. W. (1983) *J. Biol. Chem.* **258,** 1637–1644.

Carson, M. A., Tsai, M.-J., Conneely, O. M., Maxwell, B. L., Clark, J. H., Dobson, A. D. W., Elbrecht, A., Toft, D. O., Schrader, W. T., and O'Malley, B. W. (1987) *Mol. Endo.* **1,** 791–801.

Compton, J. G., Schrader, W. T., and O'Malley, B. W. (1983) *Proc. Natl. Acad. Sci. USA* **80,** 16–20.

Conneely, O. M., Sullivan, W. P., Toft, D. O., Birnbaumer, M., Cook, R. G., Maxwell, B. L., Zarucki-Schultz, T., Greene, G. L., Schrader, W. T., and O'Malley, B. W. (1986) *Science* **233,** 767–770.

Conneely, O. M., Maxwell, B. L., Toft, D. O., Schrader, W. T., and O'Malley, B. W. (1987) *Biochem. Biophys. Res. Commun.* **149,** 493–501.

Coty, W. A., Schrader, W. T., and O'Malley, B. W. (1979) *J. Steroid Biochem.* **10,** 1–12.

Denner, L. A., Weigel, N. L., Schrader, W. T., and O'Malley, B. W. (1987) *Anal. Biochem.* **161,** 291–299.

Dougherty, J. J., Puri, R. K., and Toft, D. O. (1984) *J. Biol. Chem.* **259,** 8004–8009.

Ennis, B. W., Stumpf, W. E., Gasc, J.-M., and Baulieu, E.-E. (1986) *Endocrinology* **119,** 2066–2075.

Hutchens, T. W., Gibbons, W. E., and Besch, P. K. (1984) *J. Chromatograph* **297,** 283–299.

Isola, J., Ylikomi, T., and Tuohimaa, P. (1986) *Histochemistry* **86,** 53–58.

Jantzen, H.-M., Strahle, U., Glass, B., Stewart, F., Schmid, W., Boshart, M., Miksicek, R., and Schutz, G. (1987) *Cell* **49,** 29–38.

Klock, G., Strahle, U., and Schutz, G. (1987) *Nature* **329,** 734–736.

Laemmli, U. K. (1970) *Nature* **227,** 680–685.

Pavlik, E. J., van Nagell, J. R., Jr., Donaldson, E., and Hanson, M. (1982) *J. Steroid Biochem.* **17,** 553–558.

Peleg, S., Schrader, W. T. and O'Malley, B. W. (1988) *Biochemistry* **27,** 358–367.

Riabowol, K. T., Mizzen, L. A., and Welch, W. J. (1988) *Science* **242,** 433–436.

Renoir, J.-M., Mester, J., Buchou, T., Catelli, M. G., Tuohimaa, P., Binart, N., Joab, I., Radanyi, C., and Baulieu, E.-E. (1984) *Biochem. J.* **217,** 685–692.

Schneider, C., Newman, R. Q., Sutherland, D. R., Asser, U., and Greaves, M. F. (1982) *J. Biol. Chem.* **257,** 10766–10769.

Schrader, W. T. and O'Malley, B. W. (1972) *J Biol. Chem.* **247,** 51–59.

Strahle, U., Kloch, G., and Schutz, G. (1987) *Proc. Natl. Acad. Sci. USA* **84,** 7871–7875.

Sullivan, W. P., Beito, T. G., Proper, J., Krco, C. J., and Toft, D. O. (1986) *Endocrinology* **119,** 1549–1557.

Tora, L., Gronemeyer, H., Turcotte, B., Gaub, M.-P., and Chambon, P. (1988) *Nature* **333,** 185–188.

Tsai, S. Y., Carlstedt-Duke, J., Weigel, N. L., Dahlman, K., Gustafsson, J.-A., Tsai, M.-J., and O'Malley, B. W. (1988) *Cell* **55,** 361–369.

Vedeckis, W. V., Schrader, W. T., and O'Malley, B. W. (1980) *Biochemistry* **19,** 343–349.

Weigel, N. L., Tash, J. S., Means, A. R., Schrader, W. T., and O'Malley, B. W. (1981) *Biochem. Biophys. Res. Commun.* **102,** 513–519.

Weigel, N. L., Rodriguez, R., Carson, M. A., Birnbaumer, M. L., and Schrader, W. T. (1989) in *Laboratory Methods Manual for Hormone Action and Molecular Endocrinology* 13th edi (M.R. Hughes and B. W. O'Malley, eds.), Houston Biological Associates, Houston, pp. 2:1–2:74.

Wiehle, R. D., Hofmann, G. E., Fuchs, A., and Witliff, J. L. (1984) *J. Chromatogr.* **307,** 39–51.

Wray, W., Boulikas, T., Wray, V. P., and Hancock, R. (1981) *Anal Biochem.* **118,** 197–203.

1,25-Dihydroxyvitamin D Receptor Purification and Structural Determination

Thomas A. Brown and Hector F. DeLuca

1. Introduction

The high affinity receptor for 1,25-dihydroxyvitamin D_3 [1,25-$(OH)_2$-D_3] that has been detected in intestine, bone, and kidney is considered an essential mediator of 1,25-$(OH)_2D_3$ action (DeLuca and Schnoes, 1983). According to the accepted view of the vitamin D endocrine system, vitamin D_3 is converted to 1,25-$(OH)_2D_3$ by 25-hydroxylation in the liver followed by 1-hydroxylation in the kidney. The hormonally active 1,25-$(OH)_2D_3$ then interacts with its receptor in target tissues and causes the elevation of serum calcium and phosphorus levels (DeLuca, 1985). In the intestine, 1,25-$(OH)_2D_3$ stimulates absorption of calcium and phosphorus from the lumen to the blood. Also contributing to the rise of serum calcium, 1,25-$(OH)_2D_3$, together with parathyroid hormone, stimulates the mobilization of calcium from bone and increases the reabsorption of calcium in the distal renal tubules. One line of evidence that the 1,25-$(OH)_2D_3$ receptor is required for calcium regulation by vitamin D is that patients with a defective 1,25-$(OH)_2D_3$ receptor, such as in some cases of vitamin D-dependency rickets Type II, demonstrate target organ resistance to 1,25-$(OH)_2D_3$ (Eil et al., 1986). Also, neonatal rats do not regulate calcium homeostasis in response to 1,25-$(OH)_2D_3$ until the appearance of receptor in their intestine (Halloran and DeLuca, 1981).

Receptor Purification, vol. 2   ©1990 The Humana Press

The 1,25-$(OH)_2D_3$ receptor has been found in many tissues not previously recognized as targets for 1,25-$(OH)_2D_3$. For example, specific nuclear localization of 1,25-$(OH)_2D_3$ has been found in the parathyroid, epidermis, stomach, pituitary, brain, teeth, mammary gland, testes, ovary, and thymus, as well as in numerous cancer cell lines (Link and DeLuca, 1985). The discovery of receptor in such a wide variety of tissues (Stumpf et al., 1979; Link and DeLuca, 1985) has spurred speculation that, in addition to its classical role of regulating calcium homeostasis, vitamin D has a general function in growth and differentiation of many cell types. In several instances, evidence for receptor in nonclassical target tissues has led to practical applications of vitamin D compounds in the treatment of disease (DeLuca, 1988). For example, the discovery of receptor in skin cells initiated experiments showing that 1,25-$(OH)_2D_3$ induces differentiation and suppression of proliferation of keratinocytes (Smith et al., 1986). These observations are now being used to develop a treatment of psoriasis and contact dermatitis by oral administration of 1,25-$(OH)_2D_3$ (DeLuca, 1988). Also, the presence of 1,25-$(OH)_2D_3$ receptor in many cancer cell lines prompted investigations into the effect of 1,25-$(OH)_2D_3$ on tumor growth (Eisman, 1984; Tanaka et al., 1982). As a result, it is now known that 1,25-$(OH)_2D_3$ and some analogs, such as 24-homo-1,25-$(OH)_2D_3$, will suppress growth and cause differentiation of certain cell lines (Ostrem and DeLuca, 1987). It is hoped that these investigations will lead to the treatment of certain cancers with vitamin D analogs (Eisman et al., 1987).

The action of 1,25-$(OH)_2D_3$ at target tissues is thought to occur through the 1,25-$(OH)_2D_3$ receptor by the classic steroid hormone mechanism (Link and DeLuca, 1985; Walters, 1985). By analogy to other steroid hormones, 1,25-$(OH)_2D_3$ diffuses to the nucleus and interacts with its receptor. The 1,25-$(OH)_2D_3$ receptor complex then presumably binds to specific DNA sequences near target genes and modulates the transcription of these genes. Although the available data indicates that the steroid hormone model applies to the 1,25-$(OH)_2D_3$ system, this does not provide a complete description of the mechanism of action. Unanswered questions include: How does the binding of steroid to its receptor change the in vitro and in vivo DNA affinity of receptor? Also, what specific DNA sequences are required for receptor action? Is the binding of receptor to DNA sufficient for changes in transcription, or are auxiliary factors necessary for action? One way these basic questions will be addressed is by the purification and detailed biochemical characterization of the 1,25-$(OH)_2D_3$ receptor. This paper will review the progress that has been made in receptor purification and characterization. It will also describe in detail the procedures for 1,25-$(OH)_2D_3$ receptor purification.

2. Receptor Properties

The 1,25-$(OH)_2D_3$ receptor was initially identified by its high affinity for 1,25-$(OH)_2D_3$. The equilibrium dissociation constant of the hormone–receptor complex has been determined by Scatchard analysis in crude preparations to be 10^{-10}–$10^{-11}M$ (Link and DeLuca, 1985). The ligand requirements for recognition by receptor have been extensively studied by competitive binding assays with analogs and metabolites of 1,25-$(OH)_2D_3$. The 1-hydroxyl and the 25-hydroxyl groups are very important to binding, as evidenced by the fact that receptor has 20 × lower affinity for 25-OH-D_3 and 40 × lower affinity for 1-OH-D_3 than for 1,25-$(OH)_2D_3$ (Link and De-Luca, 1988). Metabolic alterations in the sidechain of 1,25-$(OH)_2D_3$, such as in 24-OH-D, 26-OH-D, and 24-oxo-D, diminish receptor affinity significantly (Link and DeLuca, 1985).

The receptor is a trace protein. Even in the intestine, the chief source of receptor, it is present only as 0.001% of the soluble protein. Although there are conflicting reports, it appears that the receptor is predominantly nuclear. Following homogenization in low salt buffer, up to 90% of the unoccupied receptor is found with the nuclear fraction. The receptor is then readily extracted from the nuclear fraction by homogenization in high salt (Walters et al., 1980). Since these homogenization techniques require the destruction of the cell, strictly interpreted, they show only the in vitro localization of receptor. Immunocytochemical techniques should provide stronger evidence for the in vivo localization. Clemens et al. (1988) have immunocytochemical evidence indicating that the unoccupied receptor is, indeed, mainly nuclear. However, as the authors mention, the absence of detectable receptor from the cytoplasm is not yet conclusive because of the possible loss of immunoreactivity during the long fixation that was used.

The mol wt of the 1,25-$(OH)_2D_3$ receptor has been determined by Na-DodSO$_4$/PAGE for several species. Simpson and DeLuca (1982) purified the chicken intestinal receptor to a single 63kDa band; however, the protein band was not directly identified as the 1,25-$(OH)_2D_3$ receptor. Dame et al. (1985) demonstrated by renaturation from NaDodSO$_4$-polyacrylamide gel preparations and immunoblotting that pig intestinal receptor migrates as a 55kDa protein. After monoclonal antibodies became available, the receptor mol wt for various species was readily determined by immunoblotting.

The observed mol wt of receptor ranges from 60kDa for chicken to 52kDa for human (Pike, 1985). These values are in approx agreement with previous data derived from gel filtration and sucrose density gradients (Link and DeLuca, 1985).

The isoelectric point of receptor is approx 6, as determined by two-dimensional electrophoresis (Simpson et al., 1983), chromatofocusing (Simpson and DeLuca, 1982), and two-dimensional electrophoresis with immunoblotting (Pierce et al., 1987). Simpson et al. (1983) showed that highly purified chicken receptor migrated as a smear in the isoelectric focusing dimension of 2-D electrophoresis. They suggested that the smear represented multiple pI forms of receptor because of posttranslational phosphorylation. Pike and Sleator (1985) demonstrated direct evidence of receptor phosphorylation in mouse 3T6 fibroblasts by immunoprecipitation of [^{32}P]-orthophosphate labeled receptor. Also, 3 pI forms of pig receptor were detected by immunoblotting of 2-D polyacrylamide gels (Pierce et al., 1987). It is not yet clear if the multiple pI forms of chicken and pig receptor observed by 2-D electrophoresis represent different amounts of phosphate incorporation.

The unoccupied receptor is very unstable in tissue homogenates (Mellon et al., 1980; Walters et al., 1982). The 1,25-$(OH)_2D_3$ binding activity is rapidly lost at high temperature (25 and 37°C) (Mellon et al., 1980). In addition, the receptor will not bind steroid after exposure to pH conditions outside of the range pH 6–10 (Mellon et al., 1980). The receptor is also subject to rapid inactivation by sulfhydryl reagents, such as N-ethylmaleimide (Coty, 1980; Wecksler et al., 1980). Conditions attempting to maximize the stability of receptor preparations have been developed during the early attempts at purification. McCain et al. (1978) noted that receptor, although rapidly inactivated without ligand at 0°C, is quite stable even for 48 h if the 1,25-$(OH)_2D_3$-receptor complex is formed. Above 4°C, the receptor is unstable even in the presence of hormone (McCain et al., 1978). Metal chelating agents such as EDTA have also improved receptor yields (Walters et al., 1982). As would be suspected from the susceptability to N-ethylmaleimide, sulfhydryl protecting reagents stabilize receptor preparations (Mellon et al., 1980; Walters et al., 1982). Dithiothreitol was demonstrated particularly effective, whereas monothioglycerol was less effective (Mellon et al., 1980).

In summary, successful purification protocols generally include the following precautions against receptor inactivation:

1. All procedures are performed at 0–4°C;
2. Buffers are pH 7–8 if possible;

3. The tissue, especially intestinal mucosa, is extensively washed prior to homogenization (Kream et al., 1976);
4. 5 mM DTT is added to all buffers;
5. EDTA is added to all buffers; and
6. 1,25-(OH)$_2$D$_3$ is added to the soluble receptor at the beginning of procedure if this is compatible with the experimental goals.

Protease inhibitors have been used to stabilize receptor preparations with varied results. It has not been clearly demonstrated that proteolysis is the major reason for receptor instability; however, it is at least partially responsible under certain conditions. L-1-Tosylamido-2-phenylethyl chloromethyl ketone (TPCK) and N^α-p-tosylamido-2-chloromethyl ketone (TLCK) actually inhibit receptor yield possibly by binding directly to the receptor (Mellon et al., 1980; Norman et al., 1983). Phenylmethylsulfonyl fluoride (PMSF) has been reported to stabilize the free receptor at higher temperatures, but the effect is less pronounced at lower temperatures or with occupied receptor (Norman et al., 1983). The increased yield observed with EDTA may be the result of inhibition of Ca-dependent proteases, but its effect has also been explained on the basis of its chelating properties affecting the extractability of receptor from the nucleus (Walters et al., 1982). Although protease inhibitors have not generally been critical to receptor isolation, recent work with HL-60 cells in our laboratory emphasizes the caution one should use when investigating different sources of receptor (Inaba and DeLuca, 1989). The use of the serine protease inhibitor diisopropylfluorophosphate (DFP) was essential to isolation of receptor from ATCC HL-60 cells, although not necessary for isolation from LG HL-60 cells. Interestingly, other serine protease inhibitors such as PMSF, benzamidine, and aprotinin were not effective in increasing receptor yields from HL-60 cells.

3. Receptor Purification

The first significant purification of receptor was from chicken intestinal mucosa. Pike and Haussler (1979) made an important advance in receptor purification by their use of ligand affinity chromatography on DNA-cellulose and blue dextran-Sepharose. They obtained semipure receptor from vitamin D-deficient chickens by a procedure that included Polymin P precipitation followed by sequential chromatography on DNA-cellulose, Sephacryl S-200, blue dextran-Sepharose, DNA-cellulose, and heparin-Sepharose. The isolated receptor was purified a reported 86,000-fold and contained several protein bands on NaDodSO$_4$-polyacrylamide gels. Because of the low abundance of receptor and the low yield during purification, 350 rachitic chickens were used to obtain only 26 μg of semi-pure receptor. Forty percent of the 26 μg was used for the electrophoretic analysis, limiting the amount of receptor available for further studies.

Modification in receptor purification enabled our laboratory to purify receptor to apparent homogeneity (Simpson and DeLuca, 1982). Simpson and DeLuca (1982) demonstrated that adult vitamin D-replete chicken du-odena was a suitable source for receptor, thereby eliminating the cumber-some task of maintaining large numbers of vitamin D-deficient chickens. The other major improvement introduced was the isolation of intestinal nuclei in a low salt buffer before the extraction of receptor by high salt. The receptor preparation isolated from the washed nuclear fraction was more stable and fivefold purified vs whole cell extract. The purification proce-dure was further refined by Simpson et al. (1983) and will be described in detail later in this chapter. In short, a five-step purification scheme resulted in 29 μg of apparently homogeneous $1,25\text{-}(OH)_2D_3$ receptor (5800-fold pu-rified from nuclear extract, 8% yield). Note that, since the chickens were vitamin D replete, some of the receptor was occupied by endogenous $1,25\text{-}(OH)_2D_3$ causing the fold purification to be an underestimate.

Despite the progress in receptor purification made by Pike and Haus-sler (1979) and Simpson et al. (1983), the low net yields of the purification procedures made the isolation of the amounts of pure receptor necessary for physical or biochemical studies prohibitively difficult. Both groups sought to overcome the difficulties in studying such a low abundance pro-tein by developing monoclonal antibodies. Monoclonal antibodies had previously proved effective in the study of other trace proteins such as es-trogen receptor. In addition, the hybridoma technology for the generation of monoclonal antibodies required moderate amounts of only semipure receptor (Greene et al., 1980). In order to obtain the quantities of semipure receptor needed for monoclonal antibody generation, Pike et al. (1983) modified and scaled up their purification procedure for chicken receptor to obtain 412 μg of 13% pure receptor. This was of sufficient amount and purity to inject Lewis rats, obtain a titer against receptor, and produce four monoclonal antibodies. Our laboratory, however, investigated different sources for receptor. Young pig intestine was found more convenient be-cause the tissue is easier to obtain in large quantities given the large size of the animal. In addition, nuclear extract from young pig intestine contains at least twice the receptor per mg protein compared to chicken intestine. The 5-step purification scheme used for pig intestine will be described here in detail. In short, the scheme included nuclear extract preparation, DNA-cellulose chromatography, ammonium sulfate precipitation, gel filtration chromatography, and DEAE-Sepharose chromatography. The isolated receptor was used to generate 24 different monoclonal antibodies, of which, all but two crossreact with various species including pig, human, rat, monkey, and chicken (Dame et al., 1986).

Although the classical purification protocols for pig and chicken receptor were valuable in providing antigen for the generation of monoclonal antibodies, they remained very laborious. The small quantity and questionable purity of the isolated receptor still hindered physical and biochemical studies. To obtain larger quantities of purified receptor, greatly improved purification procedures using the high specificity of the anti-1,25-$(OH)_2D_3$ receptor monoclonal antibodies were developed (Brown et al., 1988; Pike et al., 1987). As will be described in detail, a one-step, immunoaffinity column was used to highly purify receptor from pig nuclear extract (Brown et al., 1988). Homogeneous receptor was then easily isolated by preparative $NaDodSO_4$/PAGE. The ease and high yield (>45%) of this procedure now make available quantities of receptor that are of sufficient purity for biochemical studies. For example, receptor isolated in this way has been peptide mapped and partially sequenced (Brown et al., 1988).

4. Purification Protocol
of Chicken Intestinal 1,25-$(OH)_2D_3$ Receptor

The purification protocol of Simpson et al. (1983) involved the preparation of a nuclear extract and ammonium sulfate precipitation, followed by chromatography on DNA-cellulose, hydroxylapatite, Sephacryl S-200, and DEAE-cellulose. The relevant data from the purification scheme is shown in Table 1.

4.1. Nuclear Extract Preparation
and Ammonium Sulfate Precipitation

Duodena were isolated from vitamin D-replete chickens and rinsed with TED (50 mM Tris-HCl, pH 7.4, 1.5 mM EDTA, 5 mM DTT). The mucosa was scraped free from the serosa, rinsed with 3 vol of TED and homogenized in 3 vol of TED with short bursts from a Polytron homogenizer (Brinkmann Instruments, Westbury, NJ). The homogenate was centrifuged (4000g, 10 min) to obtain a crude nuclear pellet. The nuclear pellet was thoroughly washed 3× by resuspension in 4 vol of TED and centrifugation (4000g, 10 min). The washed nuclear pellet was homogenized with 2 vol of TEDKMg (TED + 300 mM KCl + 10 mM MgCl$_2$) using a Polytron

Table 1
Receptor Purification Scheme

Step	Protein, mg	Receptor, dpm × 10⁻⁶	Sp Act., dpm × 10⁻³/mg	Purification, ×-fold	Yield, %
Nuclear extract	2100	9.00	4.3	1	100
$(NH_4)_2SO_4$	640				
DNA-cellulose	6.8	3.00	441.2	103	33
Hydroxyl-apatite	2.0	1.80	900.1	214	20
Sephacryl S-200	0.34	1.22	3614.0	861	14
DEAE-cellulose	0.029	0.72	24,828.4	5774	8

homogenizer. The homogenate was centrifuged at 22,000g for 60 min and the resulting supernatant (crude nuclear extract) was labeled overnight with 4 nM 1,25-$(OH)_2$-[26,27-³H]D_3 at 0°C. The labeled nuclear extract was precipitated by addition of ice-cold saturated ammonium sulfate to a final concentration of 38% (v/v). After 10–30 min, the suspension was centrifuged at 9000g for 10 min. The pellet was rinsed with ice-cold distilled water and resuspended in sufficient TED so that the ionic strength of the solution was equal to that of 50 mM KCl.

4.2. DNA–Cellulose Chromatography

DNA-cellulose, prepared as described by Alberts and Herrick (1971), was added to the ammonium sulfate precipitated nuclear extract solution to a final concentration of 10% (v/v). The suspension was swirled for 30 min, centrifuged at 100g for 5 min, and rinsed once with $TEDK_{50}$ (TED + 50 mM KCl). The DNA–cellulose was then packed into a column and rinsed with 2 column vol of $TEDK_{50}$. Using a linear gradient of 0.1–0.4M KCl in TED, receptor was eluted from the DNA–cellulose at about 0.2M KCl.

4.3. Hydroxylapatite Chromatography

The post-DNA–cellulose material was adjusted to an ionic strength of about 0.2M KCl and applied to a small column of hydroxylapatite at 0.5 mL/min. The column was washed with 2 column vol of TED and 2 vol of TED + 0.5% Triton X-100. The receptor was eluted with a linear gradient of 50–500 mM KH_2PO_4 in TED.

4.4. Sephacryl S-200 Chromatography

A column (2×76 cm) of Sephacryl S-200 was equilibrated in $TEDK_{50}$. The receptor fraction was loaded and chromatographed at 0.5 mL/min.

4.5. DEAE–Cellulose Chromatography

A DEAE–cellulose column was equilibrated in $TEDK_{50}$ + 0.5% Triton X-100. The receptor fraction was applied, loaded, and the column washed with 2 vol of $TEDK_{50}$ + 0.5% Triton X-100. The column was eluted with a linear gradient of 50–500 mM KCl with the receptor eluting at 0.15–0.20 M KCl.

4.6. Properties of the Purified Chicken Receptor

The isolated receptor preparation was purified 5800-fold with an 8% yield (Table 1). $NaDodSO_4$/PAGE showed only one protein band was present but the low amount of receptor available prohibited more rigorous evaluations of purity. Two-dimensional electrophoresis of the isolated material characterized the isolated protein as $63,000 \pm 3900$ mol wt with a pI of 6.0–6.3 (Fig. 1).

5. Purification Protocol
for Porcine Intestinal 1,25-$(OH)_2D_3$ Receptor

Pig receptor was purified by a 5-step scheme consisting of nuclear extract preparation, DNA-cellulose chromatography, ammonium sulfate precipitation, gel filtration chromatography, and DEAE-Sepharose chromatography (Dame et al., 1986). The results of the purification scheme are summarized in Table 2.

5.1. Nuclear Extract Preparation

The small intestine was removed from young pigs (18–40 kg) and washed extensively in TEDNa (TED + 150 mM NaCl). Washing was continued until the wash buffer was clear. The mucosa was then scraped from the serosa, rinsed two times with 3 vol of TEDNa, and rinsed one time with 3 vol TED. One settled vol of mucosa was added to 2 vol of TED and homogenized with a Polytron homogenizer. The homogenate was centrifuged at 3000g for 30 min to obtain a crude nuclear pellet. The pellet was washed $3 \times$ with TED by resuspending the pellet and centrifuging at 3000g for 10 min. The washed nuclear pellet was then homogenized in 2 vol of TEDKMg + 1 mM PMSF using a Polytron homogenizer. The homogenate

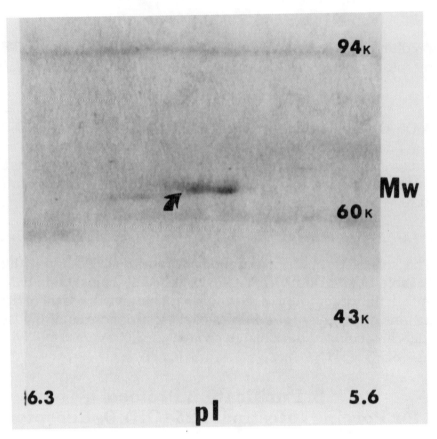

Fig. 1. Two-dimensional electrophoresis of purified chick intestinal receptor of Simpson et al. (1983). The major polypeptide (indicated by arrow) has a mol wt of 64,000kDa. The polypeptide runs as a smear in the isoelectric focusing dimension at approx pH 6.0. Internal molecular mass standards are phosphorylase a (94,000), catalase (60,000), and actin (43,000).

was centrifuged at 27,000g for 120 min. The supernatant (nuclear extract) contained the 1,25-(OH)$_2$D$_3$ receptor and was frozen in liquid nitrogen and stored at –70°C. Typically, the small intestine from 6 pigs yields 5 L of scraped mucosa and 12 L of crude nuclear extract (approx 700 fmol/mg protein; 5 mg protein/mL).

5.2. DNA-Cellulose Chromatography

The crude nuclear extract was diluted to 0.075M KCl with TED and labeled overnight at 2 nM 1,25-(OH)$_2$-[26,27-^{3}H]D$_3$. DNA–cellulose prepared according to Litman (1968) was added to the labeled receptor at about 10% (v/v). The suspension was occasionally swirled for 1 h and then centrifuged (200g, 5 min). The DNA–cellulose was washed twice with 10 vol of TEDK$_{75}$ (TED + 75 mM KCl) and then poured into a column. The

Table 2
Purification of the Porcine 1,25-(OH)$_2$D$_3$ Receptor Protein

	Sp act. pmol/mg	Purification, ×–fold	Yield, %	Purity, %
Nuclear extract (10)[a]	0.635	1	100	0.0035
DNA-cellulose (10)	47.2	74	51	0.26
(NH$_4$)$_2$SO$_4$ ppt (10) and size-exclusion				
HPLC (25)	629	990	27	3.8
DEAE-Sepharose (6)	4170	6600	23	24

[a]Number of batches or columns.

column was washed with TEDK$_{75}$ at 5 mL/min and eluted with TEDK$_{300}$ (TED + 300 mM KCl) at 1.7 mL/min (Fig. 2A). The peak of 1,25-(OH)$_2$D$_3$ binding activity was pooled and solid ammonium sulfate was added to 40% saturation. After 1 h, the suspension was centrifuged at 7000g for 15 min. The ammonium sulfate pellet could be conveniently stored at –70°C.

5.3. Gel Filtration Chromatography

The ammonium sulfate pellet was resuspended to 3 mg protein/mL in 0.02 M Na$_2$HPO$_4$ (pH 6.8) with 0.2M KCl and 5 mM DTT. The protein sample was loaded onto TSK gel filtration column type TSK3000SW (64 × 2.5 cm) with a guard column (10 × 2.5 cm). The HPLC gel filtration was run at 4 mL/min in 0.02M Na$_2$HPO$_4$ (pH 6.8) with 0.2M KCl (Fig. 2B).

5.4. DEAE-Sepharose

The pooled receptor fraction from gel filtration was diluted twofold with 8.1 mM Na$_2$HPO$_4$/1.5 mM KH$_2$PO$_4$ (pH 7.4) and made 5 mM in DTT. The receptor was then loaded at 0.2 mL/min onto a DEAE-Sepharose column. Receptor was eluted with a linear gradient of 50–500 mM NaCl in 8.1 mM Na$_2$HPO$_4$/1.5 mM KH$_2$PO$_4$ (pH 7.4) + 5 mM DTT (Fig. 2C).

5.5. Properties of the Purified Pig Receptor

The isolated receptor was purified 6600-fold with 23% yield (Table 2). The preparation contained several proteins as judged by NaDodSO$_4$/ PAGE and silver stain (Fig. 3), but was of sufficient purity to inject mice for the generation of monoclonal antibodies. It also permitted deduction of the mol wt of the porcine receptor at 55kDa.

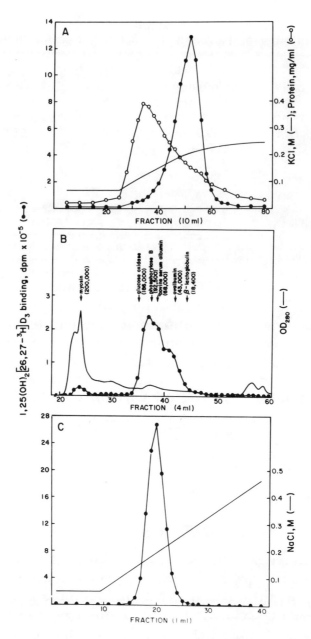

Fig. 2. Purification steps of porcine intestinal 1,25-(OH)$_2$D$_3$ receptor by Dame et al. (1986). **A.** DNA-cellulose chromatography. Aliquots were assayed for 1,25-(OH)$_2$-[26,27-^{3}H]D$_3$ (●), protein (o), and KCl concentration (-). **B.** HPLC gel filtration column chromatography. Aliquots were assayed for 1,25-(OH)$_2$-[26,27-^{3}H]D$_3$ (●) and absorbance at 280 nm (-). The arrows show the elution positions of protein standards. **C.** DEAE-Sepharose chromatography. Aliquots were assayed for 1,25-(OH)$_2$-[26,27-^{3}H]D$_3$ (●) and NaCl (-) by conductivity measurements.

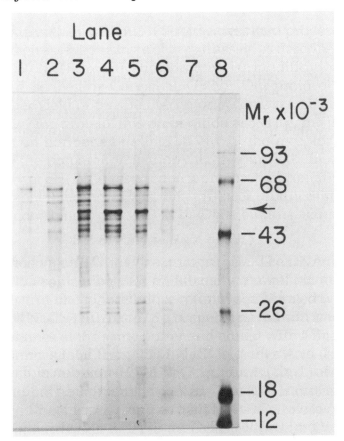

Fig. 3. NaDodSO$_4$/PAGE with silver staining of purified porcine receptor (Dame et al., 1986). The position of the 1,25-(OH)$_2$D$_3$ receptor band is designated by the arrow. Lanes 1–6 are aliquots of fractions 14, 16, 18, 20, 22, and 24, respectively, from the DEAE-Sepharose column (*see* Fig. 2C). Lane 7 is a buffer blank. Lane 8 is molecular mass standards.

6. Immunoaffinity Purification
of Porcine 1,25-(OH)$_2$D$_3$ Receptor

6.1. Preparation of Immunoaffinity Resin

Cyanogen bromide activation of Sepharose was done as described by Kohn and Wilchek (1982), using 9 mg of CNBr per 1 g drained Sepharose. Monoclonal antibody XVIE10B6A5 was coupled to the activated resin by overnight incubation at 4°C in phosphate buffered saline (1.5 mM KH$_2$PO$_4$, 8.1 mM Na$_2$HPO$_4$, pH 8.0, 137 mM NaCl, 2.7 mM KCl, 0.02% NaN$_3$). Approximately 1 mg of antibody was coupled per mL of resin. The XVIE10 B6A5-Sepharose was stored in TED with 0.025% sodium azide.

6.2. Immunoaffinity Chromatography

Pig nuclear extract was prepared as described above. The nuclear extract was thawed, centrifuged at 27,000g for 1 h, and incubated overnight with 5.0 nM 1,25-(OH)$_2$-[26,27-^{3}H]D$_3$ (2 Ci/mmol). A 20 mL precolumn of Sepharose Cl-4B linked in series to a 1.8 mL column of XVIE10B6A5-Sepharose was equilibrated in TED. Labeled nuclear extract (250 mL) was loaded onto the columns by a peristaltic pump at 1.7 mL/min. After the extract was loaded, the precolumn was removed. The antibody column was then washed with the following buffers at 0.5 mL/min: 10 mL TEDK$_{400}$ (TED + 400 mM KCl + 0.05% Tween-20); 30 mL TED; 30 mL TEDK$_{400}$ and 15 mL 50 mM 3(cyclohexylamino)-1-propanesulfonic acid (Caps) (pH 9.8) + 5 mM DTT. The receptor was eluted at 0.15 mL/min using 50 mM Caps (pH 11.2) with 5 mM DTT as elution solvent. To assay hormone binding, the eluted fractions were immediately diluted in 10 vol of TED with 1.25 mg/mL β-lactoglobulin as a carrier protein. Fractions were assayed for protein bound 1,25-(OH)$_2$-[26,27-^{3}H]D$_3$ by a hydroxylapatite binding assay (Dame et al., 1985).

Figure 4 shows the protein bound 1,25-(OH)$_2$D$_3$ present in the fractions collected during loading, washing, and elution of the XVIE10B6A5-Sepharose column. Greater than 90% of the steroid binding activity in the nuclear extract was retained on the column during loading and during the salt and detergent washes. The receptor was eluted at high pH in a sharp peak with an overall yield of 45% hydroxylapatite precipitable binding activity. Stripping the column with strong denaturing reagents, 2% NaDodSO$_4$ and 6 M urea, demonstrated that no detectable receptor remained bound to the column after the high pH elution. Other elution conditions, such as pH 5, were tried, but the isolated receptor did not retain its activity. In addition, low pH or strong denaturants removed some of the antibody from the column, adding a contaminant to the preparation.

6.3. Preparative Electrophoresis

Approximately 100 µg of immunoaffinity purified receptor was added to 1 mg of β-lactoglobulin and trichloroacetic acid/deoxycholate precipitated (Mahuran et al., 1983). The sample was subjected to electrophoresis on a 9% polyacrylamide gel. After electrophoresis, the gels were rinsed with water and stained with ice-cold 0.25M KCl + 1 mM DTT. The gel was rinsed and destained in ice-cold water containing 1 mM DTT until the protein band was visible in oblique lighting (Hager and Burgess, 1980). The receptor band was cut out and electroeluted essentially as described by

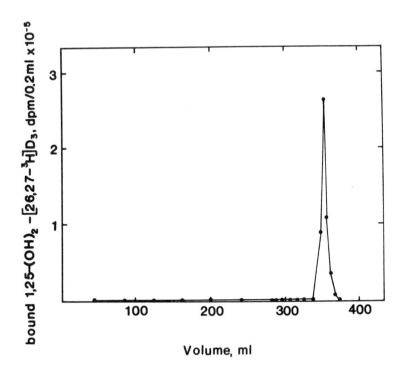

Fig. 4. Immunoaffinity chromatography of porcine 1,25-(OH)$_2$D$_3$ receptor (Brown et al., 1988). Aliquots of fractions collected during loading, washing, and elution were assayed for bound 1,25-(OH)$_2$-[26,27-^{3}H]D$_3$ by a hydroxylapatite binding assay.

Hunkapiller et al. (1983), except the Laemmli buffer system (0.025M Tris, 0.192M glycine, pH 8., 0.1% NaDodSO$_4$) was used. The elution was complete after 4 h at 50 V.

6.4. Properties of the Immunopurified Pig Receptor

The receptor eluted from the immunoaffinity column was highly purified, as judged by NaDodSO$_4$/PAGE with silver staining (Fig. 5A). As shown in lanes 3–6, the receptor fractions contained one predominant protein band at 55,000kDa. There are oftentimes minor contaminants at 200,000 and 45,000kDa. Immunoblotting of the eluted receptor with antireceptor antibody IVG8C11 shows that the 55,000kDa protein is, indeed, the 1,25-(OH)$_2$D$_3$ receptor (Fig. 5B).

Preparative electrophoresis typically gave 80–100% recovery and the electroeluted receptor migrated as a single sharp band on NaDodSO$_4$/PAGE (Fig. 5C).

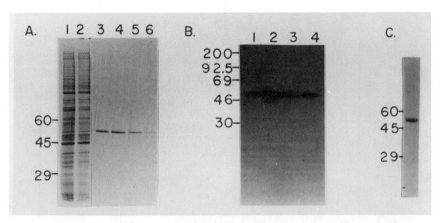

Fig. 5. NaDodSO$_4$/PAGE of 1,25-(OH)$_2$D$_3$ receptor (Brown et al., 1988). **A.** Silver staining of fractions from immunoaffinity chromatography. Lane 1, pig nuclear extract; lane 2, column wash-through; lanes 3–6, peak receptor fractions eluted at pH 11.2. **B.** Immunoblotting with monoclonal antibody IVG8C11 of receptor fractions from immunoaffinity chromatography. Lanes 1–4, fractions of receptor eluted at pH 11.2. **C.** Electroeluted receptor visualized by Coomassie Blue stain.

7. Impact of Receptor Purification

The purification of the 1,25-(OH)$_2$D$_3$ receptor has led to the production of anti-1, 25-(OH)$_2$D$_3$ receptor monoclonal antibodies and the detailed physical characterization of the receptor (Pike, 1985; Dame et al., 1985, 1986; DeLuca, 1988). As mentioned previously, immunoblotting of tissue extracts from various species identified the receptor as a 52,000–60,000 kDa protein (Dame et al., 1985; Pierce et al., 1987; Pike, 1985). In addition, immunoprecipitation of receptor from [^{35}S]-methionine- and [^{32}P]-orthophosphate-labeled 3T6 cells demonstrated phosphorylation of the receptor (Pike and Sleator, 1985). Monoclonal antibodies have also been used in competitive radioligand immunoassay to quantitate receptor that cannot be detected by the standard 1,25-(OH)$_2$D$_3$ binding assay, such as 1,25-(OH)$_2$D$_3$-bound receptor or nonbinding forms of receptor (Dokoh et al., 1984; Sandgren and DeLuca, 1988). Although this technique is not widely useful (Pike, 1985), it was important in demonstrating that some patients with vitamin D$_3$-dependent rickets Type II contain defective receptor that cannot be detected by a steroid binding assay (Pike et al., 1984).

Monoclonal antibodies were essential to the cloning of the 1,25-(OH)$_2$D$_3$ receptor cDNA. The antibodies were used to screen a chicken intestinal (McDonnell et al., 1987) and a rat kidney (Burmester et al., 1988a) λgtll cDNA expression library. Full length receptor cDNA has since been isolated (Baker et al., 1988; Burmester et al., 1988b). From these clones the

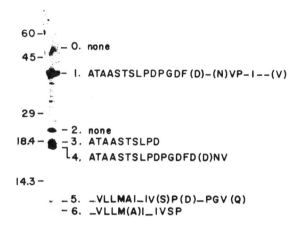

Fig. 6. Amino-terminal sequences of 1,25-(OH)$_2$D$_3$ receptor peptides (Brown et al., 1988). The amino acid sequences determined as described by Brown et al. (1988) are shown on the right in the standard 1-letter code next to the corresponding peptide band. Parentheses indicate uncertainty in the amino acid assignment, and a dash indicates that a determination could not be made at the given position. The positions of mol mass standards (kDa) are shown on the left.

entire amino acid sequence of receptor has been deduced. Analysis of the 1,25-(OH)$_2$D$_3$ receptor primary sequence confirms that the receptor is a member of the steroid hormone-thyroid hormone receptor family (Burmester et al., 1988a; McDonnell et al., 1988). The complete sequence provides a reference point for studying the receptor structure and function. For example, sequence analysis and biochemical studies have identified a 70 amino acid cysteine rich region as the probable DNA binding domain (Burmester et al., 1988b; McDonnell et al., 1988).

The immunoaffinity purification procedure described provided purified receptor for biochemical analysis. For example, the receptor was cleaved with *Staphylococcal aureus* nuclease and the N-terminal amino acid sequences of the peptides were determined (Fig. 6) (Brown et al., 1988). The amino acid sequences determined by protein sequencing were identical to the sequences deduced from the receptor cDNA clones, thereby confirming the identity of the isolated clones (Burmester et al., 1988a,b). This peptide map should be useful in the localization of covalent modifications, such as phosphorylation, to specific regions of the protein.

The pure, inactive 1,25-(OH)$_2$D$_3$ receptor protein isolated by preparative NaDodSO$_4$/PAGE following immunoaffinity chromatography has also been useful in receptor studies. The electroeluted receptor is currently being injected into rabbits for the generation of polyclonal antibodies.

Polyclonal antibodies are expected to be valuable for immunocytochemical studies. This inactive receptor is also being used as an antigen in a radioimmunoassay. The radioimmunoassay provides a sensitive assay that detects unoccupied, occupied, and nonbinding $1,25\text{-}(OH)_2D_3$ receptor for the study of receptor in various tissues (Sandgren and DeLuca, 1988).

8. Summary

The successful purification of $1,25\text{-}(OH)_2D_3$ receptor has already revealed much about its structure and function. Although it has been postulated that steroid receptors function through protein–DNA and protein–protein interactions, this has not yet been demonstrated for the $1,25\text{-}(OH)_2D_3$ receptor. A major focus of future work will be the characterization of the interaction of immunoaffinity purified receptor with specific DNA sequences. The calcium binding protein gene that has recently been isolated may be a suitable model system to study receptor binding to specific DNA sequences (Darwish et al., 1987; Perret et al., 1988). Also, the protein–protein interactions of purified receptor can now be investigated.

The entire primary amino acid sequence of the $1,25\text{-}(OH)_2D_3$ receptor has been determined through the efforts of receptor purification and molecular cloning. The overexpression of receptor cDNA promises to overcome the low abundance of receptor protein available in target tissues and allow detailed three-dimensional structure analysis by X-ray crystallography.

Acknowledgments

This work was supported by a program project grant no. DK-14881 from the National Institutes of Health and by the Harry Steenbock Research Fund of the Wisconsin Alumni Research Foundation.

References

Alberts, B. and Herrick, G. (1971) *Methods Enzymol.* **21,** 198–217.

Baker, A. R., McDonnell, D. P., Hughes, M., Crisp, T. M., Mangelsdorf, D. J., Haussler, M. R., Pike, J. W., Shine, J., and O'Malley, B. W. (1988) *Proc. Natl. Acad. Sci. USA* **85,** 3294–3298.

Brown, T. A., Prahl, J. M., and DeLuca, H. F. (1988) *Proc. Natl. Acad. Sci. USA* **85,** 2454–2458.

Burmester, J. K., Maeda, N., and DeLuca, H. F. (1988a) *Proc. Natl. Acad. Sci. USA* **85,** 1005–1009.

Burmester, J., Wiese, R. J., Maeda, N., and DeLuca, H. F. (1988b) *Proc. Natl. Acad. Sci. USA* **85**, 9499– 9502.

Clemens, T. L., Garret, K. P., Zhou, X-Y, Pike, J. W., Haussler, M. R., and Dempster, D. W. (1988) *Endocrinology* **122**, 1224–1230.

Coty, W. A. (1980) *Biochem. Biophys. Res. Commun.* **93**, 285–292.

Dame, M. C., Pierce, E. A., and DeLuca, H. F. (1985) *Proc. Natl. Acad. Sci. USA* **82**, 7825–7829.

Dame, M. C., Pierce, E. A., Prahl, J. M., Hayes, C. E., and DeLuca, H. F. (1986) *Biochemistry* **25**, 4523–4534.

Darwish, H. M., Krisinger, J., Strom, M., and DeLuca, H. F. (1987) *Proc. Natl. Acad. Sci. USA* **84**, 6108–6111.

DeLuca, H. F. (1985) in *Calcium in Biological Systems* (R. P. Rubin, G. B. Weiss, and J. W. Putney Jr., eds.), Plenum, New York.

DeLuca, H. F. (1988) *FASEB J.* **3**, 224–236.

DeLuca, H. F. and Schnoes, H. K. (1983) *Ann. Rev. Biochem.* **52**, 411–439.

Dokoh, S., Haussler, M. R., and Pike, J. W. (1984) *Biochem. J.* **221**, 129–136.

Eil, C., Liberman, U. A., and Marx, S. J. (1986) *Adv. Exp. Med. Biol.* **196**, 407–422.

Eisman, J. A., Barkla, D. H., and Tutton, P. J. M. (1987) *Cancer Res.* **47**, 21–24.

Eisman, J. A. and Frampton, R. J. (1984) in *Endocrine Control of Bone and Calcium Metabolism* (C. V. Colins, J. T. Pott Jr., and T. Fujita, eds.), Elsevier, Amsterdam.

Greene, G. L., Fitch, F. W., and Jensen, E. V. (1980) *Proc. Natl. Acad. Sci. USA* **77**, 157–161.

Hager, D. A. and Burgess, R. R. (1980) *Anal. Biochem.* **109**, 76–86.

Halloran, B. P. and DeLuca, H. F. (1981) *J. Biol. Chem.* **256**, 7338–7342.

Hunkapiller, M. W., Lujan, E., Ostranoer, F., and Hood, L. E. (1983) *Methods Enzymol.* **91**, 227–236.

Inaba, M. and DeLuca, H. F. (1989) *Biochim. Biophys. Acta* **1010**, 20–27.

Kohn, J. and Wilchek, M. (1982) *Biochem. Biophys. Res. Commun.* **107**, 878–884.

Kream, B. E., Reynolds, R. D., Knutson, J. C., Eisman, J. A., and DeLuca, H. F. (1976) *Arch. Biochem. Biophys.* **176**, 779–787.

Link, R. and DeLuca, H. F. (1985) in *The Receptors*, vol. 2 (P. M. Conn, ed.), Academic, New York.

Link, R. and DeLuca, H. F. (1988) *Steroids* **51**, 583–598.

Mahuran, D., Clements, P., Carrella, M., and Strasberg, P. M. (1983) *Anal. Biochem.* **129**, 513–516.

McCain, T. A., Haussler, M. R., Okrent, D., and Hughes, M. R. (1978) *FEBS Lett.* **86**, 65–70.

McDonnell, D. P., Mangelsdorf, D. J., Pike, J. W., Haussler, M. R., and O'Malley, B. W. (1987) *Science* **235**, 1214–1217.

McDonnell, D. P., Pike, J. W., and O'Malley, B. W. (1988) *J. Steroid Biochem.* **30**, 41–46.

Mellon, W. S., Franceschi, R. T., and DeLuca, H. F. (1980) *Arch. Biochem. Biophys.* **202**, 83–92.

Norman, A. W., Hunziker, W., Walters, M. R., and Bishop, J. E. (1983) *J. Biol. Chem.* **258**, 12876–12880.

Ostrem, V. K., and DeLuca, H. F. (1987) *Steroids* **49**, 73–102.

Perret, C., Lomri, N., Gouhier, N., Auffray, C., and Thomasset, M. (1988) *Eur. J. Biochem.* **172**, 43–51.

Pierce, E. A., Dame, M. C., and DeLuca, H. F. (1987) *J. Biol. Chem.* **262**, 17092–17099.

Pike, J. W. (1985) in *Vitamin D, A Chemical, Biochemical and Clinical Update* (A. W. Norman, K. Schaefer, H. G. Grigoleit, and D. V. Herratin, eds.), Walter de Gruyter, Berlin.

Pike, J. W., Dokoh, S., Haussler, M. R., Liberman, U. A., Marx, S. J., and Eil, C. (1984) *Science* **224**, 879–881.

Pike, J. W. and Haussler, M. R. (1979) *Proc. Natl. Acad. Sci. USA* **76,** 5485–5489.

Pike, J. W., Marion, S. L., Donaldson, C. A., and Haussler, M. R. (1983) *J. Biol. Chem.* **258,** 1289–1296.

Pike, J. W. and Sleator, N. M. (1985) *Biochem. Biophys. Res. Commun.* **131,** 378–385.

Pike, J. W., Sleator, N. M., and Haussler, M. R. (1987) *J. Biol. Chem.* **262,** 1305–1311.

Sandgren, M. and DeLuca, H. F. (1988) *FASEB J.* **2,** No. 5, pg. A1098. Abstract No. 4659.

Simpson, R. U. and DeLuca, H. F. (1982) *Proc. Natl. Acad. Sci. USA* **79,** 16–20.

Simpson, R. U., Hamstra, A., Kendrick, N. C., and DeLuca, H. F. (1983) *Biochemistry* **22,** 2586–2594.

Smith, E. L., Walworth, N. C., and Holick, M. F. (1986) *J. Invest. Dermatol.* **86,** 709–712.

Stumpf, W. E., Sar, M., Reid, F. A., Tanaka, Y., and DeLuca, H. F. (1979) *Science* **206,** 1188–1190.

Tanaka, H., Abe, E., Miyaura, C., Kuribayashi, T., Konno, K., Nishii, Y., and Suda, T. (1982) *Biochem. J.* **204,** 713–719.

Walters, M. R. (1985) Steroid hormone receptors and the nucleus. *Endocr. Rev.* **6,** 512–543.

Walters, M. R., Hunziker, W., Konami, D., and Norman, A. W. (1982) *J. Recept. Res.* **2,** 331–346.

Walters, M. R., Hunziker, W., and Norman, A. W. (1980) *J. Biol. Chem.* **255,** 6799–6805.

Wecksler, W. R., Ross, F. P., Moson, R. S., Posen, S., and Norman, A. W. (1980) *J. Clin. Endocrinol. Metab.* **50,** 152–157.

Retinoic Acid-Binding Protein

Michimasa Kato, Masataka Okuno,
and Yasutoshi Muto

1. Introduction

The identification of retinoic acid-binding proteins in intracellular compartments offered an alternative approach to understanding retinoic acid action by presenting a retinoic acid–protein interaction followed by the hormone-like actions, e.g., the augumentation of specific RNA synthesis. With this view, recent interests have focused on the intranuclear functions of retinoic acid-binding protein (Takase et al., 1986; Barkai and Sherman, 1987).

Although the intranuclear location of the specific [^{3}H]retinoic acid-cellular retinoic acid-binding protein complexes might have been anticipated the special role(s) of cellular retinoic acid-binding protein (Wiggert et al., 1977; Jetten and Jetten, 1979; Sani and Donovan, 1979; Mehta et al., 1982), no interaction of the complex with nuclear acceptor site has been reported. A recent in vivo study (Takase et al., 1986) suggested that cellular retinoic acid-binding protein itself may not bind nuclear sites but may be able to deliver the retinoic acid to nuclear sites, proposing a possible delivery system for retinoic acid between cytoplasm and nucleus. This unique transfer mechanism of retinoid has been also proposed in the action of retinol, for which cellular retinol-binding protein is a candidate for the shuttle (Liau et al.,1981; Barkai and Sherman, 1987), although it has not been known in the action of steroid hormones. Thus, the evidence available now suggests that the mecha-

Receptor Purification, vol. 2   ©1990 The Humana Press

nism by which retinoic acid affects the specific gene expression may be facilitated by appropriate retinoic acid-binding protein. The protein binds retinoic acid with high affinity and specificity, and should be able to migrate to the intranuclear compartment. Of retinoic acid-binding proteins, only cellular retinoic acid-binding protein is now available as a homogeneous preparation from target tissues, e.g., rat testis (Ong and Chytil, 1978; Kato et al., 1984), bovine retina (Saari et al., 1978a), and human placenta (Okuno et al., 1987), and its complementary DNA and gene have been cloned (Shubeita et al., 1987; Wei et al., 1987).

Thus, a chapter on the purification of retinoic acid-binding protein needs entail only the information about isolation and characterization of cellular retinoic acid-binding protein. However, there is no guarantee that the above well-characterized cellular retinoic acid-binding protein is always homologous to the other numerous retinoic acid-binding proteins that have been determined as retinoic acid-binding activity in a number of tissues and cells, especially of the neoplastic and experimental tumor origins (*see review;* Chytil and Ong, 1984). Many unpublished observations suggest the possibility that several retinoic acid-binding proteins besides cellular retinoic acid-binding protein exist in certain organs and tumor tissues. Recently, Hashimoto et al. (1988) reported the novel retinoic acid-binding protein, retinoid specific binding protein, that was expected to be the retinoic acid receptor. Although the relation between this novel protein and other candidates for retinoic acid receptor(s) (Petkovich et al., 1981; Giguere et al., 1987; Brand et al., 1988) is not yet known, the report may encourage future experiments designed to find other unknown retinoic acid-binding proteins.

This chapter aims at giving appropriate improvements on each step of conventional purification methods for cellular retinoic acid-binding protein, attempting to broaden the category and to improve the yield of cellular retinoic acid-binding proteins. By way of illustration, this chapter refers to the methodology for purifying human placental cellular retinoic acid-binding protein (Okuno et al., 1987).

2. Purification of Human Placental Cellular Retinoic Acid-Binding Protein

Cellular retinoic acid-binding protein (CRABP) was purified from a soluble acetone powder extract of human placentae by a series of purification procedures.

The cellular retinol-binding protein (CRBP) was also isolated, since the level and location were known in the rat placenta, (Kato et al., 1985). As shown in Table 1, CRABP was purified 75,400-fold, based on total soluble acetone powder extract from 2 kg of placentae.

2.1. Preparation of Placenta Extract

The placentae were dried by lyophilization and were homogenized with 4 L of ice cold acetone, using an ultradispenser (Polytron type). The dried powder was rehomogenized with 4 L of 50 mM Tris-HCl buffer, pH 8.4, containing 4 mM EDTA, 0.004% NaN_3, and 12 mM monothioglycerol. The homogenate was centrifuged at 20,000g for 60 min to remove debris. The resulting supernatant was filtrated through glass fiber filter paper (pore size, 10 μm). To this soluble acetone powder extract was added all-trans [15-^{14}C] retinoic acid diluted with cold retinoic acid (8.5 μCi/μmol) to give a final concentration of 15 μM, and incubated at 4°C for 16 h.

Taking all our experiments into consideration, the preparation of acetone powder from lyophilized tissue is likely to produce a good solubilization of certain proteins that interact with artificial micelles in the conventional extract from untreated tissue. Although a delipidization effect on the solubilization of CRABP is not properly demonstrated, the preparation of an acetone powder may have a preventive effect on the formation of high-mol-wt lipid-protein aggregate that has been reported to interact with CRBP in the rat liver homogenate (Sklan et al., 1982).

To obtain a clear acetone powder extract, present study employed a filtration method using glass fiber filter paper. However, this procedure could not handle a large quantity of homogenate. In general preparation of tissue protein, an extract obtained by ultracentrifugation of homogenate contains numerous undesirable components, including insoluble microparticles, lipids, and substances with high polarity. These components are very harmful to chromatography media, and affect the elution profile of the target protein. Moreover, the large vol of highly concentrated extracts compel the researcher to follow many cumbersome steps in the initial stage of purification.

Recently, we attempted to improve the process of large-scale preparation of relatively small proteins. The mol wt of bovine CRABP is 15,460 daltons from its amino acid sequence (Sundelin et al., 1985a). The highest value (16,300 kDa) has been estimated in the study of bovine eye CRABP (Saari et al., 1978a). In a pilot study of purification, we found that CRABP in certain liver extract could readily pass through a XM300 diaflo mem-

Table 1
Steps in the Purification of CRABP from Human Placenta

Procedure	Protein (mg)[a]	CRABP (mg)	CRBP (mg)	Purification (-fold)		Recovery (%)	
				CRABP	CRBP	CRABP	CRBP
Soluble acetone powder extract	181.000	2.40[b]	8.72[b]				
Sephadex G-50, (I)	6.000	2.06[b]	7.12[b]	26	25	86	82
Sephadex G-50, (II)	975	1.59[b]	6.73[b]	123	143	66	77
DEAE-Cellulose, pH 6.4	111	1.09[b]	5.01[b]	740	940	45	57
DEAE-Cellulose, pH 8.4	22.4	0.46[b]	4.04[b]	1,550	3,740	19	46
Sephadex G-50, (III)	9.9	0.41[b]	3.59[b]	3,120	7,530	17	41
SP-Sephadex, pH 4.9	0.32 (CRAB)	0.32[c]	2.70[c]	75,400	20,700	13	31
	2.7 (CRBP)						

[a]Estimated according to the method of Lowry et al. (1951).
[b]Estimated by a gel filtration binding assay (*see* text).
[c]Assumed since no protein other than CRABP or CRBP was demonstrable.

brane of Amicon (unpublished observations). Although the highest mol wt of filtered protein was approx 150,000kDa, only a trace amount of albumin (66,000kDa) was found to pass through a XM300 diaflo membrane. Thus, it should be noted that ultrafiltration is affected by the size and shape of a protein and by many other factors, e.g., the concentration of the sample. Fortunately, the pilot study resulted in remarkable purification of CRABP, and eliminated numerous other components.

Prior to the above ultrafiltration, tissue extract was treated with CDR (Cell Debris Remover: Whatman) to obtain a clear solution. For the purification of low-mol-wt retinoid-binding protein, the protocol below may be helpful in a large-scale preparation:

1. Homogenization of a tissue in four tissue volumes (V/W) of 10 mM Tris-HCl buffer, pH 7.6, containing 4 mM EDTA, 0.05% NaN$_3$, and 0.01% FOY®) (gabexate mesilate: Ono Pharmaceutical Co. Ltd., Osaka, Japan);
2. Centrifugation at 20,000g for 10 min;
3. Removal of both debris and undesirable components by adding 1:10 original tissue weight (W/W) of CDR to the homogenate (stirring for 60 min), and by subsequent centrifugation at 20,000g for 10 min;
4. Filtration of the supernatant through the 1:20 original tissue weight (W/W) of CDR that is sandwiched between two filter papers and packed in a Buchner funnel;
5. Ultrafiltration of the resulting solution by XM300 diaflo membrane under a weak (0.1 kg/cm^2) pressure of nitrogen gas;
6. Concentration of the filtrate by a YM5 diaflo membrane.

In the above treatments, the minimal requirement of CDR should be determined by a pilot test. The above condition has been determined for the isolation of normal liver CRBP. On the other hand, a step by XM300 diaflo membrane is not correct for the fractionation of middle size proteins (roughly, 50,000–150,000 kDa), which may suffer a partial filtration.

This new method may significantly enhance the chemical economics of a large-scale procedure, eliminating most costly and cumbersome steps. Most research laboratories have general limitation to the consumption of costly [^{3}H]retinoic acid. Actually, in many cases, reasonable expense may not allow the complete saturation of a large vol of whole homogenate with undiluted [^{3}H]retinoic acid. Thus, the appropriate treatment of starting material is the most important step in the isolation and detection of retinoic acid-binding proteins.

2.2. Fractionation Steps

Fractionation procedures were carried out at 4°C. The fractions from each chromatography were assayed for protein-bound retinol by measuring fluorescence with excitation at 350 nm and emission at 475 nm and for protein-bound [^{14}C]retinoic acid by determination of radioactivity to identify the elution portions containing binding proteins. Except in the final purification procedure, CRBP and CRABP were eluted together and were not separated from each other during fractionation procedures.

2.2.1. Gel Filtration

The placenta extract was submitted to gel chromatography on a Sephadex G-50 (medium) column (10×125 cm) in 50 mM Tris-HCl buffer, pH8.4, containing 4 mM EDTA, 0.004% NaN$_3$, and 25 mM NaCl. Fractions of 18 mL each were collected at a flowrate of 240 mL/h. Eight column runs were repeated to process the whole preparation. Fractions containing both CRBP and CRABP were combined and concentrated by ultrafiltration using a YM5 membrane in an Amicon cell. The solution was then reapplied to the same column and eluted as described above (Fig.1). Monitoring the fluorescence of protein-bound retinol and the radioactivity of protein-bound [^{14}C]retinoic acid, a prominent peak centered at 5094 mL (fraction 283) was observed.

For the purpose of CRABP purification, Sephadex G-50 is one of the best choices for gel filtration. However, different chromatography media with suitable mol wt spectra should be employed when samples contain unknown binding proteins.

2.2.2. Anion-Exchange Chromatography

The sample from the gel filtration (Fig.1) was dialyzed against 15 mM imidazole-acetate buffer, pH 6.4, containing 0.004% NaN$_3$, and was applied to a column (3.2×40 cm) of DEAE-cellulose equilibrated with the above buffer. The column was eluted with a linear gradient of imidazole-acetate buffer from 15 to 200 mM at pH 6.4 (total vol, 1600 mL). Fractions of 15 mL each were collected at a flowrate of 40 mL/h. Three sharp peaks of fluorescence and/or ^{14}C radioactivity were observed in this column (Fig. 2). The first peak was centered at 1005 mL (fraction 168), the second peak was at 1185 mL (fraction 186) and the third peak was at 1335 mL (fraction 201). Retinol-binding protein (RBP) should be suspected of binding retinoids. Since the size of RBP (approx 21,00 kDa: Kanai et al., 1968) is similar to that of CRABP and CRBP (15,460 and 15,700 kDa, respectively; Sundelin et al., 1985a,b), it is conceivable that RBP has been fractionated concurrently with CRABP and CRBP in such a large-scale

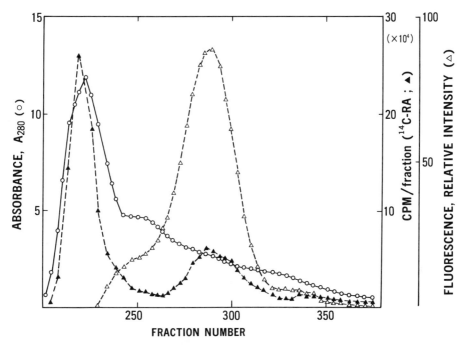

Fig. 1. Second chromatography on Sephadex G-50 of the combined binding protein-containing fractions obtained by eight runs of the first chromatography on Sephadex G-50. Fractions 250–320 containing both CRABP and CRBP were pooled. Absorbance was monitored at 280 nm (O). CRABP was monitored by radioactivity of protein-bound [^{14}C]retinoic acid (^{14}C-RA; ▲). CRBP was monitored by fluorescence of protein-bound retonol (Δ; excitation at 350 nm and emission at 475 nm).

gel filtration procedure on Sephadex G-50 (Fig. 1). On analysis of fluorescence spectra and immunological reactivities, the second and third peaks of fluorescence and/or radioactivity were clearly attributed to retinol bound to RBP. The excitation spectrum of the retinol bound to RBP exhibits a peak at 330 nm, whereas that of retinoids bound to CRABP and CRBP have a peak at 350 nm. Several assay systems for human RBP are now commercially available. RBP occurs as a free form in the human placenta. RBP circulates in plasma as a protein-protein complex with transthyretin, and very little RBP is normally present in the free state (Goodman, 1984). In the study in human, Sklan et al. (1985) reported that a significant amount of free RBP is present in the maternal and fetal circulation. Their results may be relevant to our observation. Thus, it should be noted that RBP can bind with not only retinol but also retinoic acid when excess retinoic acid is added in vitro (Heller and Horwitz, 1973). The unknown broad retinoic acid-binding activity has been also observed in the fractions between 106 to 130 (Fig. 2), but it still remains to be characterized.

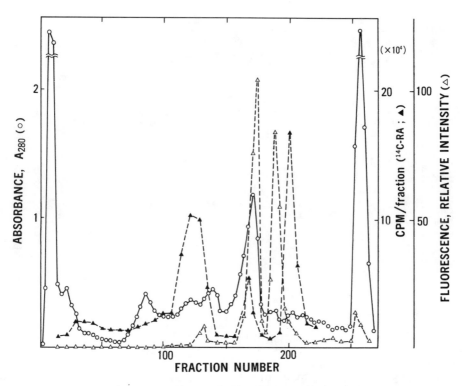

Fig. 2. Chromatography on DEAE-cellulose of the sample from the second chromatography on Sephadex G-50 (Fig. 1). The column was eluted with a linear gradient of imidazole-acetate buffer from 15 to 200 mM, pH 6.4 (total volume, 1600 mL). Three prominent peaks of fluorescence and/or radioactivity were observed on this column. Following analyses of fluorescence spectra and immunological reactivities, the first peak (fraction 157–179), containing CRABP and CRBP, was pooled. The second (fraction 180–192) and the third (fraction 193–210) peaks were attributed to the retinoids bound to RBP (*see* text). Absorbance was monitored at 280 nm (O). CRABP was monitored by radioactivity (^{14}C-RA; ▲) and CRBP was monitored by fluorescence (Δ).

Pooled and concentrated fractions containing both CRBP and CRABP (the first sharp peak in Fig. 2) were then dialyzed against 50 mM Tris-HCl buffer, pH 8.4. containing 4 mM EDTA and 0. 004% NaNa$_3$, and applied to a column (1.5×15 cm) of DEAE-cellulose equilibrated with the above buffer, and eluted with a linear gradient of NaCl from 0 to 150 mM in the same buffer (total vol, 200 mL). Fractions of 3 mL each were collected at a flowrate of 10 mL/h. On this column, CRABP and CRBP were eluted together in a single peak, centered at 40 mL of a 200 mL gradient (data not shown). As shown in Table 1, anion-exchange chromatography at pH 8.4 produced satisfactory results in the purification of CRBP, and is also suitable for CRABP purification. Although the reason is unclear, an alkaline condition may have a considerable effect on the reduced recovery of CRABP bound to anion-exchange media. Saari et al. (1978b) reported the tendency of CRABP to form insoluble aggregates and to lose binding activity during purification and storage. The chromatography on the anion exchange medium at pH 8.4 is likely to accelerate such a tendency, leading to the lower purification fold, in comparison with the similar step at pH 6.4 (Table 1).

2.2.3. Rechromatography on Sephadex G-50

The resulting sample from anion-exchange chromatography at pH 8.4 was then submitted to gel chromatography on a smaller column (2 × 95 cm) of Sephadex G-50 (superfine) eluted with 50 mM Tris-HCl buffer, containing 4 mM EDTA, 0. 004% NaN_3, and 25 mM NaCl at a flow-rate of 6 mL/h. All three peaks of fluorescence, radioactivity, and absorbance at 350 nm shared the same elution volume centered at 134 mL (data not shown). Absorbance at 350 nm of this peak was somewhat higher than that at 280 nm, indicating that the fraction was abundant in cellular retinoid-binding proteins. When a large-scale gel filtration step was assigned to the initial stage of purification, the addition of another gel filtration step, near to the final stage, produced a good effect on the advance of purification fold without significant loss of target protein(s).

2.2.4. Cation-Exchange Chromatography

Pooled and concentrated fractions from preparative gel filtration were dialyzed against 10 mM sodium acetate buffer, pH 4.9. The final step of chromatography was performed on a column (1.5 × 30 cm) of SP-Sephadex equilibrated with the above buffer. The dialyzed sample was applied to the column, and eluted with a linear gradient of sodium acetate buffer from 10 to 130 mM at pH 4.9 (200 mL total). As shown in Fig. 3, two peaks of absorbance at 350 nm were observed in an elution vol centered at 69 mL (fraction 23) and 120 mL (fraction 40). The first peak corresponded to the peak of fluorescence of protein-bound retinol, and the other overlapped with the peak of radioactivity of bound [^{14}C]retinoic acid, indicating the complete separation of CRABP from CRBP.

3. Characterization of Cellular Retinoic Acid-Binding Protein

3.1. Gel Electrophoresis

The purity of CRABP was assessed electrophoretically by three different systems (Fig. 4). Polyacrylamide slab gel electrophoresis was performed using 10% gel at pH 8.8, in order to demonstrate a clear separation of CRABP and CRBP from each other (Fig. 4A). By illuminating

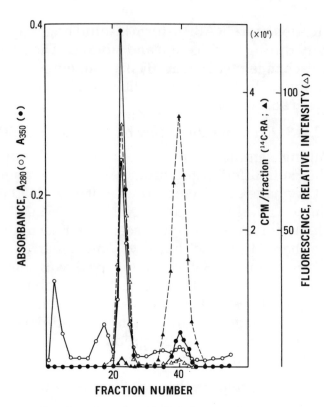

Fig. 3. Final chromatography on SP-Sephadex resulting in pure CRABP and CRBP. The column was eluted with a linear gradient of a sodium acetate buffer from 10 to 130 mM, pH 4.9 (total vol, 200 mL). Fractions 20–25 were pooled and contained pure CRBP; fractions 38–42 were pooled to yield pure CRABP. Absorbance was monitored at 280 nm (O) and 350 nm (●). CRABP was monitored by radioactivity (^{14}C-RA; ▲) and CRBP was monitored by fluorescence (Δ).

the unstained gel with long-wave ultraviolet light in the dark, a single yellow-green band was observed, which corresponded exactly to the single stained band of CRBP (cols. 1 and 2 in Fig. 4A).

Polyacrylamide disk gel electrophoresis in 0.1% sodium dodecyl sulfate (SDS) was also performed using 13% gel at pH 8.8 (Fig. 4B). The mol wt of CRABP was calculated based on a comparison of its migration rate through the gel with those of protein standards: ovotransferrin (76,000–77,000kDa), bovine serum albumin (66,200kDa), ovalbumin (45,000kDa), bovine chymotrypsinogen A (25,700kDa), horse myoglobin (17,200kDa), and horse cytochrome C (12,300kDa). The apparent mol wt of CRABP was calculated to be 14,600.

Analytical isoelectric focusing electrophoresis was perfomed on an LKB Ampholine® polyacrylamide gel electrophoresis plate, pH 3.5–9.5. The isoelectric point of the protein was calculated based on a comparison

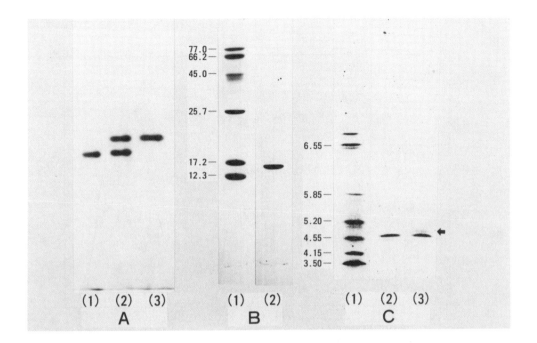

Fig. 4. Electrophoretic analyses of purified CRABP by different systems. Anode to the lowest. Ten µg of purified protein were analyzed in each gel. (**A**) Polyacrylamide slab gel electrophoresis of CRBP (columns 1 and 2) and CRABP (columns 2 and 3) showing clear separation from each other, (**B**) SDS-polyacrylamide disk gel electrophoresis of mol wt markers (gel 1) and CRABP (gel 2) suggesting size homology (approx 14.6 kDa). (**C**) Analytical isoelectric focusing in the pH interval of 3.5–9.5 performed on pI markers (gel 1) and on (gels 2 and 3) in the presence (gel 2) and absence (gel 3) of excess retinoic acid (*see* text). Note the disappearance of the faint band (arrow in gel 3) by a preincubation of the sample with excess retinoic acid (gel 2).

of its migration rate with those of protein markers: human carbonic anhydrase B (pI 6.55), bovine carbonic anhydrase B (pI 5.85), β-lactoglobulin A (pI 5.20), soybean trypsin inhibitor (pI 4.55), and glucose oxidase (pI 4.15). Purified CRABP showed two bands with isoelectric pH value of 4.78 and 4.82 (column 3 in Fig. 4C). The faint band at pI 4.82 (arrow in Fig. 4C) disappeared when the CRABP sample was incubated with excess retinoic acid before electrophoresis. This observation indicates that the faint band is apo-CRABP, and that the binding of retinoic acid to the binding protein may alter the protein charge. A similar change of charge occurs in human CRBP, as described by Fex and Johannesson (1982). The isoelectric point of human holo-CRABP (pH 4.78) is similar to the value of rat CRABP (Ross et al., 1980).

3.2. Absorption Spectrum

The absorption spectra of purified CRABP and CRBP as shown in Fig. 5, clearly indicate the presence of the respective bound ligands with the absorption characteristics of retinoic acid (left) and retinol (right).

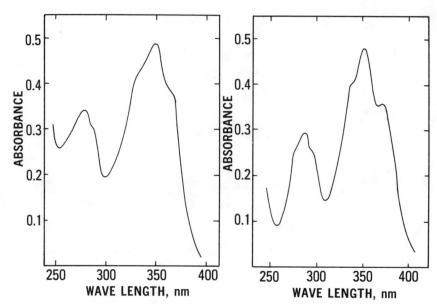

Fig. 5. Absorption spectra of CRABP bound with retinoic acid (left) and CRBP bound with retinol (right), at a protein concentration of $10^{-5}M$ in 20 mM Tris-HCl buffer (pH 8.4).

CRABP displayed two absorption peaks, with maxima at 283 and 350 nm. The former represented protein absorbance and the latter was owing to protein-bound retinoic acid. The ratio of A_{350} to A_{280} was 1.62. This absorbance ratio is somewhat lower than the value of rat CRABP, as reported by Ong and Chytil (1978). This discrepancy may be explained by the presence of apo-CRABP, which was detected on isoelectric focusing electrophoresis (Fig. 4C). The absorption spectrum of CRBP was similar to that of CRABP.

3.3. Fluorescence Spectrum

The fluorescence excitation and emission spectra of retinoic acid bound to CRABP and retinol bound to CRBP are shown in Fig. 6. The excitation spectrum of CRABP showed a major peak at 350 nm, a second peak at 370 nm, and a small shoulder at 290 nm. The emission spectrum of CRABP showed a single peak at 475 nm (Fig. 6A). Both the excitation and emission spectra of CRBP were very similar to those of CRABP (Fig. 6B).

3.4. Quantitation of Retinoid-Binding Activity

The retinoid-binding activities of CRABP and CRBP were assayed by high performance size exclusion chromatography according to the

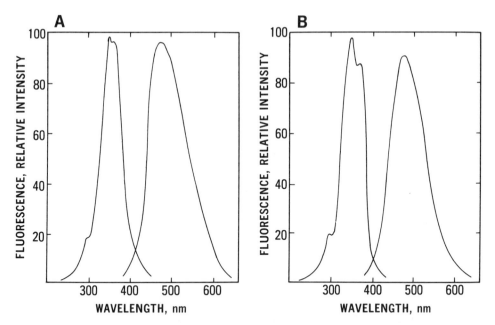

Fig. 6. Uncorrected fluorescence spectra of retinoic acid bound to CRABP (A) and retinol bound to CRBP (B), $5 \times 10^{-6}M$ in 20 mM Tris-HCl buffer (pH 8.4). The excitation spectra of CRABP (A, left) and CRBP (B, left) were determined measuring emission at 475 nm; the emission spectra of CRABP (A, right) and CRBP (B, right) were determined with excitation at 350 nm.

method of Allegretto et al. (1983). Generally, 1 mL of aliquot from each purification step was incubated with 80 nM [^{3}H]-retinoic acid or [^{3}H] retinol in the presence or absence of a 200-fold molar excess of respective retinoid. The samples were then analyzed by an HPLC system consisting of a TSK G 3000SW column (Toso Co., Tokyo, Japan), a CCPM pump (Toso Co.), a Toso UV-8010 HPLC detector set at 280 nm, and a Chromatocorder 11 integrator (System Instruments Co., Tokyo, Japan). Elution was perfomed with 200 mM phosphate buffer, pH 7.2, at a flowrate of 1 mL/min. Sixty fractions were collected for a total period of 30 min, and radioactivity in each fraction was measured with a Beckman LS-7500 liquid scintillation counter.

4. Simulation for Future Studies

Since 1973, when Bashor et al. first detected a retinol-binding activity in rat tissue, retinoid-binding activities have been assayed in a variety of tissue and cell extracts by sucrose gradient analysis. Subsequently, this idea hatched similar methods, including gel filtration (Saari et al., 1978b), agarose gel electrophoresis (Huber et al., 1978), high pressure size exclusion chromatography (Allegretto et al., 1983), as well as batch-assays by charcoal-coated dextran (Trown et al., 1980). These methods certainly contributed to the detection of conspicuous retinoid-binding proteins, i.e., CRBP and CRABP. However, special attention should be paid to the detection of unknown (novel) retinoid-binding activity.

The use of radioactive retinoids with the highest specific activity is essential to obtain high sensitivity of the assay. Further, it is desirable that the sample should be submitted to a pilot study. In such a relatively large-scale pilot study, the unlabeled tissue extract should be fraction-ated, first by a conventional gel filtration media, e.g., Sephacryl S 300 (Pharmacia). The rapid method of quantitating retinoid-binding activity shall be suitable for the assay of many fractions. For such a purpose, dex-tran-coated charcoal method (Trown et al., 1980) or filter binding assay (Hashimoto et al., 1988) shall be available to analyze a large number of samples at the same time. The appropriate fractions, corresponding to a retinoid-binding activity, are fractionated by the certain ion-exchange chromatography. The former gel-filtration step aims at the exclusion of high-mol-wt lipid-protein aggregates from specific retinoid-binding proteins. As mentioned, the aggregates might have an obstructive effect on the solubilization and binding activity of a true retinoid-binding protein. A calibrated gel filtration column can give the elution vol of known retinoid-associated proteins, including CRBP, CRABP, RBP, al-bumin, and RBP bound to transthyretin. It should be noticed that the unknown retinoid-binding protein might be overlapped with above proteins. However, the next chromatography on ion-exchange medium may produce a good chance to find a novel retinoid-binding protein.

5. Conclusion

To provide a satisfactory explanation of the molecular mechanism of action of retinoic acid, a specific experimental approach will have to be provided to the common problem that is shared with other vitamins and hormones, namely, the mechanism of control of gene expression. Of the compounds that affect gene expression, only retinoids have a well-defined unique specific cytosolic binding protein. In this respect, exhau-stive and more extensive researches are desirable to explore every intra-cellular binding activity of specific retinoic acid-binding proteins. Thus far, the conspicuous retinoic acid-binding activities of relatively small proteins have been assumed to be that of CRABP by many investigators. However, the prominent retinoic acid-binding activity of CRABP may conceal other minor activities. In this chapter, the authors proposed possible ways to find novel retinoic acid-binding protein(s).

Abbreviations

CRABP	cellular retinoic acid-binding protein
CRBP	cellular retinol-binding protein
RBP	retinol-binding protein
HPLC	high performance liquid chromatography
pI	isoelectric point

References

Allegretto, E. A., Kelly, M. A., Donaldson, C. A., Levine, N., Pike, J. W., and Haussler, M. R. (1983) *Biochem. Biophys. Res. Commun.* **116,** 75–81.

Barkai, U. and Sherman, M. I. (1987) *J. Cell Biol.* **104,** 671–678.

Bashor, M. M., Toft, D. O., and Chytil, F. (1973) *Proc. Natl. Acad. Sci. USA,* **70,** 3483–3487.

Brand, N., Petkovich, M., Krust, A., Chambon, P., The, de, Marchio, A., Tiollais, P., and Dejean, A. (1988) *Nature* **332,** 850–853.

Chytil, F. and Ong, D. E. (1984) in *The Retinoid* vol 2 (M. B. Sporn, A. B. Roberts, and D. S. Goodman, eds.) Academic, London, pp. 89–123.

Fex, G. and Johannesson, G. (1982) *Cancer Res.* **44,** 3029–3032.

Giguere, V., Ong. E. S., Segui, P., and Evans, R. M. (1987) *Nature* **330,** 624–629.

Goodman, D. S. (1984) in *The Retinoid* vol 2 (Sporn, M. B., Roberts, A. B. and Goodman, D. S., eds.) Academic Press, London, pp. 41–88.

Hashimoto, Y., Kagechika, H., Kawachi, E., and Shudo, K. (1988) *Jpn. J. Cancer Res.* (Gann) **79,** 473–483.

Heller, J. and Horwitz, J. (1973) *J. Biol. Chem.* **248,** 6308–6316.

Huber, P. R., Geyer, E., Kung, W., Matter, A., Torhorst, J., and Eppenberger, U. (1978) *J. Natl. Cancer Inst.* **61,** 1375–1378.

Jetten, A. M. and Jetten, M. E. R. (1979)*Nature* **278,** 180– 182.

Kanai, M., Raz, A., and Goodman, D. S. (196?) *J. Clin. Invest.* **47,** 2025–2044

Kato, M., Kato, K., and Goodman, D. S. (1984) *J. Cell Biol.* **98,** 1696–1704.

Kato, M., Kato, K., and Goodman, D. S. (1985) *Lab. Invest.* **52,** 475–484.

Liau, G., Ong, D. E., and Chytil, F. (1981) *J. Cell Biol.* **91,** 63–68.

Lowry, O. H., Rosenbrough, N. J., Farr, A. L., and Randall, R. J. (1951) *J. Biol. Chem.* **193,** 265–275.

Okuno, M., Kato, M., Moriwaki, H., Kanai, M., and Muto, Y. (1987)*Biochim. Biophys. Acta* **923,** 116–124.

Ong, D. E. and Chytil, F. (1978) *J. Biol. Chem.* **253,** 4551–4554.

Petkovich, M., Brand, N. J., Krust, A., and Chambon, P. (1987) *Nature* **330,** 444–450.

Ross, A. C., Adachi, N., and Goodman, D. S. (1980) *J. Lipid Res.* **21,** 100– 109.

Saari, J. C., Futterman, S., and Bredberg, L. (1978) *J. Biol. Chem.* **253,** 6432–6436.

Saari, J. C., Futterman, S., Stubbs, G. W., Hefferman, J. T., Bredberg, L., Chan, K. Y., and Albert, D. M. (1978b) *Invest. Ophthalmol. Vis. Sci.* **17,** 988–992.

Sani, B. and Donovan, M. (1979) *Cancer Res.* **39,** 2492–2496.

Shubeita, H. E., Sambrook, J. F., and McCormick, A. M. (1987) *Proc. Natl. Acad. Sci. USA* **84,** 5645–5649.

Sklan, D., Blaner, W. S., Adachi, N., Smith, J. E., and Goodman, D. S. (1982) *Arch. Biochem. Biophys.* **214,** 35–44.

Sklan, D., Shalit, I., Lasebnik, N., Spirer, Z., and Weisman, Y. (1985) *Brit. J. Nutr.* **54,** 577–583.

Sundelin, J., Anundi, H., Trägardh, L., Eriksson, U., Lind, P., Ronne, H., Peterson, P. A., and Rask, L. (1985a) *J. Biol. Chem.* **260,** 6483–6493.

Sundelin, J., Das, S. R., Eriksson, U., Rask, L., and Peterson, P. A. (1985b)*J. Biol. Chem.* **260,** 6494–6499.

Takase, S., Ong., D. E., and Chytil, F. (1986) *Arch. Biochem. Biophys.* **247,** 328–334.

Trown, P. W., Palleroni, A. V., Bohoslawec, O., Richelo, B. N., Halpern, J. M., Gizzi, N., Geiger, R., Lewinski, C., Marchlin, L. J., Jetten, A., and Jetten, M. E. R. (1980) *Cancer Res.* **40,** 212–220.

Wei, L. N., Mertz, J. R., Goodman, D. S., and Chi, M. (1987)*Molec. Endocrin.* **1,** 526–534.

Wiggert, B., Russell, P., Lewis, M., and Chader, G. (1977) *Biochem. Biophys. Res. Commun.* **79,** 218–225.

Receptor Purification of the Thyroid Hormone Receptor

Kazuo Ichikawa and Leslie J. DeGroot

1. Introduction

Partial purification of nuclear thyroid hormone receptor has been reported from several laboratories (Table 1). However, complete purification of the receptor has not yet been achieved. Difficulty in purification of the thyroid hormone receptor is a result of

1. Its extremely low abundance in target organs;
2. Rapid loss of hormone binding activity during purification; and
3. Ineffectiveness of affinity purification methods.

The following is a description of a method for purification of nuclear thyroid hormone receptors from rat liver. Nuclei are prepared from 4–5 kg rat liver using a heavy sucrose-ultracentrifugation method. After solubilizing the receptor in 0.3M KCl, the receptor is purified by sequentially using hydroxylapatite column chromatography, ammonium sulfate precipitation, Sephadex G-150 gel filtration, DNA-cellulose column chromatography, DEAE-Sephadex column chromatography and heparin-Sepharose column chromatography. This method provides a receptor preparation with approx 200 pmol 3,5,3'-triiodo-L-thyronine (T_3) binding capacity in 0.22 mg total protein. Assuming that one T_3 molecule binds to the 49,000-mol wt unit of the receptor, we obtain 6.4–14.7 µg of receptor protein with 4.2–4.9% purity.

Receptor Purification, vol. 2   ©1990 The Humana Press

Table 1
Previous Reports of Nuclear Thyroid Hormone Receptor Purification[a]

Author	Specific activity, pmol T_3/mg protein	Purity, %	Columns used
Latham et al. (1976)	1.55	0.008	QAE-Sephadex
Silva et al. (1977)	46	0.2	DEAE-Sephadex
Toressani et al. (1978)	104.4	0.5	Sephadex G-100 DEAE-Sephadex DNA-Sepharose
Apriletti et al. (1981)	247	1.2	Sephadex G-100 T_3-affinity DEAE-Sephadex
Apriletti et al. (1988)	287	1.4	Heparin-Sepharose DEAE-Sephadex Phospho-Ultrogel Size exclusion HPLC HIC HPLC

[a]Purity of nuclear thyroid hormone receptor was calculated assuming that mol wt of the receptor is 49,000 and one receptor molecule binds one T_3 molecule. HPLC, high-performance liquid chromatography. HIC, hydrophobic interaction chromatography.

2. Receptor Assay

2.1. Principles of the Assay

Receptor assay is performed by in vitro measurement of ^{125}I-labeled 3,5,3'-triiodo-L-thyronine (T_3) binding. Dowex 1×8, Cl⁻, 200- to 400-mesh anion exchange resin (BioRad, Richmond, CA) is used to separate receptor bound T_3 from free T_3 based on its ability to selectively adsorb free T_3. However, since Dowex resin adsorbs the receptor bound T_3 as well as free T_3 under certain circumstances, resin should be used in the presence of more than $0.2M$ KCl or NaCl at pH 8.0 with 30–50 µg heated nuclear extract protein. This is especially important when purified receptor is used. Nitrocellulose filter binding assay (Inoue et al., 1983) gives good linearity between concentration of receptor and binding activity at low levels of protein concentration, as is expected from the principles of this method in which bound hormone is retained on a filter. However, it is still necessary to add enough protein in the binding assay mixture in order to prevent loss of the purified receptor during incubation. The Dowex resin test gives identical linearity when heated nuclear extract is added. Because of the simplicity of the resin test, limited protein retaining capacity of the filter paper, and inaccuracy in counting bound hormone on filter papers, the resin test is routinely used.

2.2. Pretreatment of Dowex Resin and Preparation of Heated Nuclear Extract

Dowex resin should be treated as follows before use. Five hundred g of the resin is soaked in $0.1N$ HCl and heated to 90°C for 20 min. The resin is washed with ~9 L of distilled water until the pH of the resin suspension becomes neutral. Resin is dried in an oven and kept dry until used.

Heated nuclear extract is prepared by incubating nuclear extract at 100°C for 15 min; this is followed by centrifugation at $1000g$ for 15 min to remove any precipitate. This material has no specific T_3 binding activity in itself and serves to reduce nonspecific binding of nuclear receptor protein to resin beads.

2.3. Assay Methods

Receptor preparation is incubated in 0.3 mL of buffer D (*see* Section 4.1.) with $[^{125}I]T_3$ in the presence of 50 µg heated nuclear extract protein for 2 h at 22°C. Plastic tubes are usually used for the incubation. Incubation temperature of 22°C is optimum for T_3 binding, and binding reaches a plateau after 60–90 min. T_3 binding activity is stable for up to 4 h. After incubation, assay tubes are immediately cooled in an ice bath and 0.3 mL of cold buffer D containing 40 mg of Dowex 1×8, Cl⁻, 200- to 400-mesh anion exchange resin is added. This is incubated with intermittent vortexing for 5 min, followed by centrifugation. Radioactivity in 0.3 mL of the supernatant is counted for determination of receptor bound $[^{125}I]T_3$.

For screening of column fractions, 50 pM $[^{125}I]T_3$ and receptor preparation (1 µL for hydroxylapatite column; 5 µL for Sephadex G-150, DNA-cellulose, and DEAE-Sephadex A-50; and 3 µL for heparin-Sepharose column) are used. For determination of maximal T_3 binding capacity, 5 nM $[^{125}I]T_3$ is used. Duplicate tubes containing 0.3 µM unlabeled T_3 are included for determination of nonspecific $[^{125}I]T_3$ binding. Specific receptor-bound $[^{125}I]T_3$ is calculated by subtracting nonspecific $[^{125}I]T_3$ binding from total $[^{125}I]T_3$ binding.

3. Protein Assay

Protein content is estimated by a Coomassie Blue G-250 dye binding method (Bradford, 1976) using a bovine serum albumin standard. The dye binding protein assay method, although sensitive, can give different results with various proteins and may bias the extent of apparent purification. Accuracy of protein determination is critical in assessing receptor purification. The color development of sequentially purified receptor preparations, bovine serum albumin, and bovine gamma globulin, were compared using three different assays based on different mechanisms (Bradford, which is used in the present study, Lowry [Lowry et al., 1951], and Biuret [Gornall et al., 1949]). Bovine serum albumin produced a higher color development than did bovine gamma globulin in the Bradford method. The Bradford method, using bovine serum albumin standard, agreed with the other methods, whereas the same method, using bovine gamma globulin standard, did not agree with the other methods. Purified receptor preparations are more like bovine serum albumin than bovine gamma globulin in terms of their color yield in response to the dye in the Bradford method. Because of procedural simplicity and higher sensitivity, this method is used for determination of protein content of receptor preparations.

4. Preparation of Nuclear Extract

4.1. Buffers

1. Buffer A: 0.32M sucrose/3 mM $MgCl_2$/2 mM EDTA;
2. Buffer B: 2.8M sucrose/1 mM $MgCl_2$;
3. Buffer C: 0.32M sucrose/1 mM $MgCl_2$/20 mM Tris-HCl (pH 7.4)/1 mM dithiothreitol/1 mM phenylmethylsulfonyl fluoride;
4. Buffer D: 0.3M KCl/1 mM $MgCl_2$/10 mM potassium phosphate (pH 8.0)/1 mM dithiothreitol.

One hundred mM phenylmethylsulfonyl fluoride in isopropanol stored at 0°C should be warmed up to dissolve the crystals and is added to the buffer before use. One hundred mM dithiothreitol in H_2O stored at −20°C should be thawed and is added to the buffer before use.

4.2. Methods

Figure 1 shows the outline of the methods. Immediately after killing rats, the livers are excised and quickly frozen in liquid nitrogen. Frozen liver is kept at −80°C until used. Quickly frozen liver is available from

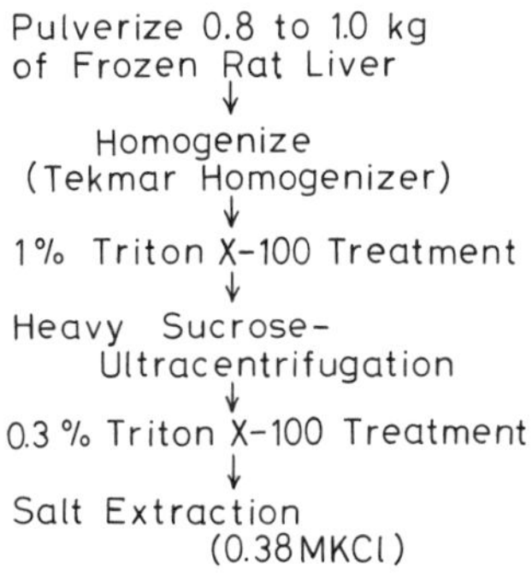

Pulverize 0.8 to 1.0 kg
of Frozen Rat Liver
↓
Homogenize
(Tekmar Homogenizer)
↓
1% Triton X-100 Treatment
↓
Heavy Sucrose–
Ultracentrifugation
↓
0.3% Triton X-100 Treatment
↓
Salt Extraction
(0.38 M KCl)

Fig. 1. Outline of the method for large scale preparation of nuclear extract from rat liver.

Rockland Co. (Gilbertsville, PA). Frozen liver (140 g) is pulverized and 300 mL of buffer A warmed to 37°C is added with stirring. The tissue is thawed in 1 min and reaches a final temperature of 5°C. The following procedures are performed at 0–4°C. The tissue is settled and the supernatant is discarded. After another wash with 300 mL of cold buffer A, tissue is divided into two parts. Two hundred mL of cold buffer A with 1 mM phenylmethylsulfonyl fluoride, 100 nM pepstatin A, 1 mM dithiothreitol, and 1 mM spermine is added to each portion. Addition of protease inhibitors is essential to prevent proteolysis of the receptor. Addition of 0.2 μg/mL of soybean trypsin inhibitor is also effective when combined with phenylmethylsulfonyl fluoride and pepstatin A. Each part of the tissue is homogenized on ice for 20–30 s using a Tekmar homogenizer (Cincinnati, OH) at submaximal speed; this is followed by immediate centrifugation at 1000g for 10 min. The supernatant is discarded. The pellet from two parts is combined and is mixed with 200 mL of ice cold buffer A and shaken vigorously. After centrifugation at 1000g for 10 min, the supernatant is discarded. The pellet is mixed with 200 mL of cold buffer A containing 1% Triton X-100, 10 mM Tris-HCl (pH 7.4), 1 mM phenylmethylsulfonyl fluoride, 100 nM pepstatin A, and 1 mM dithiothreitol, and this is shaken vigorously. After centrifugation at 1000g for 10 min, the supernatant is discarded. The pellet is mixed with 200 mL of buffer A containing 10 mM Tris-HCl (pH 7.4) and centrifuged as above. The supernatant is discarded and the pellet obtained from 140 g of tissue is mixed with 100 mL of buffer B and is shaken vigorously. This mixture is centrifuged at 30,000 rpm for 30 min using a type 30 rotor (Beckman Instruments, Palo Alto, CA). After discarding the supernatant, the pellet is mixed with buffer C (25 mL for 70 g tissue equivalent nuclei) containing 0.3% Triton X-100, vortexed, and is incubated for 10 min. After centrifugation at 1000g for 10 min, the supernatant is discarded. The nuclear pellet is dispersed by vortexing in buffer C and centrifuged at 1000g for 10 min. The nuclear pellet thus obtained is incubated with buffer D at a ratio of 1 mL for 3 g tissue equivalent nuclei in the presence of 0.2 mM phenylmethylsulfonyl fluoride for 1 h with vigorous dispersion using a vortex mixer at 10-min intervals. Samples are centrifuged at 30,000 rpm for 120 min using a type 30 rotor. The supernatant (nuclear extract) is snap-frozen in a dry ice/acetone bath and kept at –70°C for up to 2 wk. This method allows preparation of nuclear extract from 0.8–1.0 kg of rat liver in 1 d.

5. Purification of Thyroid Hormone Receptor

5.1. Buffers

1. Buffer E: 0.9M KCl/0.2 mM EDTA/1 mM MgCl$_2$/10 mM potassium phosphate (pH 7.7)/1 mM dithiothreitol/0.1 mM phenylmethylsulfonyl fluoride;
2. Buffer F: 0.3M KCl/2 mM EDTA/20 mM potassium phosphate (pH 8.0)/10 mM 2-mercaptoethanol;
3. Buffer G: 0.05M KCl/2 mM EDTA/20 mM potassium phosphate (pH 8.0)/10% glycerol/10 mM 2-mercaptoethanol;
4. Buffer H: 20 mM potassium phosphate (pH 8.0)/2 mM EDTA/20% glycerol/10 mM 2-mercaptoethanol;
5. Buffer I: 0.05M NaCl/20 mM potassium phosphate (pH 8.0)/2 mM EDTA.

Dithiothreitol and phenylmethylsulfonyl fluoride should be added before use as described in Section 4.1. Two-mercaptoethanol is also added before use.

5.2. Methods

All glassware is siliconized using Sigmacoat (Sigma, St. Louis, MO). Hydroxylapatite gel is prepared as described by Muench (1971) (*see* Appendix). Calf thymus DNA-cellulose is prepared as described by Alberts and Herrick (1974). In order to obtain cellulose with higher amount of DNA attached, this procedure is repeated twice with the same cellulose.

Figure 2 shows the outline of the purification methods. All procedures are performed at 0–4°C. Nuclear extract prepared from 4–5 kg rat liver is adjusted to 0.9M KCl by adding solid KCl or 3M KCl solution. This is applied to a hydroxylapatite column (200-mL column vol) equilibrated with buffer E. After the column is washed with 600 mL of the same buffer, the receptor is eluted by a linear gradient of 10–200 mM potassium phos-

Nuclear Extract from
4 to 5 kg Rat Liver
↓
Hydroxylapatite Column
Phosphate Gradient
↓
43 % Ammonium Sulfate
Precipitation
↓
Sephadex G-150 Column
↓
DNA-Cellulose Column
KCl Gradient
↓
DEAE-Sephadex Column
KCl Gradient
↓
Heparin-Sepharose Column
Pyr 5'-P Elution

Fig. 2. Outline of the method for purifying rat hepatic nuclear thyroid hormone receptor.

phate (pH 7.7) in a total vol of 800 mL. Flowrate of the column is 30–35 mL/cm²/h. Active fractions are pooled.

The receptor eluted from hydroxylapatite column is precipitated by adding 0.75 vol of saturated ammonium sulfate (pH adjusted to 7.4 by adding 10N NaOH) with 5 mM dithiothreitol and 2 mM EDTA to a final concentration of 43% saturation. After gentle stirring in an ice bath for 30 min, precipitate is collected by centrifugation at 12,000g for 30 min and dissolved in buffer F to a total vol of 15–20 mL. The concentrated receptor is applied to a Sephadex G-150 (Pharmacia, Piscataway, NJ) gel filtration column (internal diameter of 2.5 cm with column vol of 500 mL) equilibrated with buffer F. Column flow is maintained at 6–9 mL/cm²/h. Flow pressure should not exceed 36 cm H₂O height. Active fractions are pooled.*

After gel filtration, active fractions are diluted 6 times with buffer H, and applied to a DNA-cellulose column (50–75 mg calf thymus DNA attached to 50 mL of cellulose) equilibrated with buffer G. After washing the column with 200 mL of buffer G, the receptor is eluted by a linear gradient of 0.05–1.0M KCl in a total vol of 200 mL. After DNA-cellulose col-

---

*Typical profile of Sephadex G-150 column chromatography is shown in Fig. 3B. First T₃ binding peak at void volume coelutes with major protein peak. Second T₃ binding peak with sufficient purity should be pooled for further purification step. It is advised not to include fractions after tube 60, since it contains proteolytic T₃ binding fragment of the receptor.

Usually a single large mol wt-protein peak is eluted separately from the second major T₃ binding peak. However, when proteolysis has occurred, another protein peak appears overlapping the second T₃ binding receptor peak and the T₃ binding peak becomes broadly distributed at fractions after 60 that contain the truncated receptors. This results in apparently low purification and low recovery of the receptor at this step. In order to avoid these problems, it is important to work quickly, keep the temperature low, and to include protease inhibitors at the homogenization step and Triton X-100 treatment of crude nuclei, since proteolytic enzymes are liberated at these steps.

umn chromatography, two peaks of T_3 binding activity are obtained, suggesting at least two forms of the receptors with different affinities for DNA are present. Second peak fractions are pooled and diluted 10 times with buffer H and applied to a DEAE-Sephadex A-50 (Pharmacia) column (50 mL column vol). The operative procedures for DEAE-Sephadex A-50 column are the same as those for DNA-cellulose column chromatography.

Active fractions from DEAE-Sephadex A-50 column are diluted 6 times with buffer H and applied to a 10-mL heparin-Sepharose (Pierce, Rockland, IL) column equilibrated with buffer I. The column is washed with 30 mL of buffer I and the same buffer containing 10 mM pyridoxal 5'-phosphate is used for the elution of the receptors. After passing 7 mL of the buffer at an elution rate of 2 mL/h, the column is stopped for 9 h. Elution is then continued slowly using the same buffer. Alternatively, salt elution is used for heparin-Sepharose column chromatography.

5.3. Results and Characterization of the Purified Receptor

Column profiles of the purification steps are shown in Fig. 3. As shown in Table 2, the final receptor preparation provides a specific activity of T_3 binding of 904 pmol per mg protein. Purified receptor has a sedimentation coefficient of 3.4 S and Stokes radius of 3.4 nm. From these values, a mol wt of 49,000 is calculated. Affinity constant of T_3 binding is 6.0×10^9 L/mol. Molecular size, affinity constant for T_3, and relative affinities for iodothyronine analogs are identical to those obtained using nuclear extract, indicating that the final product is representative of the original nuclear thyroid hormone receptors and that the receptor is purified without changing its character (Ichikawa and DeGroot, 1987a).

Protein staining of sodium dodecylsulfate-polyacrylamide gel electrophoresis at various stages is shown in Fig. 4A. Among several bands seen in the sample after purification, a band having a mol wt of 49,000 is apparently the thyroid hormone receptor. The following facts support this conclusion:

1. The mol wt is the same as that calculated from sedimentation coefficient and Stokes radius.
2. This band is covalently photoaffinity labeled with [^{125}I]-thyroxine, and the label is displacable by excess unlabeled T_3 (Fig. 4B). Specific labeling is shifted to bands of mol wt 27,000 and 36,000 when the sample is treated with trypsin, or endogenous protease, respectively. The shift of the mol wt is identical to that determined using sedimentation coefficient and Stokes radius. Previous studies (Ichikawa and DeGroot, 1986)

Table 2
Results of Nuclear Thyroid Hormone Receptor Purification[a]

Procedure	T_3 binding capacity, pmol of T_3 per mg of protein	Receptor, µg	Total protein, mg
Nuclear extract	1.4 ± 0.3	255 ± 33	3650 ± 417
Hydroxylapatite	5.7 ± 1.7	227 ± 46	847 ± 222
Sephadex G-150	31.1 ± 3.4	119 ± 27	78.1 ± 18.5
DNA-cellulose	412 ± 96	37.8 ± 8.0	1.94 ± 0.56
DEAE-Sephadex A-50	604 ± 86	23.7 ± 6.1	0.81 ± 0.23
Heparin-Sepharose	904 ± 58	9.7 ± 3.4	0.22 ± 0.07

[a]T_3 binding capacity was measured by incubating the receptor preparation with 5 nM [^{125}I]T_3 in the absence or presence of 0.3 µM unlabeled T_3. Specifically bound T_3 was calculated. Amounts of receptor were calculated assuming that the mol wt of the receptor is 49,000 and the receptor has a single T_3 binding site. Results are mean ± SD of five separate purifications.

revealed the structure of the receptor being as indicated in Fig. 5, where globular T_3- and core histone-binding fragment of the receptor (mol wt 26,000) separable from DNA-binding domain by tryptic digestion is shown.

3. In lane 6 of Fig. 4A, 6 µg protein of sample with 4.2% purity assessed by T_3 binding assay, which contains about 0.24 µg of the receptor protein, was applied. Since 0.2 µg each of mol wt markers were applied, similarity of the staining intensities between 45,000-mol-wt-standard protein and 49,000-mol-wt-band suggests that the amount of the 49,000-mol-wt-protein estimated from protein staining corresponds well to that estimated from T_3 binding assay. This is the first demonstration of the receptor band on electrophoresis.

5.4. Alternative Methods

It was reported that calf thymus H1 histone inhibits the binding of the receptor to calf thymus DNA-cellulose. This inhibition is rather selective to the receptor among other DNA binding proteins in nuclear extract. Elution of the receptor attached to DNA-cellulose column by H1 histone resulted in about fivefold purification (Ichikawa et al., 1987c). In combination with subsequent DEAE-Sephadex chromatography, which removes H1 histone from the receptor, this method of elution seems well worth trying.

Recently, Apriletti et al. (1988) reported large-scale purification of the rat hepatic nuclear thyroid hormone receptor. Their method of nuclear preparation does not include heavy sucrose-ultracentrifugation and allows one to prepare crude nuclear extract from 2 kg rat liver/d. In their report, phospho-Ultrogel chromatography of the receptor using polyanion

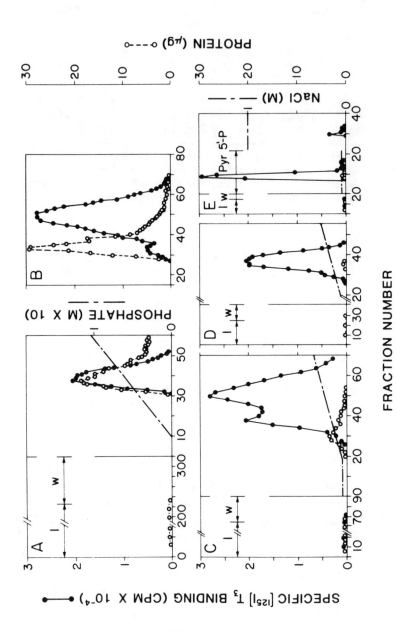

FRACTION NUMBER

Fig. 3. Sequential column chromatography of nuclear thyroid hormone receptor. **A** Hydroxylapatite column chromatography; 6.8-mL fractions were collected during the phosphate gradient. **B** Sephadex G-150 column chromatography; 5-mL fractions were collected. **C** DNA-cellulose column chromatography; 2-mL fractions were collected during the KCl gradient. **D** DEAE-Sephadex A-50 column chromatography; 2-mL fractions were collected during the KCl gradient. **E** Heparin-Sepharose column chromatography; 1-mL fractions were collected during elution with pyridoxal 5′-phosphate (Pyr 5′-P) or with 1M NaCl; 8-mL fractions were collected during sample loading (l) and washing (w). Aliquots (1 μL for A, 5 μL for B–D, and 3μL for E) of fractions were assayed for specific [125I]T$_3$ binding (●—●) and protein content (○—○). Total [125I]T$_3$ used was 30,000 cpm.

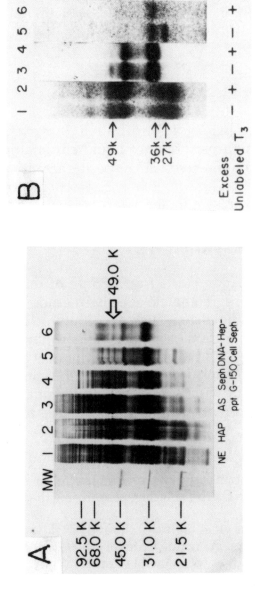

Fig. 4. A Sodium dodecylsulfate-polyacrylamide gel electrophoresis analysis at various stages of nuclear thyroid hormone receptor purification. Samples after salt extraction of nuclei (lane 1), hydroxylapatite column (lane 2), ammonium sulfate precipitation (lane 3), Sephadex G-150 column (lane 4), DNA-cellulose column (lane 5), and heparin-Sepharose column (lane 6)were analyzed by sodium dodecylsulfate-10% polyacrylamide gel electrophoresis; this was followed by silver staining. Amounts of protein were 20 μg for lanes 1–3, 10 μg for lanes 4 and 5, and 6 μg for lane 6. MW, mol wt markers. **B** Photoaffinity labeling of the nuclear thyroid hormone receptor. Samples after salt extraction of nuclei (lanes 1 and 2) and heparin-Sepharose column (lanes 3 and 4) were photoaffinity-labeled with 0.5 nM [^{125}I]-thyroxine in the absence (lanes 1, 3, 5, and 7) or presence (lanes 2, 4, 6, and 8) of 1 μM unlabeled T₃. In lanes 5 and 6, photoaffinity labeling was performed on the receptor purified up to the DNA-cellulose column step. Four μg of trypsin per mg protein was then added and incubated at 10°C for 30 min. The reaction was stopped by adding a fivefold excess (by weight) of soybean trypsin inhibitor and cooling in ice. In lanes 7 and 8 nuclear extract was prepared in the absence of protease inhibitors and purified. Small mol wt T₃ binding activity was separately pooled after Sephadex G-150 column chromatography and further purified by DNA-cellulose column chromatography. This sample was used for affinity labeling. Samples were applied on sodium dodecylsulfate-10% polyacrylamide gel electrophoresis and autoradiographed. Amounts of protein applied were 200 μg for lanes 1 and 2, 6 μg for lanes 3 and 4, and 100 μg for lanes 5–8. Exposure time was 60 d for lanes 1 and 2, 29 d for lanes 5–8, and 8 d for lanes 3 and 4.

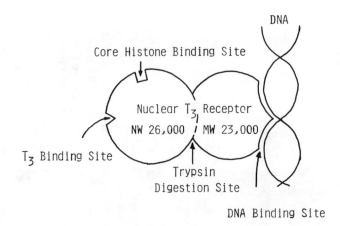

Fig. 5. Structure of the nuclear thyroid hormone receptor.

dextran sulfate for the elution resulted in 5.2-fold purification. Size exclusion and hydrophobic interaction high-performance liquid chromatography (HPLC) were also reported to be effective. Final specific activity of the receptor was 287 pmol/mg protein (Table 1). It may be worth incorporating some of their columns to the present method.

6. Discussion

Considering the fact that purification starts from 255 µg of the receptor protein per 4–5 Kg rat liver, we regret that no effective method providing substantial purification in a single step, such as affinity column chromatography, is available for purification of the thyroid hormone receptor. Therefore, we developed a method in which a large amount of pure nuclei are prepared and the receptor is purified to 4.2–4.9% purity using conventional chromatographic methods without appreciable degradation or alteration in characteristics of the receptor. Critical improvements in the column procedures are as follows:

1. A high KCl concentration used in the hydroxylapatite column allows us to remove half of nonreceptor protein at the time of sample application and washing, and also results in a sharp T_3 binding peak during the elution with phosphate gradient.
2. Hydroxylapatite prepared in this way gives high capacity and fast flow that allows us to process a large amount of starting material (1400–1700 mL of nuclear extract) quickly, and gives reproducible results in comparison to hydroxylapatite gel prepared otherwise.
3. Pyridoxal 5'-phosphate elution of the receptor from the heparin-Sepharose column gives a higher purification step than salt elution.

Another problem in purification of the receptor is the rapid loss of T_3 binding activity during purification. The effects of glycerol and phosphate* to prevent receptor degradation are utilized in the present method. Since the receptor irreversibly loses its T_3 binding activity at pH below 6.0 or above 10.0, a neutral pH is maintained during the purification. Another factor contributing to the apparent loss of T_3 binding activity resides in the T_3 binding assay. When a small amount of protein is used in the binding assay, as after substantial purification of the receptor, an appreciable amount of the receptor is adsorbed onto the resin. This is effectively avoided by adding 30–50 µg heated nuclear extract per 0.3 mL assay solution. Alternatively, one can add 100 µg core histone per 0.3 mL assay solution.**

By repeated injections of the purified receptor preparations in combination with complete or incomplete Freund's adjuvant, a specific immune response against the nuclear thyroid hormone receptor was elicited (Ichikawa and DeGroot, 1987b), demonstrating the feasibility of obtaining monoclonal antibodies against nuclear thyroid hormone receptor using the receptor preparation obtained by the method described.

Receptor preparations with the highest purity shown here still require about 20-fold further purification to homogeneity. Tremendous effort and cost will be required to achieve this. Recently, however, nucleotide sequence homology between the human glucocorticoid receptor and the v-erb A oncogene prompted the finding that human and chicken c-erb A product may be the thyroid hormone receptor (Weinberger et al., 1986 Sap et al., 1986). These products bind T_3 and regulate the expression of the hormone responsive gene. Now it is possible to produce functional receptors (c-erb A) using expression vectors and it may not be necessary to purify the native receptor to study its structure and function. However, subsequent studies suggest that there are multiple forms of the c-erb A (thyroid hormone receptor) within the same specie and even within the same organ. In addition, none of the c-erb A gene so far cloned seems to be expressed

---

*Among several compounds (dithiothreitol, thioglycerol, nitrogen atmosphere, high ionic strength, spermine, sodium molybdate, and so on) examined, glycerol and phosphate were found significantly effective in protecting the loss of T_3 binding activity of the receptor during the storage in ice. Protective effect of glycerol was also shown by heat inactivation study of the receptor. Half-lives of T_3 binding activity of the receptor at 43°C were 20, 45, and 91 min in the presence of 0, 20, and 30% of glycerol. Removal of glycerol by dialysis shortened the half-life to 19 min.

**Among other methods for separation of bound and free hormones, the nitrocellulose filter binding assay (Inoue et al., 1983) also gives accurate determination of the receptor at lower protein content, as is expected from the principles of this method, in which bound hormone is retained on the filter. However, the Dowex resin gives identical accuracy at low protein content when heated nuclear extract is added. Because of the simplicity of the resin test, limited protein-retaining capacity of the filter paper, Dowex resin is recommended.

in liver, which is one of the best target organs of thyroid hormone. It still needs to be clarified whether all the c-erb A products are the functional receptors or not and whether all the functional thyroid hormone receptors belong to the c-erb A gene products.

7. Appendix

7.1. Preparation of Hydroxylapatite Gel

Method for preparation of hydroxylapatite gel (Muench, 1971) is described here, since success of the present purification is dependent on how the hydroxylapatite gel is prepared.

Five hundred mM CaCl$_2$ (1500 mL) is added dropwise into 1800 mL 0.5M sodium phosphate buffer, pH 6.7, with gentle stirring in a 4-L beaker at room temperature. After all of the calcium chloride have been added, the suspension is stirred for 1 h. The supernatant is decanted after the calcium orthophosphate is settled. The calcium orthophosphate is suspended in 3-L H$_2$O, allowed to settle, and the supernatant is discarded. The calcium orthophosphate is transferred to a 6-L Erlenmeyer flask with 3-L H$_2$O. After addition of 1 mL of 1% phenolphthalein in ethanol, enough concentrated NH$_4$OH is added to maintain the red color of phenolphthalein while heating to 90°C with stirring. Concentrated NH$_4$OH is added as required. The suspension is kept at 90°C for 30 min to make hydroxylapatite. The hydroxylapatite is allowed to settle, and the supernatant decanted while still hot. The hydroxylapatite is washed at least 7 times with 3-L of 5 mM sodium phosphate, pH 6.7. The fine materials are removed at the same time. Gel (350 mL) as packed in column is obtained by this method. The hydroxylapatite is stored at 4°C.

References

Apriletti, J. W., Eberhardt, N. L., Latham, K. R., and Baxter, J. D. (1981) *J. Biol. Chem.* **256,** 12094.
Apriletti, J. W., Baxter, J. D., and Lavin, T. N. (1988) *J. Biol. Chem.* **263,** 9409.
Alberts, B. and Herrick, G. (1974) *Methods Enzymol.* **21,** 198.
Bradford, M. M. (1976) *Anal. Biochem.* **72,** 248.
Gornall, A. G., Bardawill, C. J., and David, M. M. (1949) *J. Biol. Chem.* **177,** 751.
Ichikawa, K. and DeGroot, L. J. (1986) *J. Biol. Chem.* **261,** 16540.
Ichikawa, K. and DeGroot, L. J. (1987a) *Proc. Natl. Acad. Sci. USA* **84,** 3420.
Ichikawa, K. and DeGroot, L. J. (1987b) *Biochem. Biophys. Res. Commun.* **144,** 178.
Ichikawa, K., Bentley, S., Fee, M., and DeGroot, L. J. (1987c) *Endocrinology* **121,** 893.
Inoue, A., Yamakawa, J., Yukioka, M., and Morisawa, S. (1983) *Anal. Biochem.* **134,** 176.
Latham, K. R., Ring, J. C., and Baxter, J. D. (1976) *J. Biol. Chem.* **251,** 7388.

Lowry, O. H., Rosebrough, N. J., Farr, A. L., and Randall, R. J. (1951)*J. Biol. Chem.* **193,** 265.
Muench, K. H. (1971) *Procedures of Nucleic Acid Research* **2,** 515.
Sap, J., Munoz, A., Damm, K., Goldberg, Y., Ghysdael, J., Leutz, A., Beug, H., and Vennstrom, B. (1986) *Nature* **324,** 635.
Silva, E. S., Astier, H., Thakare, V., Schwartz, H. L., and Oppenheimer, J. H. (1977) *J. Biol. Chem.* **252,** 6799.
Torresani, J. and Anselmet, A. (1978) *Biochem. Biophys. Res. Commun.* **81,** 147.
Weinberger, C., Thompson, C. C., Ong, E. S., Lebo, R., Gruol, D. J., and Evans, R. M. (1986) *Nature* **324,** 641.

The Vasopressin V1 Receptor

*Purification, Characterization,
and Analysis of V1 Receptor Interaction(s)
with Guanine Nucleotide-Binding Proteins*

Jordan B. Fishman and Burton F. Dickey

1. Introduction

Vasopressin, a neurohypophyseal nonapeptide, regulates a number of homeostatic functions in humans, including glucose metabolism (Lynch et al., 1985), antidiuresis (de Sousa and Grosso, 1981), calcium flux (Lynch et al., 1985; Blackmore et al., 1985), and vasoconstriction (Schmid et al., 1974), as well as stimulating prostaglandin production (Grillone et al., 1988), and acting as a neurotransmitter in the central nervous system (Tanaka et al., 1977; van Leeuwen et al., 1987). Two receptor subtypes have been characterized in great detail, the V2 receptor, found only in kidney epithelial cells, and the V1 receptor, which is expressed in virtually every other cell type in the human body. These receptors work through the second messengers cyclic AMP (V2 receptors; Ausiello and Orloff, 1982; Orloff and Handler, 1962) and inositol trisphosphate and diacylglycerol (V1 receptor; Wallace and Fain, 1985; Berridge and Irvine, 1984; Guillon et al., 1986). It has also been suggested that a third subtype of vasopressin receptor may exist in the central nervous system, however, such evidence is quite limited.

Receptor Purification, vol. 2   ©1990 The Humana Press

Whereas the binding of vasopressin agonists and antagonists to receptors on the surface of liver and smooth muscle cells from rat and calf has been well characterized (Fahrenholz et al., 1984; Cantau et al., 1980; Fishman et al., 1985; Penit et al., 1983), little information existed until recently about the molecular properties of this receptor. This is owing in large part to the inability of the solubilized vasopressin receptor to bind ligand (Boer et al., 1983; Bojanic and Fain, 1986; Guillon et al., 1980). This forced investigators to prebind radiolabeled ligand to receptor prior to solubilization in order to study the hydrodynamic properties of the receptor (Bojanic and Fain, 1986; Boer et al., 1983). Such experiments yielded a crude estimate of the mol wt of the receptor (258,000) that was far in excess of the value obtained by photoaffinity labeling techniques (Boer and Fahrenholz, 1985) or radiation inactivation (Crause et al., 1984) of approx 70,000. It was suggested that the large mol wt was a result of the ternary complex formed by the receptor, agonist, and the heterotrimeric G-protein (DeLean et al., 1980). Purification of the V1 receptor utilizing prebound ligand as a means of detection was virtually impossible because of the rapid rate of dissociation of the bound ligand. An alternative approach was taken by Fahrenholz and coworkers (Boer and Fahrenholz, 1985), who developed a photoreactive analog of vasopressin, but it bound with a low specific activity and little specificity.

It became apparent that the purification of the V1 receptor would require development of a functional reconstitution system. The solubilized receptor would be inserted into a phospholipid vesicle, which would hopefully stabilize the receptor in a ligand-binding conformation. The solubilized receptor could then be subjected to a series of separation steps whose efficacy would be determined by reconstitution of aliquots of the recovered fractions. Such a scheme was utilized in the purification of the V1 receptor (Dickey et al., 1987; Fishman et al., 1987), and is described below. In addition, we were able to utilize the reconstitution procedure to analyze the interaction of various V1 receptors with guanine nucleotide binding proteins (G-proteins), with the results summarized below.

2. Solubilization and Reconstitution of the V1 Receptor

The inability of the solubilized V1 receptor to bind ligand precluded the use of affinity steps that utilize agonists or antagonists linked to solid-phase supports in the purification scheme. Such steps can often result in several hundred-fold or more purification of receptor molecules, and, thus, can greatly simplify the purification procedure. Since conventional column chromatography systems such as lectins, size exclusion, and ion-exchange typically yield purifications on the order of severalfold to perhaps 100-fold at best, we realized early on that the purification of the vasopressin V1 receptor would require a substantial number of chromatographic steps. Each of these steps would also require a large number of reconstitution events to monitor the location and extent of purification, if any, of the V1 receptor. We therefore felt it necessary to carefully choose the type of reconstitution procedure we would utilize in order to make the purification as manageable as possible.

Liver was deemed to be the tissue of choice for purification of the vasopressin receptor because of its homogeneous nature (>90% hepatocytes) and rat was chosen because the livers can be excised quickly from live animals. The first critical decision was the choice of detergent. Earlier hydrodynamic studies had used Triton X-100 to solubilize receptors, as well as digitonin, and more recently lysolecithin. We eliminated Triton as a possible detergent for our studies (as well as Lubrol and Nonidet P-40) because of difficulties in its removal during the production of phospholipid vesicles, stemming mainly from its low critical micelle concentration (CMC; 0.2–0.3 mM). Digitonin was not used, in part because of the insoluble nature of this compound in aqueous solutions, and lysolecithin was not deemed suitable because it is a natural byproduct of many cellular reactions and is rapidly metabolized by membrane-bound enzymes. We ultimately decided to utilize CHAPS (3-[(3-cholamidopropyl) dimethylammoniol]-1-propanesulfonate), a zwitterionic detergent, with a CMC of 9 mM. The zwitterionic nature of this detergent also makes it flexible for use in a wide variety of ion-exhange applications as well as being soluble over a wide pH range.

The importance of the CMC becomes clear when one reflects on our need for a rapid, easily reproducible reconstitution procedure. Difficulty in the removal of detergent, as would be the case when using detergents with low CMCs, would increase the time and effort required to reconstitute the numerous column fractions. In contrast, the use of CHAPS allowed us to utilize a simple gel filtration procedure for reconstitution of our protein extracts. The solubilized proteins are mixed with a phospholipid/CHAPS solution, incubated together for several minutes, and chromatographed over a coarse Sephadex G-50 column. The phospholipid vesicles elute in the void vol as opalescent structures. This procedure is rapid, and several columns can be run simultaneously, allowing for a relatively rapid analysis of receptor binding activity.

Figure 1 shows a typical solubilization curve of rat liver membranes. A large portion of the membrane protein is solubilized (approx 66%) upon

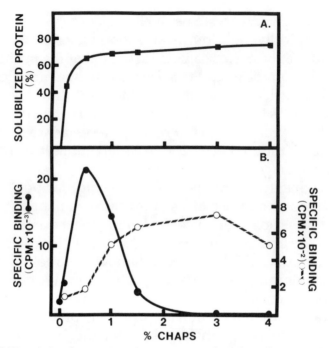

Fig. 1. Solubilization of vasopressin receptors from rat liver membranes. Membranes were treated with increasing concentrations of CHAPS as indicated, and following incubation on ice for 30 min, the membranes were centrifuged to remove insoluble material. The membranes were then assayed for [³H]-vasopressin binding activity (closed squares) by gel filtration, or the soluble material reconstituted as previously described (open circles; Dickey et al., 1987). (Reprinted from Dickey, B. F., Fine, R. E., and Navarro, J. [1987] *J. Biol. Chem.* **262,** 8738–8742.)

exposure of the membranes to 0.5% CHAPS. In contrast, almost no receptor binding activity can be reconstituted from the supernatant. Upon exposure of the membranes to 1–3% CHAPS, a substantial fraction of the receptors are solubilized along with a relatively small fraction of the remaining membrane protein. This characteristic of V1 receptors was utilized as a first purification step, whereby membranes were exposed to 0.5% CHAPS, centrifuged, and the pellet reextracted with 2–3% CHAPS. We found that such a 2-step solubilization procedure gave us approx tenfold purification of vasopressin binding activity, and, additionally, allowed for concentration of the receptors at the second solubilization step.

3. Receptor Purification and Characterization

The solubilized vasopressin receptor was subjected to a series of purification steps: size exclusion, lectin affinity, hydroxylapatite, and DEAE chromatography. Size exclusion chromatography was performed using a

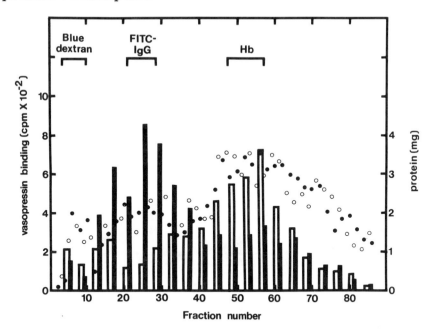

Fig. 2. Sephacryl S-200 gel filtration of CHAPS solubilized rat liver membrane proteins. Aliquots of the fractions were reconstituted and vasopressin binding assessed as previously described (Fishman et al., 1987). Open circles, protein per fraction, and open bars, vasopressin binding from membranes solubilized without prior exposure to agonist; closed circles, protein per fraction, and closed bars, vasopressin binding from membranes exposed to agonist prior to solubilization. (Reprinted from Fishman, J. B., Dickey, B. F., and Fine, R. E. [1987] *J. Biol. Chem.* **262,** 14049–14055.)

Sephacryl S-200 colulmn (2.5×75 cm) that was equilibrated with HMC buffer (30 mM HEPES, pH 7.4, 5 mM MgCl$_2$, 0.2% CHAPS). As seen in Fig. 2, the receptor was eluted as a broad peak of reconstitutable binding activity. The mol wt of the receptor complex did vary, and it was dependent on the pretreatment of the membranes prior to solubilization. If vasopressin was prebound to the membranes prior to exposure to CHAPS, a majority of the receptors eluted with an apparent mol wt in excess of 150,000. In contrast, receptors from membranes solubilized without prior exposure to vasopressin eluted with an apparent mol wt of about 70,000.

Shift in apparent mol wt of the receptor complex after exposure of the membranes to vasopressin, resulted in the elution of the receptor in fractions containing much lower amounts of protein, increasing the effective purification of the column. Fractions containing the receptor were pooled and the material subjected to lectin affinity chromatography using the lectin from *Triticum vulgaris* coupled to crosslinked agarose beads. One-fifth of the protein added to the lectin column bound, and following extensive washing, the receptor was eluted with 0.2M N-acetylglucosamine (Fig. 3).

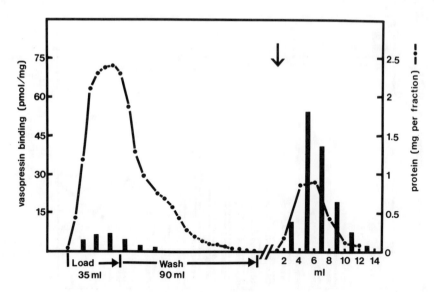

Fig. 3. Lectin affinity chromatography of solubilized membrane proteins. The fractions from the S-200 column that were enriched in vasopressin binding activity were pooled and were loaded onto a wheat germ agglutinin-Sepharose column. The column was washed exhaustively and the bound protein eluted with 0.2M N-acetylmannosamine. The fractions that were reconstituted to monitor vasopressin binding activity are indicated by the shaded bars. (Reprinted from Fishman, J. B., Dickey, B. F., and Fine, R. E. [1987] *J. Biol. Chem.* **262,** 14049–14055.)

Since previous studies had indicated that agonist binding could stabilize certain receptors with guanine nucleotide-binding proteins (G-proteins), we investigated this possiblity using guanosine 5'-thiotriphosphate (GTP-gamma-S), a nonhydrolyzable GTP analog. The interaction of receptors with G-proteins results in an increase in receptor affinity for ligand (for examples, *see* Fahrenholz et al., 1984; Lynch et al., 1985; Fishman et al., 1987; Dickey et al., 1987; Bielinski et al., 1988; for review, *see* Casey and Gilman, 1988; Birnbaumer et al., 1985). When exposed to compounds such as GTP-gamma-S, G-proteins uncouple from their associated receptors, and the receptors revert to their low affinity state. This change in affinity can be monitored through equilibrium binding assays (Fig. 4). When membranes are solubilized without prior exposure to vasopressin, the partially purified and reconstituted V1 receptors have an affinity for [^{3}H]-vasopressin of 30–40 nM. If, however, the membranes are exposed to vasopressin prior to exposure to CHAPS, the affinity of the reconstituted receptors is about 6 nM. Furthermore, if binding of the prebound material is performed in the presence of GTP-gamma-S, the affinity of the reconstituted receptors is 30 nM. Thus, it appears that agonist binding stimulates coupling of the liver V1 receptor and G-proteins. As discussed below,

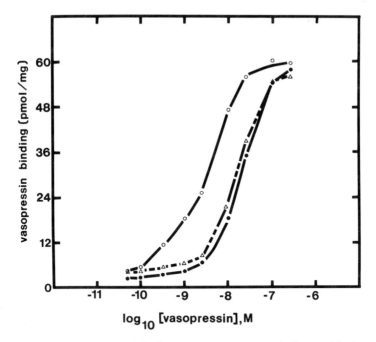

Fig. 4. Equilibrium binding of [³H]-vasopressin to partially purified and reconstituted vasopressin receptors. Vasopressin receptors were solubilized with CHAPS, subjected to size exclusion and lectin affinity chromatography, and reconstituted. The reesultant proteoliposomes were exposed to the concentrations of [³H]-vasopressin indicated in the presence or absence of 10 μM vasopressin and specific binding was determined. Open circles, membranes were exposed to vasopressin prior to solubilization; closed circles, membranes were not exposed to vasopressin prior to solubilization; triangles, membranes were exposed to vasopressin prior to solubilization, and following chromatography and reconstitution, the liposomes were exposed to 0.1 mM GTP-gamma-S prior to and during binding. (Reprinted from Fishman, J. B., Dickey, B. F., and Fine, R. E. [1987] *J. Biol. Chem.* **262**, 14049–14055.)

however, this phenomenon is not common to all vasopressin receptors. Others have noted a similar GTP-sensitive increase in the size of beta-adrenergic (Limbird et al., 1978) and pituitary dopamine D_2 receptors (Kilpatrick and Caron, 1983). In contrast, the adenosine A_1 receptor from brain (Stiles, 1985) remain associated with G-proteins following solubilization. It has also been observed that although the alpha$_2$-adrenergic receptor remains associated with a G-protein during solubilization and several chromatographic steps in the absence of ligand (Cerione et al., 1986), prebinding of agonist promotes an increase in receptor/G-protein coupling after solubilization (Smith and Limbird, 1981).

The last two steps in the purification procedure utilized the ion-exchange resins hydroxylapatite (HA) and DEAE. The material eluted from the lectin column was diluted fivefold with HMC buffer, and run over an

344 Fishman and Dickey

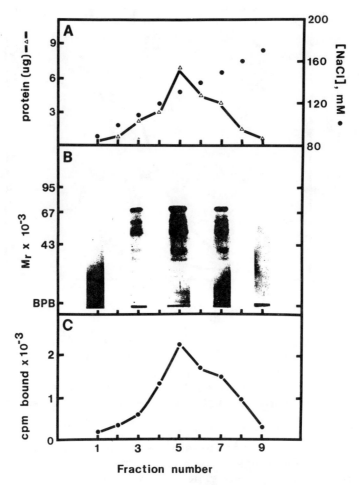

Fig. 5. Hydroxylapatite ion-exchange chromatography of partially purified vasopressin binding activity. Solubilized vasopressin receptors, prepared as described in Fig. 4, was further purified by HA ion-exchange chromatography. After loading of the sample and extensive washing to remove unbound protein, the column was washed with 80 mM NaCl followed by an 80–200 mM NaCl gradient. **A.** Protein profile and NaCl gradient profile; **B.** NaDodSO$_4$-polyacrylamide (5–12% gradient) gel electrophoresis of eluted material; **C.** [^{3}H]-vasopressin binding activity of reconstituted material. (Reprinted from Fishman, J. B., Dickey, B. F., and Fine, R. E. [1987] *J. Biol. Chem.* **262**, 14049–14055.)

HA column (Fig. 5). After exhaustive washing, the column was eluted first with 80 mM NaCl in HMC buffer, followed by a gradient elution with 80–200 mM NaCl. Reconstitutable vasopressin binding activity was eluted at a salt concentration of approx 130 mM. The receptor could also be eluted with a gradient of phosphate, however, we found this resulted in lower purification of the receptor, since it was eluted in a much broader peak.

The material eluted from the HA column was further purified by passing it over a DE-53 ion-exchange column. We had previously observed that the solubilized receptor did not bind to this ion-exchange resin. We had also found, however, that agonist prebinding had promoted receptor/G-protein coupling that persisted throughout the purification scheme (*see below*), and since DEAE-type resins are an initial step in the purification of G-proteins, we returned to this resin in an attempt to separate receptors fram their associated G-proteins. We have recently found that the best separation of receptor from associated G-protein is obtained if we increase the detergent concentration several-fold and incubate the material at room temperature in the presence of GTP. The use of GTP rather than a nonhydrolyzable substrate facilitates further assay of the G-proteins. When this material is then chromatographed over the DE-53 column, most of the reconstitutable vasopressin binding activity is found in the flowthrough. A single, silver stained protein band with a mol wt of 68,000 can be visualized following SDS-PAGE of the DE-53 purified material (Fig. 6B), and crosslinking studies using two bifunctional crosslinking reagents, sulfodisuccinimidyl tartarate and disuccinimidyl suberate, along with ^{125}I-vasopressin demonstrated specific crosslinking of bound vasopressin to a protein with a mol wt of approx 70,000 including the wt of crosslinker and vasopressin. In addition, if the SDS-PAGE was performed in the absence of reducing agents, both the crosslinked radioactivity and the silver stained protein band now migrated with an apparent mol wt of about 58,000.

Total purification of the V1 receptor, as shown in Table 1, is approx 21,000-fold from the starting membranes, which themselves are approx fourfold purified over the liver homogenate. Of particular note is the contribution of the HA purification step, which resulted in about 90-fold purification of the V1 receptor. Further characterization of the interaction of the partially purified receptor with G-proteins was performed using the HA-purified material. As shown in Table 2, crude receptor, derived from solubilized membranes without further modification, hydrolyzed approx. 3.5 pmol of [gamma-^{32}P]-GTP min^{-1} mg protein^{-1} in the presence of agonist, similar to the values obtained by Fitzgerald et al. (1986). In contrast, the highly purified V1 receptor/G-protein complex, following elution from the HA column, hydrolyzed approx. 25 nmol GTP min^{-1} mg protein^{-1} in the presence of agonist. In the absence of agonist, 40-fold less was hydrolyzed. Taken together with the equilibrium binding data in the presence and absence of GTP-gamma-S, this suggested that the liver V1 receptor couples to a G-protein following the binding of agonist. There appears to be little

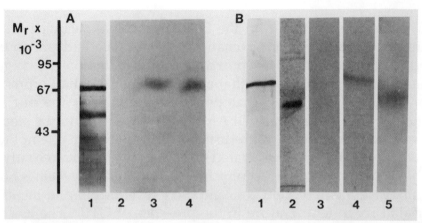

Fig. 6. NaDodSO₄-polyacrylamide gel electrophoresis of purified vasopressin receptor. **A.** Material purified as described in Figs. 1–6 was pooled, and following NaDodSO₄-polyacryamide gel electrophoresis, was silver stained (lane 1), or following reconstitution, was exposed to ^{125}I-vasopressin and crosslinked with disuccinimidyl tartarate in the presence (lane 2) or absence (lane 3) of a 1000-fold excess of unlabeled vasopressin, or crosslinked with disuccinimidyl suberate in 1% dimethyl sulfoxide. Both crosslinkers were used at a concentration of 1 mM. **B.** The HA-purified material was then further purified by DE-53 ion-exchange chromatography as described in the text. Purified receptor was then subjected to NaDodSO₄-polyacrylamide gel electrophoresis and silver-stained in the presence (lane 1) or absence (lane 2) of 2-mercaptoethanol [2-ME]. This material was also exposed to ^{125}I-vasopressin and the crosslinker disuccinimidyl suberate followed by electrophoresis in the presence of 2-ME and a 1000-fold excess of unlabeled vasopressin (lane 3), in the absence of unlabeled vasopressin (lane 4), or in the absence of both 2-ME and unlabeled vasopressin (lane 5). (Reprinted from Fishman, J. B., Dickey, B. F., and Fine, R. E. [1987] *J. Biol. Chem.* **262,** 14049–14055.)

spontaneous coupling of the receptor to G-proteins solubilized with CHAPS without the prebinding of vasopressin prior to solubilization. We have found, however, that the demonstration of a lack of spontaneous coupling of the liver V1 receptor required that careful attention be paid to the protein concentration of the membranes. If we solubilize membranes at a protein concentration of 5–10 mg/mL with 2% CHAPS, we see a substantial decrease in V1 receptor affinity in the presence of GTP-gamma-S. If, however, the protein concentration is lowered to 1 mg/mL prior to exposure to 2% CHAPS, we see no effect of GTP-gamma-S on binding. Such a critical dependence of protein concentration on subsequent GTP-gamma-S effects on vasopressin binding may explain the discrepancy in our results (Dickey et al., 1986,1987) and those of Aiyar et al. (1987), who observed that lysolecithin-solubilized vasopressin receptors are coupled to a G-protein in the absence of agonist.

The G-protein that couples to the V1 receptor appears to be pertussis toxin insensitive, as demonstrated by several laboratories (Fishman et al., 1987; Uhing et al., 1986; Lynch et al., 1986; Fitzgerald et al., 1986) and is

Table 1
Purification of the Rat Liver Vasopressin V1 Receptor[a]

Fraction	Protein, mg	Yield, %	Maximal binding, pmol/mg	Purification, –fold
Membranes	1000	100	0.55	—
0.5–2% CHAPS	48	53	4.9	9
S-200	14	31	9.6	17
Wheat germ agglutinin	2.4	26	58	106
Hydroxylapatite	0.020	12	5,300	9,636
DE-53	0.009	8	11,600	21,090

[a]Data from Fishman et al., 1987.

Table 2
Stimulation of GTP Hydrolysis by Vasopressin[a]

Reconstituted fraction	Vasopressin prebinding	Vasopressin added to assay	GTP hydrolyzed per min per mg protein
Crude receptor	–	–	0.31 ± 0.09
	–	+	0.48 ± 0.07
	+	–	0.98 ± 0.26
	+	+	3.48 ± 0.81
HA-purified receptor	–	–	4.16 ± 0.59
	–	+	28.8 ± 4.10
	+	–	610.1 ± 72.7
	+	+	$25,150 \pm 4.470$

[a]Data from Fishman et al., 1987.

therefore distinct from transducin, G_i and G_o. The lack of regulation of adenylyl cyclase also demonstrates a lack of coupling to G_s (Jard et al., 1981; Cardenus-Tanus et al., 1982).

The characteristics of the V1 receptor are quite similar to those previously described for several different receptors, most notably the beta-adrenergic receptor. The mol wts in the presence and absence of reducing agents are virtually identical, as is the low level of spontaneous coupling to their respective G-proteins. It is, therefore, a distinct possiblity that the V1 receptor is also a membrane of the ever-growing family of receptors that contain seven membrane-spanning helices. Presumptive evidence for such a structure will require full amino acid sequence, and definitive assignment of such a quarternary structure to the V1 receptor must await crystallization of the purified receptor.

4. Characterization
of the Smooth Muscle V1 Receptor

One of most important functions of vasopressin in humans is the regulation of blood pressure, through both the V2 receptor in kidney (anti-diuresis) and the V1 receptor in smooth muscle cells of arteries and arterioles (pressor response). Following characterization of the coupling of liver V1 receptors and G-proteins, we turned our attention to smooth muscle V1 receptors. Receptors were solubilized from both calf pulmonary arteries as well as primary smooth muscle cell cultures.

In contrast to the rat liver V1 receptor, the pulmonary artery smooth muscle (PASM) V1 receptors exhibited a large mol wt in the absence of agonist prebinding (Fig. 7). A majority of the receptors eluted with a mol wt of 150,000 or greater as determined by size exclusion chromatography. No difference was observed if the source of membranes was intact pulmonary artery or primary cultures of smooth muscle cells grown in the absence of vasopressin. The addition of agonist prior to solubilization had no effect on the apparent mol wt of the receptors, although a somewhat higher percentage of receptors was eluted from the column with a high (150,000+) mol wt (data not shown). The addition of GTP-gamma-S to the membranes prior to solubilization reduced the apparent size of the receptors to 60,000–70,000.

When the solubilized PASM V1 receptors were reconstituted into liposomes, they exhibited a K_d for [^{3}H]-vasopressin of 5 nM (Fig. 8). This affinity was reduced to 20–30 nM in the presence of GTP-gamma-S. Exposure of the PASM membranes to agonist prior to solubilization had no appreciable effect on the binding chartacteristics or on the ability of GTP-gamma-S to alter receptor affinity.

The effect of receptor/G-protein coupling on receptor affinity for agonist was further investigated with respect to the dissociation of pre-bound ligand (Fig. 9). Ligand dissociation from liver V1 receptors that were not exposed to agonist prior to solubilization and reconstitution was substantially faster than the dissociation of ligand from similarly treated PASM V1 receptors (k_{-1} = 0.224^{-1} vs 0.072 min^{-1}). Exposure of the rat liver membranes to agonist prior to solubilization, however, decreased the rate of ligand dissociation about threefold (k_{-1} = 0.082 min^{-1}), to about the rate observed for PASM V1 receptors. Exposure of PASM V1 receptors to GTP-gamma-S prior to solubilization increased the rate of dissociation more than threefold (k_{-1} = 0.236 min^{-1}). This suggests that the change in receptor affinity that accompanies receptor/G-protein interaction is caused, at least in part, by a change in the rate of ligand dissociation. Such an effect on

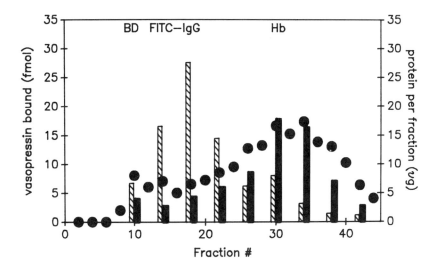

Fig. 7. Gel filtration of CHAPS solubilized PASM membrane proteins. The chromatography conditions were as described in Fig. 2. Crosshatched bars, no additions; shaded bars, membranes were incubated with 100 μM GTP-gamma-S prior to solubilization; circles, protein per fraction. The inclusion of GTP-gamma-S had no measureable effect on the protein profile. (Reprinted from Bielinski, D., McCrory, M., Cahill, M., Polgar, P., and Fishman, J. B. [1988] *Biochem. Biophys. Res. Commun.* **151,** 1293–1298.)

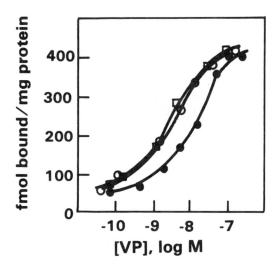

Fig. 8. Equilibrium binding of [³H]-vasopressin to CHAPS solubilized and reconstituted PASM V1 receptors. Open circles, membranes were untreated prior to solubilization and reconstitution; closed circles, membranes were exposed to 100 μM GTP-gamma-S prior to solubilization; squares, membranes were exposed to unlabeled vasopressin prior to solubilization. (Reprinted from Bielinski, D., McCrory, M., Cahill, M., Polgar, P., and Fishman, J. B. [1988] *Biochem. Biophys. Res. Commun.* **151,** 1293–1298.)

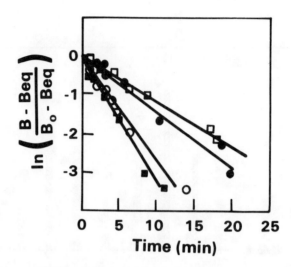

Fig. 9. Dissociation of [³H]-vasopressin bound to reconstituted PASM and rat liver V1 receptors. Approximately 77% of the total [³H]-vasopressin initially bound to PASM V1 receptors and 71% of that bound to rat liver V1 receptors dissociated as a single component that could be fit with a monoexponential function (*r* values of 0.957–0.991). B_{eq} was 0.28 and 0.33 for rat liver and PASM Vl receptors, respectively. Open circles, Liver V1 receptors; closed circles, agonist prebound liver V1 receptors; open squares, PASM receptors; closed squares, PASM V1 receptors exposed to GTP-gamma-S prior to solubilization. (Reprinted from Bielinski, D., McCrory, M., Cahill, M., Polgar, P., and Fishman, J. B. [1988] *Biochem. Biophys. Res. Commun.* **151,** 1293–1298.)

ligand dissociation is not unexpected if one considers the definition of K_d (i.e., the affinity of a receptor for ligand) = k_1 / k_{-1}. Since a ligand must bind prior to affecting a change in receptor/G-protein coupling, then only the offrate is available for modification by receptor/G-protein coupling.

5. Summary

The inability to assay soluble V1 receptors had hampered attempts at its isolation until the development of a functional reconstitution assay system. In addition, the inability of the solubilized receptor to bind ligand did not allow for the use of affinity steps, necessitating the use of multiple chromatographic steps during the V1 receptor purification. The purified rat liver V1 receptor had an apparent mol wt of 68,000 in the presence of reducing agents, 58,000 in the absence of reducing agents, and is a glycoprotein based on its binding to the lectin from *Triticulm vulgaris* (wheat germ agglutinin). The receptor does not appear to functionally couple to G-proteins in the absence of agonist, however, upon agonist binding, it stably couples with G-proteins and remains coupled during

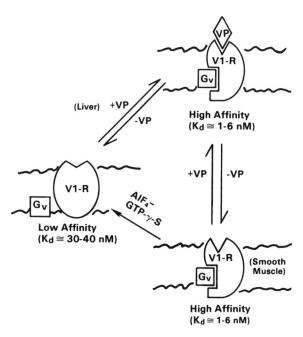

Fig. 10. Proposed scheme for the interactions of PASM and liver V1 receptors with G-proteins in the presence or absence of agonist.

several chromatographic steps. Based on its physical and kinetic characteristics, it is possible that the V1 receptor contains seven membrane spanning helices, similar to several other receptors as well as rhodopsin.

In addition to the monitoring of receptor binding activity during purification, the development of a functional reconstitution system allowed us to explore the interaction(s) of V1 receptors and G-proteins from different tissues. The use of this reconstitution system also enabled us to begin to elucidate some of the salient regulatory features of receptor/G-protein interaction and in turn signal transduction (Fig. 10).

It is apparent that the type of receptor/G-protein interactions that have been characterized fall into two categories: Those receptor systems that are not functionally coupled to G-proteins in the absence of agonist, that include the beta-adrenergic receptor from cultured heart muscle cells (Nerme et al., 1987), reticulocytes (Limbird et al., 1978) and lung (Cerione et al., 1984; Cerione et al., 1986), the pituitary dopamine D_2 receptor (Kilpatrick and Caron, 1983), and the liver V1 receptor (Dickey et al., 1987; Fishman et al., 1987); and those receptor systems that appear to be functionally coupled to G-proteins in the absence of agonist, which include the adenosine A_1 receptor from brain (Stiles, 1985), the alpha$_2$-adrenergic receptor from platelets (Cerione et al., 1986), the fMet-Leu-Phe receptor (Koo et al., 1983), and the smooth muscle V1 receptor (Bielinski et al., 1988). Such a functional difference, in particular the one observed for vasopressin receptors from liver and smooth muscle, results in populations of similar receptors with differing affinities for agonist. This allows one hormone to regulate various cellular functions according to its circulating plasma levels, without the need for multiple receptor subtypes or modification. Undoubtedly, such a variation in the ability of receptors to couple to G-proteins must be the result of slight differences in receptor structure, G-proteins, or membrane composition. It is possible that some as yet unidentified membrane or cytosolic factor can enhance or inhibit the functional interaction of receptors and G-proteins. This solubilization and reconstitution procedure should be useful in identifying such factor(s), if they exist. It will also allow for the reconstitution of PASM V1 receptors with liver G-proteins and vice versa, in an attempt to identify the regulatory component(s) of this system.

Acknowledgments

We would like to gratefully acknowledge the support of Richard E. Fine, Nancy Bucher, Javier Navarro, and Peter Polgar, as well as the expert technical assistance of Michael McCrory, Donna Bielinski, Cynthia Attisano, Hae-Yung Pyun, and Michael Cahill.

This work was supported by grants DK 39328 and BL 01530 from the USPHS. Jordan B. Fishman is a recipient of a Young Investigator's Award (N-000014-89-J-3051) from the office of Naval Research.

References

Aiyar, N., Nambi, P., Stassen, F., and Crooke, S. T. (1987) *Mol. Pharmacol.* **32,** 34–36.
Ausiello, D. A. and Orloff, J. (1982) in *Handbook of Experimental Pharmacology* vol. 58/II (Nathanson, J. A. and Kebabian, J. W., eds.), Springer-Verlag, Berlin, pp. 271–303.
Berridge, M. J. and Irvine, R. F. (1984) *Nature* (London) **312,** 315–321.
Bielinski, D., McCrory, M., Cahill, M., Polgar, P., and Fishman, J. B. (1988) *Biochem. Biophys. Res. Commun.* **151,** 1293–1298.
Birnbaumer, L., Codina, J., Mattera, R., Cerione, R. A., Hildebrandt, J. D., Sunyer, T., Rojas, F. J., Caron, M. G., Lefkowitz, R. J., and Iyengar, R. (1985) *Recent Prog. Horm. Res.* **41,** 41–99.
Blackmore, P. F., Bocckino, S. B., Waynick, L. E., and Exton, J. H. (1985) *J. Biol. Chem.* **260,** 14477–14483.
Boer, R., Crause, P., and Fahrenholz, F. (1983) *Biochem. Biophys. Res. Commun.* **116,** 91–98.
Boer, R. and Fahrenholz, F. (1985) *J. Biol. Chem.* **260,** 15051–15054.
Bojanic, D. and Fain, J. N. (1986) *Biochem. J.* **240,** 361–365.
Cantau, B. S., Keppens, S., DeWulf, H., and Jard, S. (1980) *J. Receptor Res.* **1,** 137–168.

Cardenus-Tanus, R. J., Huerta-Bahena, J., and Garcia-Sainz, J. A. (1982) *FEBS Lett.* **143**, 1–4.

Casey, P. J. and Gilman, A. G. (1988) *J. Biol. Chem.* **263**, 2577–2580.

Cerione, R. A., Codina, J., Benovic, J. L., Lefkowitz, R. J., Birnbaumer, L., and Caron, M. G. (1984) *Biochemistry* **23**, 4519–4525.

Cerione, R. A., Regan, J. W., Nakata, H., Codina, J., Benovic, J. L., Gierschik, P., Somers, R. L., Spiegel, A. M., Birnbaumer, L., Lefkowitz, R. J., and Caron, M. G. (1986) *J. Biol. Chem.* **261**, 3901–3909.

Crause, P., Boer, R., and Fahrenholz, F. (1984) *FEBS Lett.* **175**, 383–386.

Delean, A., Stadel, J. M., and Lefkowitz, R. J. (1980) *J. Biol. Chem.* **255**, 7108–7117.

de Sousa, R. C. and Grosso, A. (1981) *J. Physiol.* (Paris) **77**, 643–669.

Dickey, B. F., Fishman, J. B., Fine, R. E., and Navarro, J. (1987) *J. Biol. Chem.* **262**, 8738–8742.

Dickey, B. F., Navarro, J., Fishman, J. B., and Fine, R. E. (1986) *Fed. Proc.* **45**, 1736.

Fahrenholz, F., Boer, R., Crause, P., Fritzsch, G. and Grzonka, Z. (1984) *Eur. J. Pharmacol.* **100**, 47–58.

Fishman, J. B., Dickey B. F., and Fine, R. E. (1987) *J. Biol. Chem.* **262**, 14049–14055.

Fishman, J. B., Dickey, B. F., Bucher, N. L. R., and Fine, R. E. (1985) *J. Biol. Chem.* **260**, 12641–12646.

Fitzgerald, T. J., Uhing, R. J., and Exton, J. H. (1986) *J. Biol. Chem.* **261**, 16871–16877.

Grillone, L. R., Clark, M. A., Godfrey, R. W., Stassen, F., and Crooke, S. T. (1988) *J. Biol. Chem.* **263**, 2658–2663.

Guillon, G., Balestre, M.-N., Mouillac, B., and Devilliers, G. (1986) *FEBS Letts.* **196**, 155–159.

Guillon, G., Courand, P. O., Butlen, D., Cantau, B., and Jard, S. (1980) *Eur. J. Biochem.* **111**, 287–294.

Jard, S., Cantau, B., and Jakobs, K. H. (1981) *J. Biol. Chem.* **256**, 2603–2606.

Kilpatrick, B. F. and Caron, M. G. (1983) *J. Biol. Chem.* **258**, 13528–13534.

Koo, C., Lefkowitz, R. J., and Snyderman, R. (1983) *J. Clin. Invest.* **72**, 748–753.

Limbird, L. E., Gill, D. M., and Lefkowitz, R. J. (1978) *Proc. Natl. Acad. Sci., USA* **77**, 775–779.

Lynch, C. J., Charest, R., Blackmore, P. F., and Exton, J. H. (1985) *J. Biol. Chem.* **260**, 1593–1600.

Lynch, C. J., Prpic, V., Blackmore, P. F., and Exton, J. H. (1986) *Mol. Pharmacol.* **29**, 196–203.

Nerme, V., Severne, Y., Abrahamsson, T., and Vauquelin, G. (1987) *Mol. Pharmacol.* **30**, 1–5.

Orloff, J. and Handler, J. S. (1962) *J. Clin. Invest.* **41**, 702–709.

Penit, J., Faure, M., and Jard, S. (1983) *Am. J. Physiol.* **244**, E72–E82.

Schmid, P. G., Abboud, F. M., Wendling, M. D., Ramberg, E. S., Mark, A. L., Heistad, D. D., and Eckstein, J. W. (1974) *Am. J. Physiol.* **227**, 998–1004.

Stiles, G. L. (1985) *J. Biol. Chem.* **260**, 6728–6732.

Tanaka, M., de Kloet, E. R., de Wied, D., and Versteeg, D. H. G. (1977) *Life Sci.* **20**, 1799–1808.

Uhing, R. J., Prpic, V., Jiang, H., and Exton, J. H. (1986) *J. Biol. Chem.* **261**, 2140–2146.

van Leeuwen, F. W., van der Beek, E. M., van Heerikhuize, KJ. J., Wolters, P., van der Meulen, G., and Wan, Y.-H. (1987) *Neurosci. Lett.* **80**, 121–126.

Wallace, M. A. and Fain, J. N. (1985) *J. Biol. Chem.* **260**, 9527–9530.

Williamson, K., Dickey, B. F., Pyun, H.-Y., and Navarro, J. (1988) *Biochem.* **27**, 5371–5379.

Purification
of an Angiotensin II Receptor

*Terry S. Elton, Suzanne Oparil,
and J. Edwin Blalock*

1. Introduction

The renin–angiotensin system plays an important homeostatic role in the regulation of electrolyte and fluid balance and arterial pressure. This regulatory function is exerted through the multiple actions of the biologically active component of the renin–angiotensin system, angiotensin II (AII) (Plentl and Page, 1945; Skeggs et al., 1956). AII, an octapeptide hormone, neurotransmitter, and autocoid, stimulates aldosterone bio-synthesis and secretion by adrenal glomerulosa cells (Saltman et al., 1975; Douglas et al., 1976; Capponi and Catt, 1979), produces smooth muscle constriction (Page and Bumpus, 1977), and has positive inotropic and chronotropic effects on mammalian cardiac muscle (Kobayashi et al., 1978). In the brain, AII acts to elevate blood pressure (Lang et al., 1983) and to stimulate drinking and salt appetite (Fitzsimmons, 1979; Denton, 1982). Further, AII induces the secretion of several pituitary hormones, including vasopressin (Ramsay et al., 1978), ACTH (Ramsay et al., 1978; Reid, 1984), oxytocin (Lang et al., 1983; Keil et al., 1984), and luteinizing hormone (Steele et al., 1982,1983). In the liver, AII stimulates glycogenolysis and gluconeogenesis (Hems et al., 1978; Whitton et al., 1978; Garrison and Wagner, 1982) and angiotensinogen production (Khayyall et al., 1973). Finally, AII has multiple roles in the kidney, including modulation of glomerular filtration rate, renal plasma flow (Thurau, 1974; Navar and Rosivall, 1984), and

Receptor Purification, vol. 2   ©1990 The Humana Press

renal sodium excretion (Levens et al., 1981), and inhibition of renin release (Davis and Freeman, 1976).

Numerous studies have demonstrated that the biological responses to AII are mediated by its interaction with specific high affinity AII receptors located on the cell surfaces of the various target tissues (Mendelsohn, 1985). Once the AII receptor is occupied with AII, a receptor generated signal is transduced to intracellular effectors by guanine nucleotide-binding, "G" proteins (DeLeon et al., 1984; Pobiner et al., 1985; Lynch et al., 1986; Hausdorff et al., 1987). The AII receptor or AII receptor subtypes can interact with several distinct G proteins (DeLeon et al., 1984; Pobiner et al., 1985; Lynch et al., 1986; Hausdorff et al., 1987), thereby activating or in-activating different intracellular effectors or second messengers.

The nature of the coupling between G proteins and the AII receptor is not clear. To understand how a receptor generated signal is transduced to intracellular effectors, the AII receptor and the various G proteins must be purified and characterized. This chapter will focus on the purification and characterization of AII receptors.

2. Characterization of Angiotensin II Receptors

Extensive efforts have been made to purify and characterize the AII receptor, but progress has been relatively slow because of the difficulty of solubilizing the receptor in a stable, active state. To circumvent these problems, several laboratories have utilized both photoaffinity labeling and radioligand crosslinking techniques prior to AII receptor solubilization and characterization.

2.1. Characterization of Angiotensin II Receptors Using Photoaffinity Labels

Initial photoaffinity labeling experiments utilized ^{125}I-2-nitro-5 azido-benzoyl-angiotensin II, a photosensitive AII analog derivatized at the N terminus (Capponi and Catt, 1980). Using this photoaffinity label, Capponi and Catt (1980) demonstrated that the binding chain of the AII receptor (herein after referred to as the AII receptor) from dog adrenal and uterine membranes had a M_r of 67,000. Based on gel filtration experiments, they concluded that the AII receptor may be a dimer (M_r of 126,000), constituted of two subunits of similar mol wts.

Recent photoaffinity labeling experiments have utilized [Sar^1-(3'-^{125}I) Tyr^4-(4'-N$_3$)Phe^8] AII (IN$_3$ AII) and [Sar^1-(3'-^{125}I) Tyr^4-(4'-N$_3$) DPhe^8] AII (IN$_3$ DPhe AII), photosensitive AII analogs with an azido-phenylanine residue at the carboxyl terminus. These AII analogs are excellent photoaffinity reagents, since they permit isolation of covalently labeled AII receptors in high yields (Guillemette et al., 1986). Using the photoaffinity label, IN_3 AII, Guillemette and Escher (1983) demonstrated that the AII receptor from bovine adrenal is a 300,000 dalton protein complex that probably contains only one AII binding subunit with a M_r of 63,000. Further, Guillemette et al. (1986) demonstrated that they could photolabel an AII receptor from rat liver membranes with a M_r of 63,000, utilizing IN_3 AII (an AII agonist) or IN_3 DPhe AII (an AII antagonist). They also showed that the effective mol wt of rat liver AII receptors labeled with a reversible AII agonist (M_r of 145,000) is greater than that of AII receptors labeled with a reversible antagonist (M_r of 108,000) (Guillemette et al., 1984). These results suggest that AII agonists, but not AII antagonists, stabilize an interaction between the AII receptor and another membrane component, probably a G-protein.

Carson et al. (1987), using the photoaffinity label, IN_3 AII, analyzed the AII receptor from several species and tissues. They found that the apparent mol wt of AII receptor differed among species for a given target tissue, and among target tissues for a given species. The range of M_rs for the AII binding protein was 58,000 (cow adrenal membranes) to 92,000 (rat pituitary membranes). However, because of the breadth of the specifically labeled protein bands observed on SDS-PAGE, these M_r values are subject to error.

2.2. Characterization of Angiotensin II Receptors by Chemical Crosslinking

To identify and characterize the molecular components involved in AII binding, several laboratories have utilized chemical crosslinking reagents. Paglin and Jamieson (1982) have used the chemical crosslinker, disuccinimidyl suberate, to crosslink AII to its binding sites on rat adrenal membranes. Their results indicate that, in rat adrenal membrane fractions, a protein with an apparent M_r of 116,000 was covalently linked to ^{125}I-AII by disuccinimidyl suberate. Similarly, Rogers (1984), using various bifunctional crosslinking reagents, demonstrated that ^{125}I-AII was covalently linked to a myocardial sarcolemmal membrane component with an apparent M_r of 116,000. Laribi et al. (1987) have performed chemical crosslinking experiments on cultured mouse spinal cord cells. In contrast to the above results, they showed that ^{125}I-AII was covalently linked to a protein of apparent M_r 68,000 by SDS-PAGE. However, when the crosslinked cells were solubilized and analyzed by HPLC gel filtration, three radioactive species were identified, with apparent M_rs of 65,000, 115,000, and 185,000. Furthermore, Sen et al. (1983,1984) have demonstrated that if rabbit liver membranes are first solubilized and then crosslinked with disuccinimidyl suberate, the ^{125}I-AII labeled protein has a M_r of 68,000.

2.3. Discussion

Experiments employing both photoaffinity labeling and radioligand crosslinking techniques have demonstrated that the AII receptor from different species and tissues have apparent M_rs of 58,000–116,000, with an average M_r of 66,000. The significantly higher M_r values (116,000) for the AII binding protein were obtained by Paglin and Jamieson (1982) and Rogers (1984) after using disuccinimidyl suberate or other reagents to crosslink ^{125}I-AII to membrane preparations. Since these experiments were performed on membrane preparations, it is possible that adjacent proteins were crosslinked to the AII binding protein, leading to an overestimation of its apparent mol wt. The data of Laribi et al. (1987) and Sen et al. (1983,1984) support this explanation. Laribi et al. (1987) performed crosslinking experiments on intact cells using disuccinimidyl suberate. They demonstrated that the AII binding protein had an apparent M_r of 68,000. Sen et al. (1983,1984), also using disuccinimidyl suberate, first solubilized liver membrane preparations and then crosslinked ^{125}I-AII to the binding protein. Their results again showed that a protein with a M_r of 68,000 was specifically labeled. These crosslinking data support the photoaffinity labeling data, in that the AII binding protein has an average apparent M_r of 66,000, with some variation, depending on species and tissue. Photoaffinity labeling and chemical crosslinking techniques have been useful in the identification and partial characterization of the AII binding protein, but they have not resulted in its purification.

3. Molecular Recognition Theory

Membrane-bound receptors have generally proven difficult to purify. A novel methodology has recently been used to isolate, purify, and characterize several hormone receptors, including the corticotropin receptor (ACTH) (Bost et al., 1985a; Bost and Blalock, 1986), delta-class opiate receptor (Carr et al., 1986,1987), luteinizing hormone releasing hormone receptor (LHRHR) (Mulchahey et al., 1986), and, most recently, the fibronectin receptor (Brentani et al., 1988).

This methodology is based on a molecular recognition theory, which suggests that recognition sequences of interacting peptides can be encoded by complementary strands of DNA (Blalock and Bost, 1986). The theory states that ligands and receptor binding sites can be derived from complementary nucleic acid sequences (Fig. 1).

MOLECULAR   COMPLEMENTARITY

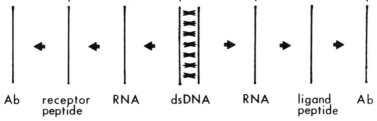

Fig. 1. Molecular complementarity. One sequence of DNA (–) encodes a peptide ligand binding site via a + strand of RNA. The complementary sequence of DNA (+) encodes a receptor binding site via a – strand of RNA. The "–" ligand peptide and "+" receptor peptide react with one another with specificity and relatively high affinity. Antibodies to the ligand and receptor peptides are idiotypic and anti-idiotypic antibodies, respectively, that may share sequence homology with their peptide immunogen's complementary partner. Antibodies (–) to the receptor peptide (+) can be used to purify that particular receptor.

This concept arose from the observation that codons for hydrophobic amino acids were complemented by those for hydrophilic amino acids and vice versa. Moreover, the average tendency of codons for uncharged (slightly hydrophilic) amino acids was to be complemented by codons for other uncharged amino acids (Blalock and Smith, 1984). One possible consequence of the aforementioned pattern was suggested by the finding that many biologically important peptides composed of 10–50 amino acids, such as hormones, can assume amphiphilic secondary structure, such as a membrane or a receptor binding site (Kaiser and Kezdy, 1984). Thus, it was theorized that two peptides that are encoded by complementary nucleic acid strands might form or impose such amphiphilic structures or conformations and bind one another with high affinity and specificity. Consistent with this idea was the finding of regions of nucleotide complementarity that were uniquely associated with receptors and their ligands (Bost et al., 1985b). For example, interleukin 2 and its receptor contain such complementary sequences, which were recently shown to represent the contact points between this ligand and its receptor (Weigent et al., 1986; Ju et al., 1987; Brandhuber et al., 1987).

In further support of the molecular recognition theory, several laboratories have demonstrated that peptides encoded by complementary strands of nucleic acids (designated "complementary peptides") have the ability to bind one another (Bost et al., 1985a; Blalock and Bost, 1986; Shai et al., 1987). For example, it was shown that complementary peptides to ACTH (Bost et al., 1985a), γ-endorphin (Carr et al., 1986), and LHRH

(Mulchahey et al., 1986) bind to their respective hormones with high affinity and specificity, suggesting that the complementary peptides resemble receptor binding sites. Additionally, it was demonstrated that antibodies directed against a peptide complementary to a particular hormone would bind to that hormone's receptor binding site (Bost et al., 1985a; Bost and Blalock, 1986; Carr et al., 1986, 1987; Gorcs et al., 1986; Mulchahey et al., 1986; Brentani et al., 1988). Finally, antibodies directed against peptides complementary to ACTH, γ-endorphin, LHRH, and fibronectin have been used to immunoaffinity purify these specific hormone receptors (Bost et al., 1985a; Bost and Blalock, 1986; Carr et al., 1986,1987; Mulchahey et al., 1986; Brentani et al., 1988). Thus, the molecular recognition receptor purification methodology may be applicable and practical for a wide variety of peptides and proteins whose receptors have to date been difficult to purify.

4. Purification of the Angiotensin II Receptor Using Complementary Peptide Methodology

The AII receptor is the only component of the renin–angiotensin system that has not been purified and characterized. Chemical crosslinking and photoaffinity techniques have been useful in the identification and partial characterization of the AII receptor, but they have not resulted in its purification. We have investigated the usefulness of molecular recognition methodology in the purification of the AII receptor, since antibodies to a peptide encoded by the complementary nucleic acid strand of angiotensinogen (or AII) may allow for the easy purification and characterization of this receptor.

We have taken advantage of the availability of a cDNA clone corresponding to rat angiotensinogen, thus providing the nucleotide sequence of AII (Ohkubo et al., 1983). From this, it was easy to deduce the sequence of a complementary peptide to the octapeptide, AII. It is also worth noting that it is possible to generate a complementary peptide from a primary amino acid sequence without prior knowledge of the nucleotide sequence (Bost and Blalock, 1988).

The published amino acid and mRNA sequences for rat AII are shown in Fig. 2. Also shown in Fig. 2 is the derived nucleotide sequence of the complemetary RNA and the amino acid sequence of the complementary peptide (IIA) when assigned in the 5'–3' direction and in the same reading frame. The complementary peptide IIA (Fig. 2, Lys-Gly-Val-Asp-Val-Tyr-Ala-Val-COOH) was synthesized by standard solid-phase methods and purified by HPLC. We initially investigated whether this complementary peptide would interact with AII in a specific manner by performing solid-

		1			4			8	
AII		Asp	Arg	Val	Tyr	Ile	His	Pro	Phe
mRNA	5'	GAC	CGC	GUA	UAC	AUC	CAC	CCC	UUU	3'
Complementary RNA	3'	CUG	GCG	CAU	AUG	UAG	GUG	GGG	AAA	5'

Complementary Peptide IIA		Val	Ala	Tyr	Val	Asp	Val	Gly	Lys
		8			4			1	

Fig. 2. Amino acid sequence of the complementary octapeptide IIA encoded by an RNA that is complementary to the mRNA for AII. The amino acid and nucleotide sequence for rat AII were from Ohkubo et al. (1983). Codons in the complementary RNA were assigned in the 5'–3'direction to derive the amino acid sequence of the complementary peptide IIA (*from* Elton et al., *Proc. Natl. Acad. Sci. USA* **85**, 2518–1522, 1988).

phase binding assays (Orth, 1975). We were unable to demonstrate specific binding of IIA and AII, possibly because of the restrictions imposed on the small complementary peptide bound to the solid support (data not shown).

The complementary peptide IIA was then assayed for its ability to interact with AII in a specific manner by utilizing solution phase assays (Elton et al., 1988). Specifically, IIA was tested for the capability to inhibit the binding of ^{125}I-AII to AII receptors on rat adrenal particles or purified rat adrenal plasma membranes (Douglas et al., 1978). Briefly, ^{125}I-AII (0.1 nM) was incubated for 60 min at 22°C in 0.1–0.2 mL of assay buffer (50 mM Tris•HCL, pH 7.5, 120 mM NaCl, 5 mM EDTA, 0.2% bovine serum albumin) with 100–200 µg of adrenal particles and various concentrations of unlabeled AII, an AII analog ([Sar^1, Ile^8]AII), complementary peptide IIA, or a control peptide (Ser-Thr-Thr-Thr-Asn-Tyr-Thr-COOH). The reactions were stopped by filtration on GF/B glass filters, followed by four rapid washings with 5 mL of ice-cold incubation buffer. Nonspecific binding, measured in the presence of an excess of unlabeled AII (10 µM), always represented <10% of total radioactivity bound.

Displayed in Fig. 3 is a typical ^{125}I-AII competition binding experiment using rat adrenal particles. The results show that IIA is capable of inhibiting (K_d ~5 × 10^{-8} M) the binding of ^{125}I-AII to rat adrenal AII receptors in a manner comparable to [Sar^1, Ile^8] AII and nonlabeled AII. Similar results were obtained with purified rat adrenal plasma membranes. The inhibition was dose dependent (Fig. 3) and did not occur if the particles or membranes were preincubated with IIA and subsequently washed (data not shown), indicating that IIA did not bind to the AII receptor. The inhibition of ^{125}I-AII binding appeared to be specific since control peptide did not significantly inhibit the binding of radiolabeled AII, even at concentrations of 0.1 mM (Fig. 3). Five other control peptides were also used in competi-

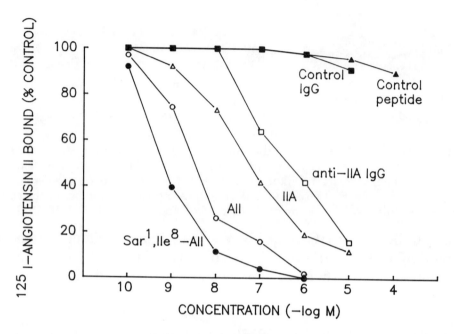

Fig. 3. Competition curves for ^{125}I-AII binding to rat adrenal particles. ^{125}I-AII (0.1 n*M*) was incubated with 100–200 µg of rat adrenal particles and increasing concentrations of unlabeled ligands for 60 min at 25°C. The amount of bound ^{125}I-AII was determined by filtration. The ligands used were [Sar^1,Ile^8]AII (●), AII (○), complementary peptide IIA(△), anti-IIA IgG(□), control IgG (■), and control peptide (▲) (*from* Elton et al., *Proc. Natl. Acad. Sci. USA* **85,** 2518–2522, 1988).

tion binding experiments and the results were identical to those shown in Fig. 3 for this control peptide. Thus, although solid phase binding was not observed, these experiments suggest that the complementary peptide IIA in solution interacts specifically with ^{125}I-AII and, thus, interferes with its ability to bind to adrenal AII receptors.

To determine if the interaction between IIA and AII was owing to the complementary peptide having a shape or conformation resembling the binding site of the AII receptor, an antibody was prepared against IIA. Since most peptides are not very immunogenic, IIA was coupled to key-hole limpet hemocyanin (KLH) protein with glutaraldehyde (6.7 m*M*) (Avarmeas and Ternyck, 1969). Three injections (250 µg each) of the peptide–KLH conjugate in 0.5 mL of complete Freund's adjuvant were administered at 2-wk intervals and the rabbits were also bled at 2-wk intervals. Multiple animals were immunized.

Antibodies to complementary peptide IIA (anti-IIA IgG) were purified by affinity chromatography. Briefly, the IIA peptide was coupled to an agarose gel bead support (Affi-Gel 10) as described by the manufacturer

(Bio-Rad). Protein A purified IgG from rabbits that had been immunized with complementary peptide IIA was then chromatographed on the IIA-Affi-Gel 10 affinity column. Antibodies (anti-IIA IgG) that bound to the column were eluted with $3M$ $MgCl_2$ and collected. The eluted material was then dialyzed against phosphate-buffered saline and concentrated in a centricon 30 microconcentrator (Amicon) to yield protein concentrations ranging from 0.5 to 3 mg/mL.

The affinity purified antibody, anti-IIA IgG, was evaluated for its ability to recognize the AII receptor. AII binding assays were performed as described earlier, except that various concentrations of anti-IIA IgG and control IgG were utilized in the assay. Figure 3 shows that affinity-purified anti-IIA IgG inhibits the binding (K_d ~4 × 10^{-7}) of radiolabeled AII to AII receptors present on rat adrenal membranes. Inhibition of ^{125}I-AII binding appeared to be specific, since control antibody (preimmune serum) did not significantly inhibit binding of the radiolabeled AII.

To investigate whether anti-IIA IgG recognizes a biologically relevant AII receptor, aldosterone secretion assays were performed. Briefly, collagenase-dispersed glomerulosa cells were prepared from rat adrenal glands, as described by Douglas et al. (1978). Adrenal cells were resuspended in medium 199 containing 0.2% bovine serum albumin. Cell viability, as monitored by trypan blue exclusion, was usually >90%. One-milliliter aliquots of the cell suspension (~10^5 cells) were incubated with AII, anti-IIA IgG, preimmune IgG, or AII with anti-IIA IgG for 2 h at 37°C under 5% CO_2/95% air. The cells were pelleted, and the aldosterone content of the media was measured by direct radioimmunoassay (Kubasik et al., 1979). The results of the aldosterone stimulation assay are shown in Table 1. When glomerulosa cells were incubated without exogenous AII, basal rates of aldosterone secretion were ~330 pg/mL. When cells were incubated with AII, there was twofold stimulation (~659 pg/mL) of aldosterone production. In contrast, when anti-IIA IgG was incubated with glomerulosa cells, aldosterone secretion was reduced below basal levels (~189 pg/mL). Similar aldosterone production results were obtained when anti-IIA IgG and AII were co-incubated with glomerulosa cells (~199 pg/mL). Control experiments utilizing preimmune serum did not inhibit the spontaneous or AII-mediated secretion of aldosterone (data not shown). Results of the aldosterone production assays and the AII binding assays suggest that anti-IIA IgG recognizes and specifically interacts with a biologically relevant AII binding site. This conclusion is also supported by our recent observation that anti-IIA IgG can reduce both the magnitude and the duration of AII induced vascular smooth muscle contractions in rabbit aortic rings (unpublished results).

Table 1
Aldosterone Production in Rat Adrenal Glomerulosa Cell[a]

Stimulus	Aldosterone production, pg/mL ($n = 4$)
None	330 ± 46
AII, 1 μM	659 ± 28
Anti-IIA IgG, 1 μM	189 ± 26
AII, 1 μM + anti-IIA IgG, 1 μM	199 ± 34

[a]Dispersed rat adrenal glomerulosa cells were suspended in serum-free medium 199 at 1×10^5 cells per tube. The cells were incubated for 2 h at 37°C with AII, control IgG, anti-IIA IgG, or AII and anti-IIA IgG. After incubation, samples were centrifuged, and the cell-free medium was assayed for aldosterone content. Each value represents the mean ±SD of four experiments, each point assayed in duplicate. Control preimmune serum did not inhibit the spontaneous or AII-mediated release of aldosterone (*from* Elton et al., *Proc. Natl. Acad. Sci. USA* **85,** 2520, 1988).

Since anti-IIA IgG appears to recognize a biologically relevant AII binding site, this antibody should be useful for immuno-purifying an AII binding protein. To investigate this possibility, anti-IIA IgG was used as an immunoadsorbent. Affinity-purified anti-IIA IgG (0.5 mg/mL) was coupled to Reacti-Gel 6X agarose beads, as described by the manufacturer (Pierce), and utilized for affinity chromatography. Illustrated in Fig. 4 is a flowchart that outlines the steps involved in the purification of the AII receptor. Initially, rat adrenal particles were prepared as described by Douglas et al. (1978). The particles (~5 mg) were resuspended in 3 mL of solubilization buffer (50 mM Tris-HCl/pH 7.5, 5 mM CHAPS, 0.32M sucrose, 30% glycerol, and 0.1 mM phenylmethylsulfonyl fluoride) and solubilized for 90 min at 4°C under constant agitation (Capponi et al., 1983). The zwitterionic detergent CHAPS (3-[3-cholamido-propyl) dimethylamino]-L-propanesulfonate) was used to solubilize the rat adrenal particles, since it is effective in solubilizing membrane proteins in a nondenatured state (Hjelmeland, 1980). CHAPS was chosen because it has been useful in solubilizing active opiate receptors (Simonds et al., 1980; Carr et al., 1986) and AII receptors (Capponi et al., 1983; Chansel et al., 1984). The concentration of CHAPS used to solubilize membrane proteins is a very important consideration. Capponi et al. (1983) demonstrated that a solubilization buffer containing 5 mM CHAPS was optimal and could extract in a nondenaturated state about 50% of the angiotensin II receptors present in membrane particles. Therefore, our solubilization buffers contained 5 mM CHAPS.

After solubilization, samples were centrifuged at 105,000g for 60 min and the supernate was diluted to a final CHAPS concentration of 0.3 mM. The diluted supernate was then chromatographed on an anti-IIA IgG affin-

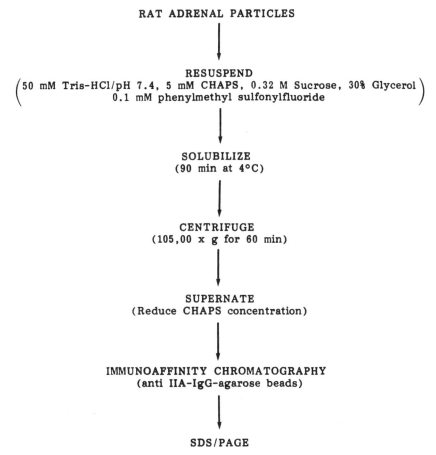

RAT ADRENAL PARTICLES

RESUSPEND
$\left(\begin{array}{c}\text{50 mM Tris-HCl/pH 7.4, 5 mM CHAPS, 0.32 M Sucrose, 30\% Glycerol} \\ \text{0.1 mM phenylmethyl sulfonylfluoride}\end{array}\right)$

SOLUBILIZE
(90 min at 4°C)

CENTRIFUGE
(105,00 x g for 60 min)

SUPERNATE
(Reduce CHAPS concentration)

IMMUNOAFFINITY CHROMATOGRAPHY
(anti IIA-IgG-agarose beads)

SDS/PAGE

Fig. 4. Immunoaffinity purification of the AII binding protein.

ity column. Protein that bound to the column was eluted with $3M$ $MgCl_2$, collected, and dialyzed overnight against H_2O. Protein samples were then concentrated in a centricon 30 microconcentrator (Amicon) and subsequently analyzed by $NaDodSO_4$/polyacrylamide slab gel electrophoresis (SDS-PAGE) as described by Laemmli (1970). After electrophoresis, gels were stained for protein by a silver-staining method as described by Merrill et al. (1981). The electrophoretic results presented in Fig. 5 show that the immunoaffinity-purified material consisted of a single protein species (lane 1 or 2). Purified protein that was treated with 2-mercaptoethanol (5%) migrated with an apparent M_r of 66,000 ± 2000 (Fig. 5, lane 2). Interestingly, in the absence of a reducing agent, the purified protein migrated faster, with a M_r of 58,000 ± 2000 (Fig. 5, lane 1). These data suggest that the immunoaffinity-purified protein contains intramolecular disulfide bridge(s), and at least in a nonstimulated state, does not interact with other proteins through intermolecular disulfide linkages.

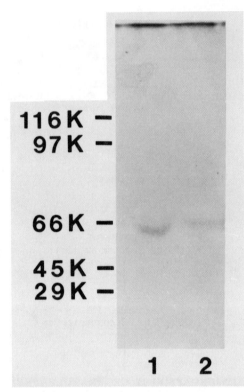

Fig. 5. SDS-polyacrylamide gel electrophoresis of immunoaffinity purified material. Immunoaffinity purified protein was electrophoresed on a 7.5% SDS-polyacrylamide gel, and subsequently silver stained (Merrill et al., 1981). Lane 1: immunoaffinity purified material without 2-mercaptoethanol. Lane 2: immunoaffinity purified material with 2-mercaptoethanol (5%). The mol wt markers were carbonic anhydrase (M_r 29,000), ovalbumin (M_r 45,000), bovine serum albumin (M_r 66,000), phosphorylase b (M_r 97.400), and β-galactosidase (M_r 116,000) (*from* Elton et al., *Proc. Natl. Acad. Sci. USA* **85**, 2518–2522, 1988).

To demonstrate that this immunoaffinity purified protein contains the AII binding site, 0.1–5 µg of this protein was incubated with ^{125}I-AII (0.2 nM) in the presence or absence of nonlabeled AII (10 µM) at 4°C for 1 h in 500 µL of phosphate buffered saline (PBS) (pH 7.4). These samples were then crosslinked with 3.0 mM disuccinimidyl suberate in dimethyl sulfide for 15 min at 25°C. The crosslinked samples were dialyzed for 18 h against PBS to remove free ^{125}I-AII, concentrated using a centricon 30 microconcen-trator (Amicon), and subjected to SDS-PAGE. Gel electrophoresis was performed on homogeneous slab gels (7.5% acrylamide) (Laemmli, 1970). After electrophoresis, gels were silver-stained (Merrill et al., 1981), dried, and finally subjected to autoradiography at –90°C. The silver-stained gel displayed in Fig. 6A again shows that the immunoaffinity-purified material consists of a single protein species (M_r of 66,000 ± 2000), even though

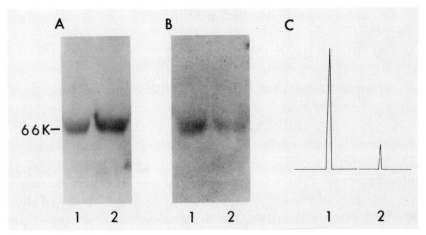

Fig. 6. Specific binding of ^{125}I-AII to immunoaffinity purified protein. Immunoaffinity purified material was labeled (^{125}I-AII) in the absence or presence of AII ($1 \times 10^{-6}M$). The material was then crosslinked with DSS and electrophoresed on a 7.5% SDS-polyacrylamide gel. A) Crosslinked immunoaffinity protein visualized by silver staining. Lane 1; protein labeled in the absence of AII. Lane 2; protein labeled in the presence of AII. B) Autoradiograph of the SDS-polyacrylamide gel in A. The autoradiograph was developed after exposure at –80°C for 10 d. C) Densitometric evaluation of the autoradiogram in B. Lane 1 is scaled to 100% absorbance (*from* Elton et al., *Proc. Natl. Acad. Sci. USA* **85,** 2518–2522, 1988).

the lanes are overloaded (lanes 1 and 2). Figure 6B shows the autoradiograph of the same gel. It is apparent that the purified protein bound ^{125}I-AII (Fig. 6B, lane 1). These results strongly suggest that this binding is specific, since the amount of radiolabeled AII binding was reduced when the protein was incubated in the presence of unlabeled AII (Fig. 6B, lane 2). Densitometric evaluation of the data in Fig. 6B showed that ^{125}I-AII binding was reduced by 80% in the presence of unlabeled AII (Fig. 6).

4. Discussion

Recently, it was hypothesized that there is a molecular recognition code in which peptide ligands and their receptor binding sites can be encoded by complementary nucleotide sequences (Bost et al., 1985a). Since antibodies to a peptide encoded by the complementary nucleic acid strand of angiotensinogen (or AII) may allow for the easy purification and characterization of the AII receptor, we have investigated the usefulness of molecular recognition methodology in the purification of the AII receptor.

In this report, we demonstrate that the complementary peptide IIA specifically inhibits ($K_d \sim 5 \times 10^{-8}M$) the binding of ^{125}I-AII with the AII receptors on rat adrenal membranes (Fig. 3). If membranes were preincubated with IIA and subsequently washed, inhibition of ^{125}I-AII was not observed, suggesting that IIA does not bind directly to the AII receptor. These results indicate that AII and its complementary peptide interact with each other, suggesting that IIA resembles the AII binding site.

If the complementary peptide IIA resembles the AII binding site, then antibodies generated against IIA should crossreact with the AII receptor. Our results show that anti-IIA IgG specifically inhibited the binding (K_d ~4 $\times 10^{-7}M$) of ^{125}I-AII to rat adrenal AII receptors (Fig. 3). Anti-IIA IgG also inhibited aldosterone secretion from freshly dispersed rat adrenal glomerulosa cells (Table 1). The inhibition of aldosterone production was not caused by toxicity of the antibody, as judged by the ability of the adrenal cells to exclude trypan blue dye. These results provide strong evidence that antibodies against the complementary peptide IIA interact specifically with the biologically relevant site, or AII receptor, involved in the regulation of aldosterone secretion. Further, results of the aldosterone production assay suggest that the complementary peptide antibody behaves like an AII antagonist.

Since anti-IIA IgG appears to recognize a biologically relevant AII receptor, this antibody should be useful for immuno-purifying the molecule. Electrophoretic results (Fig. 6) show that anti-IIA IgG affinity chromatography can purify a single protein species that has an apparent M_r of 66,000 ± 2000 under reducing conditions using 2-mercaptoethanol. Moreover, the immunoaffinity-purified protein specifically binds ^{125}I-AII, strongly suggesting that the purified material is an AII receptor.

The molecular weight of the imunoaffinity-purified AII receptor (M_r of 66,000) is in excellent agreement with most other mol wt determinations for the AII receptor (*see* Sections 2.0.–2.3.). However, the M_r of our purified rat adrenal AII receptor (66,000) was significantly smaller than that reported (116,000) by Paglin and Jamieson (1982) using the same species and tissue. This difference can best be explained, as discussed previously in Section 2.3., by attributing Paglin and Jamieson's mol wt determination to an overestimation owing to the crosslinking of a protein in close proximity to the AII binding protein.

Our electrophoresis results showed that the immunoaffinity-purified AII receptor displays an electrophoretic mobility shift after reduction with 2-mercaptoethanol, suggesting the presence of intramolecular disulfide bridge(s) in the protein. Other purified receptors, including β_1- and β_2-adrenergic receptors (Moxham and Malbon, 1985; Moxham et al., 1986), the hepatic glucagon receptor (Iwanij and Hur, 1985), the opiate receptor purified from bovine striatum (Gioannini et al., 1985), the receptors for interleukin 2 (Leonard et al., 1985) and interleukin 3 (Sorensen et al., 1986), the luteinizing hormone/human chorionic gonadotropin receptor (Wimalasena et al., 1986), and the putative AII receptor (Sen et al., 1984) display a similar mobility shift after reduction. Moxham and Malbon (1985) have shown that the integrity of one or more intramolecular disulfide bridges is essential for ligand binding, and chemical cleavage of these bonds destabilizes the binding of the β_1-adrenergic receptor to bind radioligands. Creighton (1984) has suggested that agonist binding may promote disulfide exchange reactions and/or cleavage of intramolecular disulfide bridges and transform the receptor into a structurally altered state capable of activating G proteins. Recently, it has been shown that AII receptors interact with these regulatory proteins (Pobiner et al., 1985; DeLeon et al., 1984; Lynch et al., 1984; Hausdorff et al., 1987). Taken together, these results indicate that disulfide bridges present in the AII receptor may play an important role in transforming the AII receptor into a conformation capable of activating G proteins.

5. Conclusion

This report describes the purification of an AII receptor that retains its capacity to bind ^{125}I-AII specifically. Our ability to generate microgram quantities of the purified AII binding protein will facilitate a more detailed biochemical analysis of the AII receptor. For example, we will be able to investigate whether or not the AII receptor is a dimer (Capponi and Catt, 1980) and to examine the nature of the coupling between G proteins and the AII receptor. We will also be able to investigate the basis for differences in M_rs among the various AII receptors and whether these differences are relevant to the functional diversity of AII.

Finally, this report provides strong support for the hypothesis that ligands and receptor binding sites can be derived from complementary sequences of nucleic acids. Therefore, it may be possible to purify any peptide hormone's receptor using complementary peptide methodology.

References

Avarmeas, S. and Ternyck, T. (1969) *Immunochemistry* **6**, 53–66.
Blalock, J. E. and Smith, E. M. (1984) *Biochem. Biophys. Res. Commun.* **121**, 203–207.
Blalock, J. E. and Bost, K. L. (1986) *Biochem. J.* **234**, 679–683.
Bost, K. L., Smith, E. M., and Blalock, J. E. (1985a) *Proc. Natl. Acad. Sci. USA* **82**, 1372–1375.
Bost, K. L., Smith, E. M., and Blalock, J. E. (1985b) *Biochem. Biophys. Res. Commun.* **128**, 1373–1380.
Bost, K. L. and Blalock, J. E. (1986) *Mol. Cell. Endocrin.* **44**, 1–9.

Bost, K. L. and Blalock, J. E. (1988) in *Methods in Enzymology* (P. M. Conn, ed.), vol. 168, pp. 16–28.

Brandhuber, B. J., Boone, T., Kenny, W. C., and McKay, D. B. (1987) *Science* **238**, 1707–1709.

Brentani, R. R., Ribeiro, S. F., Potocnjak, P., Pasqualini, R., Lopes, J. D., and Nakaie, C. R. (1988) *Proc. Natl. Acad. Sci. USA* **85**, 364–367.

Capponi, A. M., Birabeau, M. A., and Vallotton, M. B. (1983) *J. Rec. Res.* **3**, 289–299.

Capponi, A. M. and Catt, K. J. (1979) *J. Biol. Chem.* **254**, 5120–5127.

Capponi, A. M. and Catt, K. J. (1980) *J. Biol. Chem.* **255**, 12081–12086.

Carr, D. J. J., Bost, K. L., and Blalock, J. E. (1986) *J. Neuroimmunol.* **12**, 329–337.

Carr, D. J. J., DeCosta, B., Jacobson, A. E., Bost, K. L., Rice, K. C., and Blalock, J. E. (1987) *FEBS Lett.* **224**, 272–276.

Carson, M. C., Harper, C. M. L., Baukal, A. J., Aguilera, G., and Catt, K. J. (1987) *Mol. Endocrinology* **1**, 147–157.

Chansel, D., Ardaillou, N., and Ardaillou, R. (1984) *FEBS Letts.* **170**, 243–246.

Creighton, T. E. (1984) in *Methods in Enzymology* (J. G. Hardman and B. W. O'Malley, eds.), vol. 107, pp. 305–329.

Davis, J. O. and Freeman, R. H. (1976) *Physiol. Rev.* **56**, 1–56.

DeLeon, A., Ong, H., Guttowska, J., Schiller, P. W., and McNicoll, N. (1984) *Mol. Pharmacol.* **26**, 498–508.

Denton, D. (1982) *The Hunger for Salt*, Springer-Verlag, Berlin.

Douglas, J., Aguilera, G., Kondo, T., and Catt, K. J. (1978) *Endocrinology* **102**, 685–696.

Douglas, J., Saltman, S., Fredlund, P., Kondo, T., and Catt, K. J. (1976) *Circ. Res.* **38**, 108–112.

Elton, T. S., Dion, D. L., Bost, K. L., Oparil, S., and Blalock, J. E. (1988) *Proc. Natl. Acad. Sci. USA* **85**, 2518–2522.

Fitzsimmons, J. T. (1979) *The Physiology of Thirst and Sodium Appetite*, Cambridge University Press, Cambridge, UK.

Garrison, J. C. and Wagner, J. D. (1982) *J. Biol. Chem.* **257**, 13135–13143.

Gioannini, T. L., Howard, A. D., Hiller, J. B., and Simon, E. J. (1985) *J. Biol. Chem.* **260**, 15117–15121.

Gorcs, T. J., Gottschall, P. E., Coy, D. H., and Arimura, A. (1986) *Peptides* **7**, 1137–1145.

Guillemette, G. and Escher, E. (1983) *Biochemistry* **22**, 5591–5596.

Guillemette, G., Guillon, G., Marie, J., Pantaloni, C., Balestre, M. N., Escher, E., and Jard, S. (1984) *J. Recept. Res.* **4**, 267–281.

Guillemette, G., Guillon, G., Marie, J., Balestre, M. N., Escher, E., and Jard, S. (1986) *Mol. Pharmacol.* **30**, 544–551.

Hausdorff, W. P., Sekura, R. D., Aguilera, G., and Catt, K. J. (1987) *Endocrinology* **120**, 1668–1678.

Hems, D. A., Rodrigues, L. M., and Whitton, P. D. (1978) *Biochem. J.* **172**, 311–317.

Hjelmeland, L. M. (1980) *Proc. Natl. Acad. Sci. USA* **77**, 6368–6370.

Iwanij, V. and Hur, K. C. (1985) *Proc. Natl. Acad. Sci. USA* **82**, 325–329.

Ju, G., Collins, L., Kaffka, K. L., Tsien, W. H., Chizzonite, R., Crowl, R., Bhatt, R., and Kilian, P. L. (1987) *J. Biol. Chem.* **262**, 5723–5731.

Kaiser, E. T. and Kezdy, F. J. (1984) *Science* **223**, 249–255.

Keil, L. C., Rosella-Dampman, L. M., Emmert, S., Chee, O., and Summy-Long, J. (1984) *Brain Res.* **297**, 329–336.

Khayyall, M., MacGregor, J., Brown, J. J., Lever, A. F., and Robertson, J. I. S. (1973) *Clin. Sci.* **44**, 87–90.

Kobayashi, M., Furukawa, Y., and Chiba, S. (1978) *Eur. J. Pharmacol.* **50**, 17–25.

Kubasik, N. P., Warren, K., and Sine, H. E. (1979) *Clin. Biochem.* **12**, 59–61.

Laemmli, U. K. (1970) *Nature* **227**, 680–685.

Lang, R. E., Unger, T., Rascher, W., and Ganten, D. (1983) in *Handbook of Psychopharmacology*, vol. 16 (L. D. Iversen and S. H. Synder, eds.), Plenum, New York, pp. 307–361.

Laribi, C., Allard, M., Vincent, J. D., and Simonnet, G. (1987) *Neuropeptides* **9**, 345–356.

Leonard, W. J., Depper, J. M., Kronke, M., Robb, R. J., Waldman, T. A., and Green, W. C. (1985) *J. Biol. Chem.* **26**, 1872–1880.

Levens, N. R., Peach, M. J., Carey, P. M. (1981) *Circ. Res.* **48**, 157–167.

Lynch, C. J., Prpic, V., Blackmore, P. F., and Exton, J. H. (1986) *Mol. Pharmacol.* **29**, 196–203.

Mendelsohn, F. A. O. (1985) *J. Hypertension* **3**, 307–316.

Merrill, C. R., Goldman, D., Sedman, S. A., and Ebert, M. H. (1981) *Science* **211**, 1437–1438.

Moxham, C. P. and Malbon, C. C. (1985) *Biochemistry* **24**, 6072–6077.

Moxham, C. P., George, S. T., Graziano, M. P., Brandwein, H., and Malbon, C. C. (1986) *J. Biol. Chem.* **261**, 14562–14570.

Mulchahey, J. J., Neill, J. D, Dion, L. D., Bost, K. L., and Blalock, J. E. (1986) *Proc. Natl. Acad. Sci. USA* **83**, 9714–9718.

Navar, L. G. and Rosivall, L. (1984) *Kidney Int.* **25**, 857–868.

Ohkubo, H., Kageyama, R., Ujihar, M., Hirose, T., Inayama, S., and Nakanishi, S. (1983) *Proc. Natl. Acad. Sci. USA* **80**, 2196–2200.

Orth, D. N. (1975) in *Methods in Enzymology*, vol. 37 (J. G. Hardma and B. W. O'Malley, eds.), Academic, New York, pp. 22–38.

Page, I. H. and Bumpus, F. M. (1977) *Physiol. Rev.* **41**, 331–390.

Paglin, S. and Jamieson, J. D. (1982) *Proc. Natl. Acad. Sci. USA* **70**, 3739–3743.

Plentl, A. A. and Page, I. H. (1945) *J. Biol. Chem.* **158**, 49– 56.

Pobiner, B. F., Hewlett, E. L., and Garrison, J. C. (1985) *J. Biol. Chem.* **260**, 16200– 16209.

Ramsay, D. J., Keil, L. C., Sharpe, M. C., and Shinsako, J. (1978) *Am. J. Physiol.* **234**, R66–R71.

Reid, I. A. (1984) *Am. J. Physiol.* **246**, F533–F543.

Rogers, T. B. (1984) *J. Biol. Chem.* **259**, 8106–8114.

Saltman, S., Baukal, A., Waters S., Bumpus, F. M., and Catt, K. J. (1975) *Endocrinology* **97**, 275–282.

Sen, I., Jim, K. F., and Soffer, R. L. (1983) *Eur. J. Biochem.* **136**, 41–49.

Sen, I., Bull, H. G., and Soffer, R. L. (1984) *Proc. Natl. Acad. Sci. USA* **81**, 1679–1683.

Shai, Y., Flashner, M., and Chaiken, I. M. (1987) *Biochemistry* **26**, 669–675.

Simonds, W. F., Koski, G., Streaty, R. A., Hjelmeland, L. M., and Klee, W. A. (1980) *Proc. Natl. Acad. Sci. USA* **77**, 4623–4627.

Skeggs, L. T., Kahn, J. R., and Shumway, N. P. (1956) *J. Exp. Med.* **103**, 295–299.

Sorensen, P., Farber, N. M., and Krystal, G. (1986) *J. Biol. Chem.* **261**, 9094–9097.

Steele, M. K., McCann, S. M., and Negro-Vilar, A. (1982) *Endocrinology* **111**, 722–729.

Steele, M. K., Gallo, R. V., and Ganong, W. F. (1983) A possible role for the brain renin-angiotensin system in the regulation of LH secretion. *Am. J. Physiol.* **245**, R805– R810.

Thurau, K. (1974) in *Angiotensin* (I. H. Page and F. M., Bumpus, eds.), Springer-Verlag, Heidelberg, pp. 475–489.

Weigent, D. A., Hoeprich, P. D., Bost, K. L., Brunck, T. K., Reiher, W. E., and Blalock, J. E. (1986) *Biochem. Biophys. Res. Commun.* **139**, 367–374.

Whitton, P. D., Rodrigues, L. M., and Hems, D. A. (1978) *Biochem. J.* **176**, 893–898.

Wimalasena, J., Abel, Jr., J. A., Wiebe, J. P., and Chen, T. T. (1986) *J. Biol. Chem.* **261**, 9416– 9420

Purification of Prostaglandin E$_1$/Prostacyclin Receptors of Human Blood Platelets

Asim K. Dutta-Roy and Asru K. Sinha

1. Introduction

Activation of adenylate cyclase by prostanoids like prostaglandin E$_1$ (PGE$_1$) or prostacyclin (PGI$_2$) in human blood platelets inhibits platelet aggregation through the increase of cellular cyclic AMP level (Robinson et al., 1969; Tateson et al., 1977). It is generally accepted that the stimulation of adenylate cyclase in different target cells by various hormones and prostaglandins is initiated through the interaction of these agonists with the specific receptors on the outer surface of the membrane bilayer (Ross and Gilman, 1980; Kahn and Sinha, 1988). The occupancy of these hormone responsive receptors by the agonists subsequently activates a GTP regulatory protein that in turn activates catalytic unit of the enzyme leading to the formation of cyclic AMP from ATP (Lefkowitz et al., 1983).

In the case of human platelets, the adenylate cyclase linked PGE$_1$ and PGI$_2$ receptors have been identified with the membrane structure (Schafer et al., 1977; Siegel et al., 1979; Dutta-Roy and Sinha, 1987). PGE$_1$ receptors, which are similar to PGI$_2$ receptors, have been recently solubilized and purified to homogeneity (Dutta-Roy and Sinha, 1987). However, the role of GTP in the activation of adenylate cyclase by prostaglandins is still unclear in that the enzyme could be activated by these agonists in the absence of added GTP (Tsai and Lefkowitz, 1979; Ashby, 1986). Scatchard analysis (1949) of the binding of [^{3}H] PGE$_1$ to the platelet receptors showed the presence of one high affinity-low capacity receptor population and one low affinity-high capacity receptor population, indicating the heterogeneity of the binding sites (Schafer et al., 1977; Siegel et al., 1979; Dutta-Roy and Sinha, 1987). It is not known whether the difference in affinity is a result of the presence of two different classes of binding sites or the negative cooperativity created in the single class of receptors owing to increased concentration of the ligand. However, in the case of red cell membrane that also showed high and low affinity PGE_1 binding sites, no negative cooperativity owing to increased ligand concentrations in the assay mixture could be shown (Dutta-Roy and Sinha, 1985). Although it was postulated earlier that the activation of adenylate cyclase occurred only through the interaction of the ligand with the low affinity receptors, and the high affinity PGE_1 receptors were involved in the "desensitization" reaction (Lefkowitz et al., 1977; Siegel et al., 1979), it has been recently shown that the high affinity receptors of platelets also activate adenylate cyclase (Kahn and Sinha, 1988) and might be related to the activation of a platelet membrane cysteine protease (Dutta-Roy et al., 1986).

2. Purification of PGE_1 Receptor

2.1. Chemicals

Five:six [^{3}H]PGE_1 (specific activity: 50 Ci/mmol), unlabeled prostaglandins were obtained from New England Nuclear Corp., Boston, MA, and Sigma Chemical Co., St. Louis, MO, respectively. Triton X-100 was the product of J. T. Baker Chemical Corp., and Bio-Rad Laboratories supplied Biobeads SM-2.

2.2. Assay of Soluble PGE_1-Receptor

The binding of [^{3}H]PGE_1 to the solubilized receptors of the platelets was assayed by incubating 20–100 µg of protein in 50 mM Tris-HCl buffer, pH 7.4, containing 5 mM MgCl$_2$ in a total volume of 200 µL with 0.3 µM PGE_1 containing 0.1 µCi [^{3}H]PGE_1. Since less than 1% of the PGE_1 bound to the preparation, the concentration of the free ligand essentially remained constant throughout the incubation. Unless otherwise noted, parallel experiments were run using 50-fold excess unlabeled PGE_1 in the incubation mixture to determine the nonspecific binding. This value was subtracted from the total PGE_1 bound to calculate the specific binding. After the incubation, 1.0 mL of the above buffer with Mg^{2+} (0°C) was added to each tube, and the assay mixture was filtered under vacuum through Whatman GF/C glass fiber filters that had been presoaked with the buffer. Since detergent solubilized PGE_1 receptor preparation has a net negative charge at physiological pH, filtration of the solubilized receptor through the filter would retain the receptor protein on the filter. The receptors adsorbed on the filters were then washed with 15 mL of the buffer at 0°C.

The filter was dried and suspended in 10 mL of scintillation fluid (Amersham, ACS-11) and counted in a Beckman scintillation spectrometer (LS-8000) with 60% efficiency for ^{3}H.

2.3. Removal of Triton X-100

Since the presence of Triton X-100 in the sample inhibits the PGE_1 binding to the receptor, it was necessary to remove Triton X-100 from the protein samples using Biobeads SM-2 (Holloway, 1973). Optimum conditions for the removal of the detergent with minimum loss of protein was standardized by treating samples (0.8 mL) with different amounts of moist beads for 30 min at 4°C. The samples were then centrifuged at 35,000xg for 30 min, and the supernatants were analyzed for protein and Triton X-100 content (Garewal, 1973). Complete removal of the excess detergent was achieved using 1.1 mL of moist beads per 0.8 mL of sample. It has been found that removal of Triton X-100 increased the binding by 10–12-fold, indicating that the detergent itself is inhibitory in the ligand receptor interaction.

2.4. Preparation of Platelets

Typically, 10–12 g fresh bloody platelets (obtained from The American Red Cross) were mixed with EDTA (final concentration, 1 mM). The mixture was centrifuged at 200xg for 15 min at 23°C to remove residual erythrocytes and leukocytes from the platelet suspension. The supernatant, platelet-rich plasma, was next centrifuged at 2000xg for 20 min at 23°C. The platelet pellet thus obtained was washed three times by centrifugation after suspending the pellet in 50 mM Tris-HCl buffer, pH 7.4, containing 1.0 mM EDTA and 0.15M NaCl. The wet wt of the platelet pellet was approx 1 g/mL.

2.5. Solubilization of PGE_1 Receptor

The washed platelets were next suspended in 50 mM Tris-HCl buffer, pH 7.4 containing 1 mM EDTA, 0.15M NaCl, 1.0 mM phenylmethylsulfonyl fluoride (PMSF), 5 mM dithiothreitol (DTT), and 0.3M sucrose plus 0.05% Triton X-100 at 0°C. For each g of wet wt platelets (7.8×10^{10} cells), 1.0 mL of the above solution was used. The suspension was incubated for 60 min at 0°C with occasional mild shaking. Care was taken to avoid frothing. After incubation the detergent treated platelet suspension was centrifuged at 35,000 x g for 30 min at 0°C. The supernatant that contains the solubilized receptors was collected. The PGE$_1$ binding activity of the supernatant was determined after removing the detergent by using Biobeads SM-2, as described above.

2.6. DEAE-Cellulose Chromatography

All chromatographic procedures were carried out at 4°C. The supernatant containing the solubilized receptor (12 mL) was reduced to 6 mL by covering the samples in a dialysis bag with powdered polyethylene glycol (M_r 15,000–20,000) at 0°C and periodically removing the wet solid. The concentrated solution was clarified by centrifugation at 8000xg at 0°C and immediately applied to DEAE-cellulose column (2.5 × 60 cm) equilibrated with 10 mM Tris-HCl buffer, pH 7.4, containing 0.05% Triton X-100, 1.0 mM PMSF, 1.0 mM DTT, 5 mM MgCl$_2$, and 0.3M sucrose. The column was then eluted with stepwise increasing the concentrations of KCl in the same buffer except that the concentration of the buffer was increased to 50 mM. The eluates from the washings of the column with 75 mL each of 0, 0.1, 0.2, and 0.5M KCl in 50 mM Tris-HCl buffer did not show the presence of PGE$_1$ binding activity and were discarded.

At 0.7M KCl, the PGE$_1$ binding activity began to emerge from the column in a single peak (Fig. 1). Fractions (12–32) that contained PGE$_1$ binding protein with the highest specific activity were pooled and concentrated to 6.0 mL. Further elution of the column with 1.0, 1.25, and 1.5M KCl (75 mL each) did not show the presence of any PGE$_1$ binding activity.

2.7. Sephadex G-200 Chromatography

The concentrated fractions from the DEAE-cellulose column were applied to Sephadex G-200 column (1.0 × 40 cm) previously equilibrated with the same buffer used in the DEAE-cellulose chromatography except that KCl was omitted. The elution was also carried out with the above buffer. The eluates containing PGE$_1$ binding activity emerged in a single peak in fractions 27–35 in a total vol of 8 mL (Fig. 2). The active fractions were pooled and concentrated to 0.4 mL as described above. The purified receptor was kept at –70°C in 50 mM Tris-HCl buffer, pH 7.4, containing 0.3M sucrose and 0.05% Triton X-100.

2.8. Summary of the PGE$_1$ Receptor Purification

The purification protocol has been summarized in Table 1. By using fresh platelets approx 90% of the PGE$_1$ receptor was recovered in Triton X-100 supernatant. The procedure described above resulted in an increase

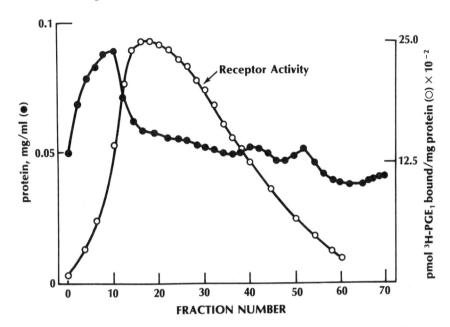

Fig. 1. DEAE-cellulose chromatography of Triton X-100 solubilzed PGE₁ receptors from platelets. Six mL of a concentrated Triton X-100 extract of platelets (25 mg of protein) was applied to a column (2.5 × 60 cm) equilibrated with 10 mM Tris-HCl buffer, pH 7.4. The column was washed with 75 mL of 50 mM Tris-HCl buffer, pH 7.4 containing 0.05% Triton X-100, 1.0 mM PMSF, 1.0 mM DTT, 5 mM MgCl₂, and 0.3M sucrose. Elution was continued with the same buffer containing 0.1, 0.2, or 0.5M KCl in successive steps (75 mL each), and the eluates were discarded. The receptor was then eluted in the same buffer system containing 0.7M KCl. Each fraction was 2.0 mL. Only those fractions that contained [³H]PGE₁ activity are shown here. Protein, ●—●; [³H]PGE₁ binding, O—O. (Reprinted from *J. Biol. Chem.* **262**, pp. 12685–12691, 1987, with the permission of the publisher.)

of specific binding of [³H]PGE₁ from 2.2 p mol/mg protein to 49,000 p mol/mg of protein that represented over 2200-fold purification of PGE₁ receptor with 30% yield.

3. Purity of PGE₁ Receptor Preparation

Alkaline gel electrophoresis of the isolated receptor, under nonreducing conditions, showed that the product was homogeneous when gels were stained either with AgNO₃ (Merril et al., 1983; Fig. 3A) or with Coomassie Brilliant Blue (Fig. 3B). In a parallel experiment, an identical gel that was not stained with the dye to avoid denaturation, was cut into slices, the binding protein from the slices was eluted, and the [³H]PGE₁ binding activity of the eluates was determined. The position of the gel slices showing the [³H]PGE₁ bindng activities corresponded exactly to the protein

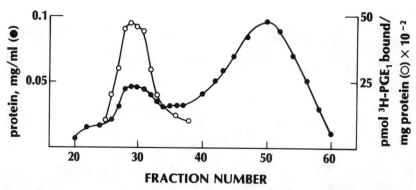

Fig. 2. Chromatography of PGE$_1$ binding proteins from DEAE-cellulose column on Sephadex-G200. The concentrated (6.0 mL, 0.5 mg of protein) active fractions from the DEAE-cellulose column was applied to a Sephadex G-200 column (1.0 × 40 cm), equilibrated, and eluted with the same buffer system used in the elution of the DEAE-cellulose column, except that KCl was omitted. Each fraction was 1.0 mL. Protein, ●—●; [³H]PGE$_1$ binding, ○—○. (Reprinted with permission of the publisher, *J. Biol. Chem.* **262**, pp. 12685–12691, 1987.)

Table 1
Summary of the Purification of PGE$_1$ Receptor

Step	Total specific binding of PGE$_1$	Total Protein	Specific activity	Yield	purification
	pmol		pmol PGE$_1$ bound/mg protein		
Washed platelets	2442	1.11 g	2.2	100	
Triton X-100 supernatant	2357	25 mg	94.3	96.2	42.8
DEAE-cellulose	1225	0.5 mg	2450	50	1113.6
Sephadex G-200	732	0.15 mg	4900	30	2227.2

band in the gel stained by the dye, except that in the case of the unstained gel that was not fixed, the receptor protein apparently diffused more when compared to the dye stained gel (Fig. 3B).

4. Stability of the Purified Receptor

The purified platelet PGE$_1$ receptor is unstable (half-life about 6 h) in dilute solution (<50 µg/mL) even at 0°C. However, the purified receptor at a concentration of 1.0 mg/mL or more when kept at –70°C in 56 mM Tris-HCl buffer, pH 7.4, containing 0.3M sucrose and 0.05% Triton X-100 found

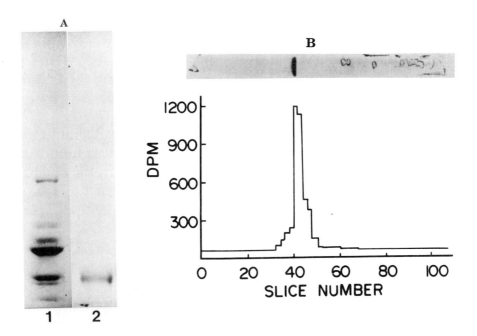

Fig. 3. Electrophoretic homogeneity and the binding of [³H]PGE₁ to the protein eluted from polyacrylamide gel electrophoresis.**A** Polyacrylamide gel electrophoresis of Triton X-100 extract from (100 μg of protein) and the purified receptor (10 μg of protein) were carried in 1.5 mm slab at pH 8.5 using Tris/borate buffer. The gels were stained with AgNO₃. Lane 1, Triton X-100 extract; Lane 2, the purified receptor. **B** Purified PGE₁ binding protein (10 μg) electrophoresed in 7.5% polyacrylamide gel and stained with Coomassie brilliant blue. In a parallel experiment, the gel was not stained for proteins but sliced (1 slice = 0.125 cm), and the proteins were eluted with 50 mM Tris-HCl buffer, pH 7.4, as described under experimental procedures. The figures shown here are typical of three different experiments. (Reprinted from *J. Biol.Chem.* **262,** pp. 2685–2691, 1987, with permission).

to have a half-life of approx 2–3 mo. The presence of detergent has a protective effect on the stability of the receptor. In the absence of the detergent, the purified receptor lost 50% of its binding activity in less than 1 mo at –70°C. It was found that during the storage of the purified receptor at –70°C, even in the presence of the detergent, protein fragments of small mol (<10,000 M_r) began to accumulate in the preparation with a concomitant loss of specific activity. These fragments, however, could be easily removed from the preparation by gel filtration in a small desalting column (PD-10, bed vol 9.0 mL, LKB). This procedure not only provided an electrophoretically homogeneous PGE₁ preparation but also restored the specific activity of the preparation to approx its original level.

5. Critical Steps in Purification

The PGE_1 receptors of human blood platelets is an unstable protein and the binding activity is rapidly lost in stored platelets. Almost 90% of the binding activity would be lost if the platelet preparation is stored for 24 h. It is, therefore, extremely important that the platelets used in the preparation of PGE_1 receptor are as fresh as possible.

The preparation of the PGE_1 receptor once started should be completed without any intermediate overnight stop. It was found that the platelet PGE_1 receptor in the Triton X-100 extract or in the eluates from the DEAE-cellulose column are rapidly destroyed, even at $-70°C$.

6. Discussion

The purified receptor showed an apparent M_r of 190,000 in SDS-polyacrylamide gel electrophoresis. A similar weight estimate of the purified protein was also obtained by Sephadex G-200 gel fitration on a calibrated column. SDS-polyacrylamide gel electrophoresis of the reduced protein showed that the receptor molecule is composed of two non-identical subunits with mol wts of 85,000 and 95,000, held together by disulfide bond(s) (Dutta-Roy and Sinha, 1987). Scatchard analysis of PGE_1 binding to the purified receptor protein showed one high affinity-low capacity binding sites population ($K_d1 = 9.8\,nM$) and one low affinity-high capacity binding sites population ($K_d2 = 0.7\,\mu M$). These K_d values compared well the K_d values of platelet membrane ($K_d1 = 6.6\,nM$, $K_d2 = 1.9\,\mu M$). Binding of $[^3H]PGE_1$ to the purified receptor was rapid and attained equilibrium within 5–6 min at 23°C. The binding of the radio ligand was also specific. Only PGI_2 at 15 μM and PGE_2 at 30 μM concentrations displaced approx 50 and 38% of the bound $[^3H]PGE_1$ from the receptor, respectively. No other prostanoids including PGA_2, PGB_1, PGD_2, $PGF_{2\alpha}$ at 15 μM displaced any bound $[^3H]PGE_1$ from the receptor. Mg^{2+} is needed for the binding of PGE_1 to its receptor; the optimum concentration of the metal ion is 5 mM. The purified receptor prepared by the method described was found to be biologically active in that the receptor could be used to reconstitute PGE_1 or PGI_2 responsive adenylate cyclase of platelets stripped of the receptors (Dutta-Roy and Sinha, 1987).

It has been discussed above that Triton X-100 not only solubilizes PGE_1 receptors from platelets, the detergent also inhibits the binding of the radio ligand to the receptor as well as protects the loss of binding activity of the isolated protein. The highly selective nature of the solubilization of the

PGE$_1$ receptors from the platelets by Triton X-100 was found to be related to the affinity for the detergent for PGE$_1$-receptors. It was found that Triton X-100 (mM) and PGE$_1$ (nM) actually compete with each other for the prostanoid receptors in platelet membranes. This competition between the detergent and the prostanoid was also found to be highly selective (unpublished observation).

Acknowledgments

We thank Jackie Assante and Janice Brewton for their assistance in the preparation of this manuscript. This work was supported, in part, by the National Institutes of Health Grant HL-41386 and by the American Heart Association (Ohio).

References

Ashby, B. (1986) *J. Cyclic Nucleotide Protein Phosphorylation Res.* **11,** 291–300.

Dutta-Roy, A. K., Ray, T. K., and Sinha, A. K. (1986) *Science* **231,** 385–389.

Dutta-Roy, A. K. and Sinha, A. K. (1985) *Biochim. Biophys. Acta.* **812,** 671–678.

Dutta-Roy, A. K. and Sinha, A. K. (1987) *J. Biol. Chem.* **262,** 12685–12691.

Garewal, H. S. (1973) *Anal Biochem.* **54,** 319–324.

Holloway, P. W. (1973) *Anal. Biochem.* **53,** 304–308.

Kahn, N. N. and Sinha, A. K. (1988) *Biochem. Biophys. Acta* **972,** 45–53.

Lefkowitz, R. J., Mullikin, D., Wood C. , Gore, T. B., and Mukherjie, C. (1977) *J. Biol. Chem.* **252,** 5295–5303.

Lefkowitz, R. J. Stadel, J. M., and Caron, M. G. (1983) *Annu. Rev. Biochem.* **52,** 159–186.

Merril, C. R., Goldman, D., Van Keuren, M. L., and Ebert, M. H. (1983) in *Electrophoresis 82: Advanced Methods. Biochemical and Clinical Application.* Proceedings of the International conference on electrophoresis, Athens, Greece. Walter De Gruyter, New York, pp. 327–342.

Robinson, G. A., Arnold, A., and Hartmann, R. E. (1969) *Pharmacol. Res. Commun.* **1,** 325–333.

Ross, E. M. and Gilman, A. G. (19800 *Annu. Rev. Biochem.* **49,** 533–564.

Scatchard, G. (1949) *Annu. NY Acad. Sci.* **51,** 660–672.

Schafer, A. I. ,Cooper, B., O'Hara, D., and Handin, R. I. (1977) *J. Biol. Chem.* **254,** 2914–2917.

Siegel, A. M., Smith, J. B., Silver, M. J., Nicolaou, K. C., and Ahern, D. (1979) *J. Clin. Invest.* **63,** 215–220.

Tateson, J. E., Moncada, S., and Vane, J.R. (1977) *Prostaglandin* **13,** 389–397.

Tsai, B. S. and Lefkowitz, R. J. (1979) *Biochim. Biophys. Acta.* **587,** 28–41.

The Asialoglycoprotein
Receptor

Richard J. Stockert, Janna C. Collins,
and Anatol G. Morell

1. Introduction

The asialoglycoprotein receptor mediates the endocytosis of galactose/N-acetylgalactosamine terminating glycoproteins by the parenchymal cells of the liver (for reviews, *see* Stockert and Morell, 1983; Schwartz, 1983). The receptor was originally solubilized from a rabbit liver particulate fraction with the nonionic detergent Triton X-100 and purified by affinity chromatography on immobilized galactose-terminating glycoproteins (Hudgin et al., 1974). Subsequently, asialoglycoprotein receptors have been purified from rat (Tanabe et al., 1979), mouse (Hong et al., 1988), and human liver (Baenziger and Maynard, 1980) by analogous procedures. No physiologic function has yet been ascribed to the receptor, but its study has provided many insights into the role of sugars as possible sites of recognition and to the intracellular pathways taken by receptor and ligand during endocytosis.

2. Preparation of the Affinity Gel

Commercial galactose/N-acetylgalactosamine affinity gels have become available, but, for large scale preparations of the receptor, it is still convenient and more economical to prepare a desialylated glycoprotein agarose conjugate.

Receptor Purification, vol. 2   ©1990 The Humana Press

The two most commonly used glycoproteins are acid$_1$-glycoprotein (orosomucoid) and fetuin. Both proteins are available from a number of commercial sources. The glycoproteins are desialylated by the selective acid hydrolysis of their sialic acid residues. A 1% solution of protein dissolved in 0.1N H$_2$SO$_4$ is incubated at 80°C for 1 h. The release of sialic acid can be monitored using a thiobarbituric acid assay (Warren, 1959); in practice, acid treatment results in the hydrolysis of over 90% of the sialyl residues. The desialylated glycoprotein solution is neutralized by the addition of NaOH and dialyzed overnight against distilled H$_2$O. This preparation can be used as is, stored frozen, or lyophilized.

Cyanogen bromide-activated Sepharose 4B can be obtained from a commercial source or prepared by the following procedure. Sepharose 4B is washed several times with 10 volumes of distilled water and the fines removed by decantation. To 50 mL of settled gel is added an equal vol of water and 50 mL of a 10% solution of cyanogen bromide. The pH of the reaction mixture, which falls as the reaction proceeds, is promptly brought to 11 and maintained there by the continuous addition of 4N NaOH until no further drop in pH is observed. The reaction is usually completed in 8–10 min at room temperature. The activated gel is collected on a Büchner funnel and washed rapidly with 1 L of cold 0.1N NaHCO$_3$ at pH 9.1. The residue, suspended in the same buffer, is brought to a final vol of 100 mL, and 5 mL of a 2% solution of desialylated orosomucoid is added. The coupling reaction is allowed to continue overnight at room temperature with gentle mixing on a rotating wheel.

On the following day, the suspension is poured into a chromatographic column and washed with several bed volumes of distilled water until the A$_{280}$ is less than 0.05. The filtrate and washings are pooled, and the unreacted glycoprotein is recovered, after dialysis, by lyophilization. It is estimated that between 0.3–0.5 mg of desialylated orosomucoid is bound per milliter of settled gel. The extent of the reaction can be monitored by the inclusion of trace amounts of radioactive glycoprotein in the coupling mixture or by simply determining the amount of unreacted desialylated orosomucoid recovered. When not in use, the gel may be stored at 4°C in the presence of 0.01% sodium azide. Such gels have been used repeatedly over a period of months without apparent loss of activity.

3. Solubilization
of the Asialoglycoprotein Receptor

In every case, the solubilization of the receptor from a crude fraction of rabbit, rat, mouse, and human liver is accomplished by the addition of

1–2% Triton X-100 at an alkaline pH. The procedure described here is used to prepare receptor from rat liver. This protocol has an additional step of cadmium precipitation that allows for the processing of larger amounts of tissue without a high speed centrifugation. All steps of the procedure are performed at 4°C. Livers (100 g) are homogenized in 800 mL of 1 mM NaHCO$_3$, pH 9.1, containing 0.5 mM CaCl$_2$, by using a Brinkmann Polytron at speed 4 for 30 s and the homogenate centrifuged at 1000g for 10 min. The supernatant is adjusted to contain 100 mM Tris-Cl, pH 7.9, 1.25M NaCl and 1% Triton X-100, and the mixture extracted for 30 min. The extract is made 50 mM in cadmium acetate, allowed to stand for an additional 15 min followed by centrifugation at 10,000g for 10 min. The pellet is suspended in 300 mL of 100 mM Tris-Cl, pH 7.9, 1.25M NaCl, 20 mM ethylene glycol bis (β-amino ethyl ether)-N, N 'tetraacetate (EGTA) followed by the subsequent addition of 600 mL of the same buffer without EGTA. Triton X-100 is added to 1% (assuming no residual detergent from previous extraction step) and the suspension is centrifuged at 10,000g for 10 min.

For small samples for which high speed centrifugation steps present no problem, the homogenate can be adjusted to contain 100 mM Tris-Cl, pH 7.9, 1.25M NaCl, and 1% Triton X-100, directly extracted for 30 min and centrifuged (100,000g for 30 min). When removing the supernatant, the surface lipid containing layer is to be avoided, since it will clog the affinity gel.

An alternative procedure, employing a liver acetone powder, has been found to be a convenient and stable starting material (Hudgin, 1974). The acetone and a Waring Blender container are cooled to –20°C. Fresh or frozen and thawed liver is chilled on ice, minced with scissors, and blended with 10 volumes of cold acetone for two 30 s periods. The suspension is immediately filtered under reduced pressure on Whatman No. 1 paper. As soon as the acetone disappears from the surface, the cake is crumbled into the blender, extracted a second time, and filtered. The cake is washed with cold acetone, crumbled, and pushed quickly through a fine wire mesh screen. The powder is allowed to air dry for 30 min, placed in a dessicator, and stored under reduced pressure at 4°C.

Although the receptor activity of extracts of the powder remains stable for several weeks, better recoveries are obtained when the powder is used within a few days after preparation. The dry powder is stirred for 30 min at 4°C with 40 volumes of a buffer containing 0.1M sodium acetate at pH 6.0, 10 mM EGTA, and 0.2M NaCl, centrifuged at 10,000g for 10 min, and the supernatant is discarded. This step is repeated with the same buffer and then with cold distilled water. The residual pellet is resuspended by homogenization in 40 volumes of an extracting buffer consisting of 10 mM

Tris-Cl at pH 7.9, 0.4M KCl, and 1% Triton X-100. After stirring for 30 min at 4°C, the crude, slightly turbid extract is centrifuged (10,000g for 10 min) and the supernatant made 50 mM in CaCl$_2$. The recovery of receptor protein ranges from 90–100% of that present in the acetone powder.

4. Affinity Purification

Sepharose 4B, to which desialylated orosomucoid (0.3–0.5 mg protein/mL gel) has been coupled is equilibrated with 10 mM Tris-Cl, pH 7.9, 50 mM CaCl$_2$, 1.25M NaCl, and 0.5% Triton X-100. The crude extract is mixed with the affinity gel in a ratio of 20 vol of extract to 1 vol of gel. The suspension is adjusted to 50 mM CaCl$_2$ and incubated at 4°C for 2 h with constant mixing. The suspension is then centrifuged at 1000g for 10 min, resuspended in half its original vol with the equilibration buffer containing only 20 mM CaCl$_2$, and poured into a chromatography column. After washing with 5 vol of the resuspension buffer receptor is eluted with 20 mM sodium acetate, pH 5.2, 1.25M NaCl, 1% Triton X-100. The major portion of the binding activity is usually recovered in the second bed vol of eluting buffer. Upon readjustment of the pH to 7.9 and addition of CaCl$_2$ to 50 mM the receptor is adsorbed onto a second, smaller affinity column approx one-tenth the size of the original bed vol. The addition of CaCl$_2$ should be made just prior to filtration. The gel is washed with 5 vol of the resuspension buffer. Triton X-100 can be removed by washing the bound receptor with the resuspension buffer minus Triton X-100 until the bulk of the detergent has been removed as is determined by monitoring the absorbance at 280 nm. The receptor is eluted in 3–4 bed volumes of a buffer containing 20 mM ammonium acetate at pH 6.4, and 0.5M NaCl.

An alternative procedure is to elute the receptor from the second column with a buffer containing Triton X-100 and remove the detergent by cadmium acetate precipitation. There are additional steps, but this second procedure results in a higher yield of receptor in a more concentrated solution. The receptor is eluted from the second column in 20 mM ammonium acetate, pH 6.4, 1.25M NaCl, and 0.5% Triton X-100 and extensively dialyzed against 20 mM ammonium acetate, pH 6.4, and 0.25% Triton X-100. Cadmium acetate is added to the dialysate to a final concentration of 20 mM and the solution is kept on ice for 30 min. The precipitated protein is recovered by centrifugation at 10,000g for 15 min, suspended in 2 mM cadmium acetate, recentrifuged, and resuspended in 20 mM ammonium acetate, pH 6.4, 1.25M NaCl. Solubilization of the receptor protein is effected by the addition of EGTA to a final concentration of 20 mM. The preparation is clarified by centrifugation at 10,000g for 10 min.

Taking advantage of the differential binding activities of the major and minor forms of the rat receptor polypeptides in the presence of reducing agent, a method has recently been described to separate these proteins (Halberg et al., 1987). A solution of receptor in 10 mL of 1.25M NaCl containing 50 mM Tris-Cl, pH 7.9, 50 mM CaCl$_2$, and 0.5% Triton X-100 is made 1M in 2-mercaptoethanol, incubated for 1 h at room temperature, and then applied to a 1.5 × 7 cm affinity column of galactose-Sepharose equilibrated in loading buffer. The column is rinsed with 100 mL of loading buffer and is eluted with 50 mL of 1.25M NaCl containing 40 mM ammonium acetate, pH 5.4, and 0.05% Triton X-100. The minor receptor species is eluted between 30–70 mL in the loading buffer rinse, whereas the major species elutes in the second and third column volumes of elution buffer.

5. Assay of Solubilized Receptor

Assay of the receptor activity in Triton X-100 extracts of homogenates and of purified protein are based on the selective precipitation of the receptor ligand complex with either ammonium sulfate (Hudgin et al., 1974) or polyethylene glycol (Stockert et al., 1974). Radiolabeled desialylated orosomucoid is commonly used ligand. Orosomucoid obtained from commercial sources or prepared from serum (Whitehead and Sammons, 1966) should be enzymatically desialylated to avoid high nonspecific precipitation. Desialylation of orosomucoid is easily accomplished by incubation with 0.025 U of immobilized *Clostridium perfringens* neuraminidase in 0.15M sodium acetate pH 5.6 for 24 h at room temperature with constant mixing. The extent of sialic acid released can be measured by the Warren procedure (1959). Desialylated orosomucoid is radiolabeled with 125-iodine by either a chloramine-T method (Greenwood, 1963) or by a solid phase lactoperoxidase–glucose oxidase system. Alternative ligands that have also been utilized with success are desialylated fetuin (Evarts, 1984; Townsend et al., 1986), galactosylated bovine serum albumin (Kawaguchi, 1981; Lee, 1982), and desialylated bovine submaxillary mucin whose carbohydrate moieties terminate in N-acetylgalactosamine (Stockert et al., 1977).

The ammonium sulfate precipitation assay first described by Hudgin et al. (1974) for extracts of rabbit liver is the most commonly used. The 0.1 mL incubation mixture containing: 25 mM Tris-Cl; pH 7.9; 0.5M NaCl; 20 mM CaCl$_2$; 0.1% (w/v) Triton X-100; 0.6% (w/v) bovine serum albumin, 1 µg ^{125}I-desialylated orosomucoid, and 1 µg purified lectin or crude extract is incubated at 25°C for 30 min and then chilled in ice. The receptor–ligand complex is precipitated by the addition of 0.5 mL of a cold, saturated solution of ammonium sulfate, adjusted to pH 7.9 with solid Tris. After 10 min at 4°C, the suspension is filtered onto Whatman GF/C disks (2.4 cm) under reduced pressure. The incubation tube is washed 3 times with 1.0 mL of 40% saturated ammonium sulfate, pH 7.9, containing 10 mM CaCl$_2$ and 0.1% bovine serum albumin (w/v). The bovine serum albumin must be free of fatty acids, since they will inhibit the assay by over 90% (Schwartz et al., 1980). The extent of nonspecific binding, i.e., blanks, is determined by inclusion of 100-fold excess of ligand or 50 mM N-acetyl galactosamine in the assay mixture. The blank value is approx 0.2% of the total radioactivity added. This value usually increases with the age of the ^{125}I- desialylated glycoprotein preparation.

The substitution of polyethylene glycol 6000 (PEG) for ammonium sulfate has resulted in the reduction of nonspecific binding. It has also made possible detection of the contribution that detergent and salt concentration have on the assay system (Grant and Kaderbhai, 1986). Purified receptor (0.1–1 µg) is added to 0.2 mL of an assay mixture containing: 50 mM Tris-Cl, pH 7.9; 20 mM CaCl$_2$; 150 mM NaCl; 0.02% Triton X-100; 0.1% bovine serum albumin; 0.1% bovine IgG; and 1 µg ^{125}I-desialylated orosomucoid. After incubation for 30 min at 22°C, 0.5 mL of a 20% (w/v) PEG solution in 50 mM Tris-Cl, pH 7.9, 10 mM CaCl$_2$, and 150 mM NaCl is added. The suspension is vortexed, allowed to stand on ice for 10 min, and is filtered on a Whatman GF/A disk under reduced pressure. The assay tube and disk is washed with an 8% (w/v) solution of PEG in assay buffer minus Triton X-100 and protein. When the appropriate blanks are subtracted, this assay is linear and proportional to protein concentration with 1 µg of pure lectin binding approx. 100 ng ASOR.

6. Assay of Receptor Activity in Cell Cultures

These assays work well for the well-differentiated hepatoblastoma cell line, HepG2 (Knowles et al., 1980; Schwartz et al., 1981) and isolated hepatocytes. Expression of the receptor by HepG2 is maximal at or just near confluence (Theilmann et al., 1983). In addition HepG2 should be grown in media containing not less than $10^{-8}M$ biotin or 10% fetal bovine serum, which is required for receptor expression (Collins et al., 1988).

Rat hepatocytes are isolated *in situ* by collagenase perfusion (Seglen, 1973). Isolated cells are suspended in Waymouth's 752/1 medium containing 5% heat-inactivated fetal bovine serum, 1.7 mM additional CaCl$_2$, 5 µg/ mL bovine insulin, 100 U/mL penicillin, 0.1 µg/mL streptomycin, and 25 mM HEPES, pH 7.2. Three mL aliquots containing 1×10^6 hepatocytes are plated on 60 mm Lux Contur dishes and cultures are maintained at 37°C in a 5% CO_2-95% air atmosphere for 1–2 h. Nonadherent cells are aspirated and 3 mL of Waymouth's medium added to each plate. Adherent mono-layer cells are maintained for 16–24 h prior to use in experiments. One hour prior to use in experiments, overnight cultured hepatocytes are washed with serum-free Waymouth's and incubated in 3 mL of medium at 37°C. This 1-h preincubation is necessary for the clearance of inhibitors present in the fetal bovine serum and in crude collagenase preparations if newly isolated cells are used (Stults and Lee, 1986).

To limit ligand binding to the cell surface, cells are chilled to 4°C. ^{125}I-Desialylated orosomucoid (1 μg) is added to the culture dish in 1 mL of Waymouth's medium and incubated at 4°C for 60 min to achieve maximal binding of ligand to surface receptor (Wolkoff et al., 1984). Nonspecific li-gand binding is quantitated from dishes that also contained 100 μg of un-labeled glycoprotein. Unbound ligand is removed from the culture dishes by four washes with 1.5 mL of cold medium, the third of which contains 0.5 m*M* *N*-acetylgalactosamine (GalNAc) (Harford et al., 1983). The amount of ligand binding to the cell surface is assessed from EGTA (20 m*M*) or GalNAc (50 m*M*)-released radioactivity at 4°C.

Internalization of surface receptor-bound ^{125}I desialylated orosomu-coid is quantitated following incubation at 37°C in pregassed and warmed media. The cells are then washed twice and incubated in 50 m*M* GalNAc or 20 m*M* EGTA for 10 min at 4°C to remove residual surface radioactivity. Hepatocytes are washed again and harvested by scraping with a rubber policeman. The fraction of initial surface-bound radioactivity that is inter-nalized is quantitated as the fraction of total radioactivity not released by EGTA or GalNAc.

Degradation is quantitated as acid soluble radioactivity in media. This precipitation procedure is usually performed in an Eppendorf tube. To an aliquot of media, an equal vol of 20% trichloroacetic acid/4% phospho-tungstic acid is added. The mixture is allowed to stand on ice for at least 20 min, and then centrifuged in an Eppendorf centrifuge at top speed for 2–3 min. An aliquot of the supernatant is removed for counting and multi-plied by the dilution factor to determine the percent of acid soluble radio-activity.

To measure the total number of receptors available inside the cell as well as on the surface, digitonin is included in the incubation at a con-centration of 0.055% (Weigel and Oka, 1983). The permeabilized cells are processed as for the determination of surface binding activity. To deter-mine the number of intracellular binding sites, cells are incubated with unlabeled desialylated orosomucoid (1 μg) at 4°C for 1 h to fill the surface sites. Once the unbound unlabeled ligand is washed away, the cells are incubated in Waymouth's containing 1 µg ^{125}I-desialylated orosomucoid and 0.055% digitonin for 1 h at 4°C.

Once internalized, the receptor–ligand complexes are localized rapidly to an acidic prelysosomal system of membranous vesicles and tubules. These complexes may dissociate because of the low pH and segregate with receptor, ultimately returning to the cell surface, whereas ligand is delivered to lysosomes where it is degraded. Alternatively, the receptor–ligand complex may return to the cell by a process termed diacytosis.

Dissociation of the internalized ^{125}I-desialylated orosomucoid receptor complex is assayed in hepatocyte lysates by differential precipitation in ammonium sulfate (Hudgin, 1974). Residual surface-bound ligand is removed by washing the cells with ice-cold 20 mM EGTA. The hepatocytes are then washed twice with 1.5 mL of cold 0.28M sucrose, 2 mM CaCl$_2$, 0.01M Tris-Cl, pH 7.6. Cells are solubilized by the addition of 1 mL/plate of an ice-cold solution containing 1% Triton X-100, 1 µg/mL of unlabeled desialylated orosomucoid, 50 mM CaCl$_2$, 0.15M NaCl, and 0.02M Tris-Cl, pH 7.6. The plates are scraped with a rubber policeman and the contents added to 1 mL of saturated ammonium sulfate (pH 7.6 with Tris) at 25°C. The precipitated receptor–ligand complex is filtered on glass fiber filters that are washed twice with 1.5 mL of 45% saturated ammonium sulfate containing 20 mM CaCl$_2$. Radioactivity associated with the filters (receptor bound ligand) and in the filtrate (free ligand) is quantitated.

Intracellular segregation of ligand is assayed from the extent of monensin-induced reassociation of receptor and ligand (Wolkoff et al., 1983). Monensin (5 mM), in absolute ethanol, is diluted 100-fold and added to the hepatocytes for 30 min at 37°C. Cell monolayers are then subjected to the solubilization-precipitation procedure to quantitate the extent of receptor–ligand reassociation in the presence of ionophore. To measure diacytosis, cells are incubated with ^{125}I-desialylated orosomucoid (1 µg) at pH 7.5 at 4°C for 60 min to saturate surface receptors. After three washes at 4°C, the medium is replaced with 1.5 mL of pregassed and warmed (37°C) media and the cells incubated at this temperature for 10 min to allow ligand internalization. Cells are then washed at 4°C, and residual surface ligand is removed by a 10 min incubation in 50 mM GalNAc or 5 mM EGTA. Cells are switched to pregassed and warmed media containing 50 mM GalNAc or 5 mM EGTA for 30 or 60 min at 37°C. Radioactivity in aliquots of medium is determined, and degradation products are quantitated as described above. Intact, ^{125}I-desialylated orosomucoid released into the medium (diacytosis) is determined as acid precipitable radioactivity.

7. Characterization

A considerable amount is now known about the structures of the receptors (for review, *see* Drickamer, 1988). Regardless of the source, as assessed by SDS-PAGE analysis, there are at least two different mol wt asialoglycoprotein receptor polypeptides. In all cases, there is a major more abundant polypeptide and at least one minor one with a lower level of expression. By either direct chemical analysis (Hudgin et al., 1974; Kawasaki and Ashwell, 1976) or a combination of metabolic labeling and changes in SDS-PAGE migration following treatment with bacterial gly-cosidase (Schwartz and Rup), it has been shown that all peptides are glycosylated. The isolated rabbit receptor consists of two molecular species, a major 40,000 and minor 48,000 dalton (Kawasaki and Ashwell, 1976). Rat and mouse receptor can be resolved into three polypeptides. The rat re-ceptor consists of a major species of 41,500 dalton and two minor ones of 49,000 and 54,000 (Drickamer, 1984; Tanabe, 1979). The amino acid sequence of the minor species are identical and appear to differ only in the extent of glycosylation (Halberg et al., 1987). No sequence information is yet available for the major 42,000 or two minor 45,000 and 51,000 dalton polypeptides isolated from mouse liver (Hong et al., 1988). By a combination of cDNA sequencing and selective peptide immunoprecipitation of receptor labeled in a hepatoma cell line, HepG2, the human receptor is proposed to consist of two polypeptides of 46,000 and 50,000 dalton (Bischloff et al., 1988; Spiess and Lodish, 1985).

The receptor exists in membranes in an oligomeric state, perhaps as high as a hexamer (Halberg et al., 1987). Crosslinking of receptor poly-peptides has not resolved whether it is present as two independent homooligomers, as has been described for the rat receptors (Halberg et al., 1987), or as a hetero-oligomers found in HepG2 cell membranes (Bischoff et al., 1988). The major and minor polypeptides are capable of binding desialylated glycoproteins, but reconstitution of the receptor endocytotic activity requires the expression of both polypeptides (McPhaul and Berg, 1986; Bischoff et al., 1988).

Acknowledgment

This work was supported by grants DK-32972 and by a grant from the Gail I. Zuckerman Foundation for Research in Childhood Liver Disease and the Foundation for the Study of Wilson's Disease.

References

Baenziger, J. V. and Maynard, Y. (1980) *J. Biol. Chem.* **255,** 4607– 4613.

Bischoff, J., LiGresco, S., Shia, M. A., and Lodish, H. F. (1988) *J. Cell Biol.* **106,** 1067–1074.

Collins, J. C., Paietta, E., Green, R., Morell, A. G., and Stockert, R. J. (1988) *J. Biol. Chem.* **263,** 11280–11283.

Drickamer, K. (1988) *J. Biol. Chem.* **263,** 9557–9560.

Drickamer, K., Mamon, J. F., Binns, G., and Leung, J. O. (1984) *J. Biol. Chem.* **259,** 770–778.

Evarts, R. P., Marsden, E., Hanna, P., Wirth, P. J., and Thorgeirsson, S. S. (1984) *Cancer Res.* **44,** 5718–5724.

Grant, D. A. W. and Kaderbhai, N. (1986) *Biochem. J.* **234,** 131–137.

Greenwood, F. C., Hunter, W. M., and Glover, J. S. (1963) *Biochem. J.* **89,** 114–123.

Halberg, D. F., Wager, R. E., Farrell, D. C., Hildreth, J. I. V., Quesenberry, M. S., Loeb, J. A., Holland, E. C., and Drickamer, K. (1987) *J. Biol. Chem.* **262,** 9828–9838.

Harford, J., Wolkoff, A. W., Ashwell, G., and Klausner, R. D. (1983) *J. Cell Biol.* **96,** 1824–1828.

Hong, W., Le, A. V., and Doyle, D. (1988) *Hepatology* **8,** 553–558.

Hudgin, R. L., Pricer, W. E., Jr., Ashwell, G., Stockert, R. J., and Morell, A. G. (1974) *J. Biol. Chem.* **249,** 5536–5543.

Kawaguchi, K., Kuhlenschmidt, M., Roseman, S., and Lee, Y. C. (1981) *J. Biol. Chem.* **256,** 2230–2234.

Kawasaki, T. and Ashwell, G. (1976) *J. Biol. Chem.* **251,** 1296–1302.

Knowles, B. B., Howe, C. C., and Aden, D. P. (1980) *Science* **209,** 497–499.

Lee, R. T. (1982) *Biochemistry* **21,** 1045–1050.

McPhaul, M. and Berg, P. (1986) *Proc. Natl. Acad. Sci. USA* **83,** 8863–8867.

Schwartz, A. L. (1983) *CRC Critical Reviews in Biochemistry* **16,** 207–233.

Schwartz, A. L., Fridovich, S. E., Knowles, B. B., and Lodish, H. F. (1981) *J. Biol. Chem.* **256,** 8878–8881.

Schwartz, A. L. and Rup, D. (1983) *J. Biol. Chem.* **258,** 11249–11255.

Schwartz, A. L., Rup, D., and Lodish, H. F. (1980) *J. Biol. Chem.* **255,** 9033–9035.

Seglen, P. O. (1973) *Exp. Cell Res.* **82,** 391–398.

Spiess, M. and Lodish, H. F. (1985) *Proc. Natl. Acad. Sci. USA* **82,** 6465–6469.

Stockert, R. J. and Morell, A. G. (1983) *Hepatology* **3,** 750–757.

Stockert, R. J., Morell, A. G., and Scheinberg, I. H. (1974) *Science* **186,** 365–366.

Stockert, R. J., Morell, A. G., and Scheinberg, I. H. (1977) *Science* **197,** 667, 668.

Stults, N. L. and Lee, C. Y. (1986) *Proc. Natl. Acad. Sci. USA* **83,** 7775–7779.

Tanabe, T., Pricer, Jr., W. E., and Ashwell, G. (1979) *J. Biol. Chem.* **254,** 1038–1043.

Theilmann, L., Teicher, L., Schildkraut, C. S., and Stockert, R. J. (1983) *Biochem. Biophys. Acta* **762,** 475–477.

Townsend, R. R., Hardy, M. R., Wong, T. C., and Lee, Y. C. (1986) *Biochemistry* **25,** 5716–5725.

Warren, L. (1959) *J. Biol. Chem.* **234,** 1971– 1975.

Weigel, P. H. and Oka, J. A. (1983) *J. Biol. Chem.* **258,** 5095–5102.

Whitehead, P. H. and Sammons, H. G. (1966) *Biochem. Biophys. Acta* **124,** 209– 211.

Wolkoff, A. W., Klausner, R. D., Ashwell, G., and Harford, J. (1984) *J. Cell. Biol.* **98,** 375–381.

Periplasmic and Membrane Receptors of Bacterial Chemotaxis

Sherry L. Mowbray

1. Introduction

Chemotaxis is the process by which cells move in response to chemicals in their environment (for reviews, *see* Berg, 1975; Koshland, 1979; Ordal, 1985). Chemotaxis in bacteria is best described as the result of two alternating behavioral states (*see* Fig. 1). In the first, the flagella (seven or so per cell for *E. coli*) are all rotating counterclockwise; as this direction corresponds to the twist of the helical units that make up the flagella, a bundle is formed that will propel the cell smoothly forward. At intervals, one or more of the flagella begin rotating clockwise; the flagellar bundle is thus disrupted, and the bacteria tumble in place. Each tumbling period serves to randomly reorient a bacterium for its next round of smooth swimming. When the composition of the medium is unchanging, the periods of smooth swimming are interrupted about once per second by a tumble; the net result is a random walk through the medium.

A primitive memory device provides the bacteria with the ability to compare the concentrations of chemicals in the medium at one time point with those a short while later. This gives them the information necessary to alter the proportion of time spent in each of these two behavioral states to their advantage. When a cell is swimming in a direction that is perceived to be favorable (e.g., when the external concentration of an attractant is

Receptor Purification, vol. 2   ©1990 The Humana Press

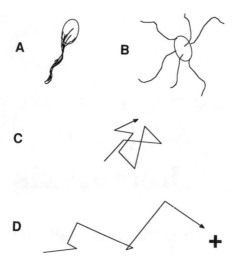

Fig. 1.   Behavioral states in gram-negative bacteria. **A.** Counterclockwise rotation generates smooth swimming. **B.** Counterclockwise rotation causes a tumble. **C.** When the bacterium is in an unchanging environment, tumbling and smooth swimming alternate at frequent intervals. **D.** When the bacterium is traveling up a gradient of attractant, smooth swimming predominates. Tumbling is more frequent when the direction of travel is down the gradient.

going up, or that of a repellent is going down), smooth swimming predominates. If the current direction of swimming is an unfavorable one (e.g. up a concentration gradient of a repellent, or down one of attractant), a tumble is more likely to occur. The sum of these two effects is that chemotactic bacteria spend much more time swimming in favorable directions than in unfavorable ones.

The receptors that recognize attractants and repellents are located on the exterior "surface" of the cell. Gram-negative bacteria actually have two types of receptors, ones that are periplasmic and ones that are membrane-bound (*see* Fig. 2). Some stimulants are bound directly by a membrane receptor, whereas others must first be activated by binding to a periplasmic protein. In both cases, it is the membrane receptor that is responsible for conveying the signal across the cytoplasmic membrane, for translating that information into a format that is "readable" by the cellular machinery, and for adapting to chemical signals that continue unabated for a period of time. The mechanism by which a signal is conveyed across the membrane has been proposed to involve a conformational change in the receptor on ligand binding (Mowbray and Koshland, 1987). The nature of the intracellular signal is largely unknown, although recent work has indicated that a

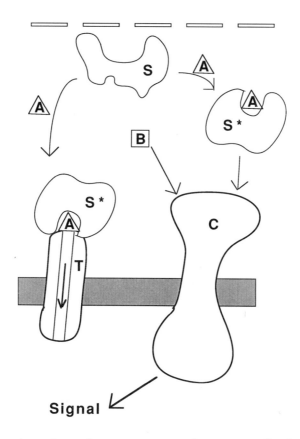

Fig. 2. Periplasmic and membrane receptors of gram-negative bacteria. Binding of a small attractant molecule (A) to a soluble periplasmic receptor (S) activates it (S*). The activated binding protein will then interact with the membrane proteins for chemotaxis (C) and transport (T). Other small molecules (B; which may be either attractants or repellents) can activate the membrane proteins for chemotaxis by direct interaction.

series of phosphorylation events is a likely possibility (Hess et al., 1987, 1988; Parkinson, 1988; Stock et al., 1988). The nature of the adaptation process is also unclear, although there is strong evidence that multiple methylation of specific internal sites of the membrane receptor is involved (Springer et al., 1979).

The types of periplasmic and membrane receptors of *E. coli* are shown in Fig. 3. Purification of many of these receptors has been sought with the goals of reconsituting the primary events of the system in vitro, and of obtaining structural data. In the following, the purification of the receptors will be described in general, followed by an example of each type (the ribose and aspartate receptors) in more detail.

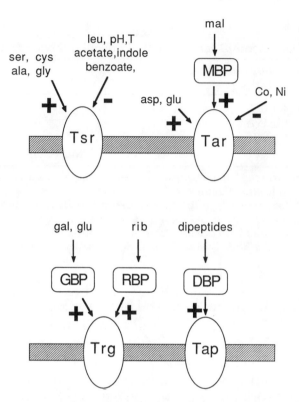

Fig. 3. Specificities of various receptors in chemotaxis of *E. coli*. The attractants are indicated by positive signs and the repellents are indicated by negative signs. The names given for the membrane receptors follow the name of the gene (e.g. Tsr is the product of the *tar* gene). The periplasmic receptor proteins are generally named for their primary ligand, and are indicated as X-binding protein (XBP). In the case of the repellents, direct binding to the membrane receptor has generally not been established, even though the behavioral links are clear.

2. Periplasmic (Soluble) Receptors

2.1. General Features

The periplasmic receptors (binding proteins) are in fact the primary receptors of various transport systems that have gained the ability to participate in chemotaxis as well (*see* Fig. 2). Not all of the transport receptors are able to perform in this fashion; the ones that can are shown in Fig. 3. Except for the matter of sharing a primary receptor in these systems, transport and chemotaxis are independent events; either can take place in the absence of the other (Ordal and Adler, 1974). The two functions appear to be separable within the binding proteins, since in some cases it has been possible to isolate mutants that have normal transport characteristics, but defective chemotaxis to the relevant compound, and vice versa (Ordal and Adler, 1974).

As a group, these transport-only and transport/chemotaxis receptors have many common features. Each is a protein of 30–50kDa with a dissociation constant for its ligands in the range of 10^{-6}–$10^{-7}M$. All are low in sulfur-containing (i.e., oxidizable) amino acids, as would be consistent with the need for stability in the often-uncontrolled environment of the periplasm.

The class seems as well to have a common structural architecture, despite the fact that sequence similarity is often vanishingly small. The proteins for which X-ray structural information is available (Gilliland and Quiocho, 1981; Mowbray and Petsko, 1983; Saper and Quiocho, 1983; Vyas et al., 1983; Pflugrath and Quiocho, 1985; Sack et al., 1989a, b; Mowbray et al., 1990) are composed of two similar domains linked by a small amount of amino acid chain (*see* Fig. 4). The binding site for ligand is located in the cleft formed between the two domains (Newcomer et al., 1979). Each domain is made up of a core of beta sheet surrounded by alpha helices; the *N*-terminus of each helix is aimed toward the cleft. A calcium binding site of unknown function has been found in the galactose/glucose-binding protein structure (Vyas et al., 1987; Mowbray et al., 1990).

Various biochemical and biophysical studies indicate that a conformational change occurs on the binding of ligand (Boos, 1971; Boos et al., 1972; Zukin et al., 1977,1979; Newcomer et al., 1981; Boos et al., 1981; Mowbray and Petsko, 1982). In addition, X-ray structural work has shown that, once bound to a periplasmic receptor, ligand molecules are almost completely sequestered from the solvent. A form of a transport-only receptor, believed to represent the more open state found prior to ligand binding, has been examined by X-ray crystallography (Sack et al., 1989a, b). The movement of the molecule on closing has been estimated to involve a 18° twist between the domains in the hinge region of the protein (Newcomer et al., 1981).

As similar approaches appear to work for the purification of many members of the binding protein class (Willis et al., 1974), general considerations will be discussed first, then an example will be given in detail for the ribose-binding protein of *E. coli*. This protein is the primary receptor for ribose in both chemotaxis and transport (Aksamit and Koshland, 1974; Galloway and Furlong, 1977; Furlong, 1982). It is unusual among the binding proteins in that it will compete for another (the galactose/glucose receptor) for the same membrane component in chemotaxis (Strange and Koshland, 1976; Kondoh et al., 1979).

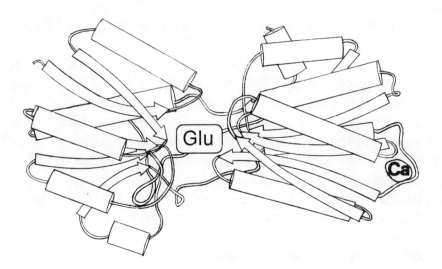

Fig. 4. Structure of the periplasmic galactose binding protein from *Salmonella typhimurium.* Derived from data presented in Mowbray et al., 1990; and Vyas et al., 1987. Like the other binding proteins studied by X-ray crystallography, this receptor has two domains consisting of beta sheet surrounded by alpha helix. The sugar ligand is located in the cleft between the two domains, and a calcium binding site of unknown function resides in a loop structure.

2.2. General Purification

2.2.1. Cell Growth

The type of approach to cell growth will depend on whether or not the protein is available on a multicopy plasmid. If such a plasmid is available, it should of course be used for the obvious reasons of yield and ease of purification. A more subtle point is the fact that because these proteins are not constitutive, induction by growth of the organism in the presence of ligand or a nonmetabolizable analog (at, e.g., 0.5% in the medium) is otherwise required. The receptor so purified thus generally contains a substantial concentration of bound ligand from this source (or, more fortuitously, as in the case of glucose introduced into the glucose/galactose receptor as the result of a contaminant during sucrose shock; Richarme and Kepes, 1974; Miller et al., 1980). The use of a multicopy plasmid removes the need to add ligand, and so may produce a product without significant amounts of "contaminating" ligand.

The choice of medium and growth conditions will depend as well on whether or not plasmid is available. Without a plasmid, a minimal medium at 30° is generally used, to avoid the suppression of the chemotaxis system observed with richer media and higher temperatures (this author, unpublished data). If the technology is locally available, the use of high yield growth techniques involving special media and pH control has proven successful in giving high cell densities (25–50 g wet wt/L of cell culture) without sacrificing production of binding protein (Willis et al., 1974). The growth of the bacteria to stationary phase is often practical, although harvest in late exponential phase may prove necessary if the observed proteolysis of the protein of interest is unacceptably high. If a multicopy plasmid is available, as illustrated for the ribose-binding protein below, a few liters of rich medium grown to stationary phase in an environmental shaker may suffice.

2.2.2. Release of Periplasmic Proteins

Release of periplasmic protein has traditionally been carried out by the method of Neu and Heppel (1965). In this method, the periplasm is loaded with sucrose, after which the bacteria are rapidly resuspended in cold deionized water containing magnesium chloride. The resulting osmotic shock disrupts the outer membrane while leaving the inner membrane intact. More recently, a method using chloroform has been applied for the same purpose (Ames et al., 1984). This latter method has proven practical in this laboratory for preparations involving up to 300 g of cells, if ample care is taken to mix completely at all steps. As a general comment, these procedures work only if the cells are used right after harvest; they should not be frozen prior to either chloroform or osmotic shock.

2.2.3. Ligand-Binding Assays and Removal of Bound Ligand

The binding of ligand may be measured in several ways as a method of locating material in the column fractions, of measuring binding affinity and of determining whether or not the material purified contains bound ligand.

The filter-binding assay is one of the most popular methods (Lever, 1972; Richarme and Kepes, 1983). An example of this type of assay is given below for the ribose receptor. Simply stated, the protein of interest is mixed with radioactive ligand at a concentration somewhat higher than the K_d. After binding has occurred, the ligand is "frozen" in by the addition of concentrated ammonium sulfate. The mixture is then washed through a nitrocellulose filter, and counted in a scintillation cocktail.

Nondenaturing isoelectricfocusing gels have also proved useful for determining the state of ligand binding, and as an assay (Boos et al., 1981; Mowbray and Petsko, 1982; Blank, 1987). In one method, ligand is introduced into one side of a prepoured gel (LKB), either by spraying or by temporarily applying a piece of thick (e.g., Whatman 3MM) filter paper previously soaked in a concentrated ligand solution. A comparison of the migration behavior on the two sides of the gel will reveal a shift in pI of several tenths of a pH unit on ligand binding. This method has been used successfully for at least the galactose/glucose, ribose, and dipeptide receptors. In another application, a gel is run on samples without added ligand. Radioactive ligand is immediately soaked into the gel for a brief (about 15 min) period, and the gel rapidly dried onto paper. Autoradiography (generally with an enhancer such as that obtained from NEN) will locate the receptor protein on the gel.

Fluorescence spectroscopy may also be used as an assay (Boos et al., 1972; Miller et al., 1980), since the environment of aromatic residues at or near the binding site is altered when the protein closes around ligand. Receptor fluorescence is compared in the presence and absence of ligand, using an emission wavelength of, e.g., 330–350 nm and varying the excitation wavelength in the range of 240–320 nm. Alternatively, excitation at 290 nm and observation of the emission in the range of 300–440 nm may be performed. Slit widths of 10 nm (excitation) and 8 nm (emission) are generally acceptable for receptor concentrations of 20–50 μg/mL.

Removal of ligand may be necessary for further work. Various approaches have been tried for this purpose, with various degrees of success. Dilution to very low protein concentrations followed by exhaustive dialysis (Richarme and Kepes, 1983), or dialysis against solutions containing guanidine hydrochloride (2–6M; Miller et al., 1980) have been used with radioactive ligand to follow the course of removal. In many cases, isoelectricfocusing gels with and without ligand as described above would be useful as a final check.

Another approach to ligand removal that should be feasible requires the use of an enzyme that will act on the ligand to chemically alter it. In this method, a standard equilibrium dialysis cell would contain the receptor protein of interest on one side of a dialysis membrane, and an enzyme that will modify its ligand on the other side. As ligand comes off the receptor, some will diffuse across the membrane before rebinding. Once modified, it will no longer be recognizable to the receptor, and the reaction should be drawn to completion in much the same way that precipitation can pull a chemical reaction to the right.

2.3. Purification
of the Ribose Receptor from E. coli

2.3.1. Materials

Sephacryl S200 and chlorobutanol (trichloromethyl-propanol) were obtained from Sigma. Spectragel DEAE was a Spectrum Industries product from Fisher Scientific. PBE 94 and Polybuffer 74 were products from Pharmacia. Nitrocellulose filters (0.45 μ, type HA) for the binding assays were from Millipore. [1-^{14}C]-D-ribose (7.9 mCi/mmol) was obtained from ICN. All other reagents were of the highest grade commercially available.

Strain MRi7 and plasmid pCMB1 (carrying the structural gene for the *E. coli* ribose binding protein) were obtained from Camilla Mauzy in the laboratory of Mark Hermodson at Purdue University.

2.3.2. Assays

For the filter binding assay (Richarme and Kepes, 1983), sample (containing 1–15 μg receptor) was mixed with buffer and ^{14}C-ribose (at a final concentration of 10 μM) to a vol of 100 μL. After 10 min on ice, the samples were mixed with 1 mL cold 90% saturated ammonium sulfate, and applied to a nitrocellulose filter. Two additional 1 mL portions of 90% ammonium sulfate were used to wash the filter, which was then counted in 10 mL of scintillation fluid.

Nondenaturing isoelectricfocusing gels were useful for evaluating purity and ligand contamination. Isoelectricfocusing was performed according to the LKB literature using prepoured narrow-range gels from pH 4–6.5. Before loading samples and placing the electrode wicks on the gel, ligand was introduced by diffusion. This was performed by applying a piece of Whatman 3MM filter paper which had been soaked in ribose solution (1 mM) directly to the gel. Diffusion was allowed to proceed for 15 min before removing the paper and running the gel.

2.3.3. Purification

The chromatography steps could safely be carried out at room temperature, but centrifugation and precipitation steps were performed on ice or at 4°C. Fractions and pools between columns were stored at –20° (or 4°C if stored for a brief time).

Mri7/pCMB1 were grown to saturation at 37°C in 4 L of Luria broth without glucose (Maniatis et al., 1982) containing 25 µg/mL ampicillin (OD_{600} approx 2.5 after overnight growth). Cells were harvested in a Sorvall GS3 rotor (8000 rpm, 10 min). After pouring off the supernatant, the pellets were resuspended in the residual medium. Chloroform (40 mL) was then added, mixed in thoroughly, and the resulting paste allowed to sit for 15 min at room temperature with frequent stirring. Then 400 mL of 10 mM TrisHCl, pH 8.0 were added and mixed in thoroughly by shaking. The solution was centrifuged in the GS3 rotor (8000 rpm, 10 min). The top layer (hereafter referred to as shock fluid) was carefully removed without disturbing either the cell pellet or the chloroform layer underneath the pellet.

The basic procedure of ammonium sulfate precipitation, followed by molecular sieve and ion-exchange chromatography were adapted from the procedure described by Willis and Furlong (1974). Solid ammonium sulfate was added to 40% saturation (in small portions while stirring on ice), followed by stirring on ice for 1 h. The solution was centrifuged in the GS3 rotor at 10K rpm for 20 min; a moderate-sized, yellowish pellet was visible. The supernatant was then brought to 100% saturation with ammonium sulfate, and stirred on ice for 2 h, followed by centrifugation (GS3 rotor at 8000 rpm for 20 min). After careful removal of all of the supernatant, the large light brown pellet was resuspended completely in 10–15 mL of 10 mM TrisHCl, pH 8.0. This mixture was clarified in a Sorvall SS34 rotor at 10,000 rpm for 10 min.

The sample was then loaded onto a Sephacryl S200 column (2.5 cm × 80 cm) in 10 mM Tris-HCl, pH 8.0. The column was developed at 20 mL/h using a peristaltic pump, collecting 144 fractions of 6 mL each. The samples were analyzed by SDS* polyacrylamide gel electrophoresis; the ribose-binding protein (M_R about 30 kDa) was the major band in its peak fractions (eluting as a monomer). Further information could be obtained from a filter binding assay using 5 µL of the appropriate fractions.

The fractions with highest proportion of ribose-binding protein were pooled and loaded onto a Spectragel DEAE column (2.5 × 15 cm) in 10 mM Tris-HCl, pH 7.0 at a flow rate of 60 mL/h, collecting 10 mL fractions. The column was washed with about 100 mL of the same buffer at the same flowrate. A gradient was applied (1.3 L total vol) from 0–250 mM NaCl in 10

---

*Abbreviations used: EDTA, ethylenediamine tetraacetic acid; PMSF, phenylmethyl sulfonyl fluoride; SDS, sodium dodecyl sulfate.

mM Tris-HCl, pH 7.0. Fractions were analyzed by SDS polyacrylamide gel electrophoresis, and the fractions with the highest proportion of RBP (eluting at about 30 mM NaCl) were pooled.

The pool from the DEAE column was dialyzed against 10 mM imidazole-HCl, pH 7.4 at 4°C. This sample was then run onto a chromatofocusing column (PBE 94, 1.5 × 25 cm) that was developed with 10 column vol of Polybuffer 74, pH 5.0. The flow rate was 10 mL 1 h; 2.5 mL fractions were collected. Fractions were chosen based on SDS-polyacrylamide and isolectric focusing gels; ribose receptor eluted with an apparent pI of 6.2.

An extinction coefficient of 0.55 mL/mg-cm (Willis and Furlong, 1974) was used to estimate the quantity of protein, after which the relevant fractions were precipitated with 100% saturated ammonium sulfate. The precipitate was washed twice with 90% saturated ammonium sulfate to remove residual ampholyte, then redissolved in 10 mM sodium phosphate, pH 7.0, to approx. 10 mg/mL of protein, followed by dialysis against the same buffer. The resulting protein was stable for extended periods at 4°C; bacterial growth was inhibited by the addition of small amounts of chlorobutanol.

The yield was approx 15–30 mg of protein at a purity of greater than 98% as judged by SDS gels. Isoelectricfocusing gels with and without ribose confirmed the purity and the fact that the protein was prepared free of endogenous ligand. The protein purified in this fashion was suitable for obtaining high-quality three-dimensional crystals (Mahendroo et al., 1990).

3. Membrane Receptors

3.1. General Features

Like the periplasmic proteins, the membrane receptors of bacterial chemotaxis form a class of similar molecules, the members of which were introduced in Fig. 3. Sequence similarities and work with mutants have suggested the same basic model for each (Boyd et al., 1981,1983; Krikos et al., 1983; Russo and Koshland, 1983; Bollinger et al., 1984). The N-terminus appears to be located on the cytoplasmic side of the bacterial inner membrane. A short hydrophobic segment (of appropriate length and character to be a helix) then crosses to the periplasm where the next 15kDa or so of the protein form a soluble ligand-binding domain. A low degree of sequence similarity in this region is consistent with the fact that differing ligands are bound to each of these receptors. A second hydrophobic region (again, probably a helix) then crosses to the cytoplasm where a second soluble portion (about 35kDa) forms the region of the protein associated most closely with intracellular signaling and adapation. As expected, this region of the protein is very well conserved among the members of the class.

The membrane receptors of chemotaxis are known to be multiply methylated and demethylated during adaptation (Van der Werf and Koshland, 1977; DeFranco and Koshland, 1980; Chelsky and Dahlquist, 1980; Boyd and Simon, 1980; Engstrom and Hazelbauer, 1980) by two cytoplasmic enzymes, a transferase (Springer and Koshland, 1977), and esterase (Stock and Koshland, 1978), both of which are encoded for by chemotaxis genes.

The membrane receptors in chemotaxis are rather reminiscent of a class of eucaryotic receptors represented by the insulin (Ullrich et al., 1985) and epidermal growth factor (Ullrich et al., 1984) receptors. In these proteins, as well, a few small hydrophobic regions are implicated in the task of carrying information across a membrane. As such, the chemotaxis receptors become even more interesting as a testing ground for ideas about the mechanics of transmembrane signaling (*for instance,* Mowbray and Koshland, 1987).

The gene for the so-called aspartate receptor (*tar*) has been cloned and sequenced in both *S. typhimurium* (Russo and Koshland, 1983) and *E. coli* (Krikos et al., 1983). In *E. coli*, this receptor directs responses from aspartate (Springer et al., 1977) by direct binding (Wang and Koshland, 1980) and from maltose after it has bound to a periplasmic maltose-binding protein (Koiwai and Hayashi, 1979; Richarme, 1982; Manson et al., 1985). In the case of the *Salmonella* receptor, the response is to aspartate alone, apparently owing to an intrinsic difference in the membrane protein (Dahl and Manson, 1985). Aspartate is bound with a K_d of 1 μM (Wang and Koshland, 1980; Russo and Koshland, 1983; Foster et al., 1985). The aspartate receptor methylated is at four sites (Kehry et al., 1983; Terwilliger and Koshland, 1984) in varying degrees in response to both maltose and aspartate (Mowbray and Koshland, 1987).

Evidence in the case of the aspartate/maltose receptor is that the protein is found in the membrane (as well as in the detergent-solubilized state) as a dimer of elongated shape (Milligan and Koshland, 1988). The available experimental data suggest that the dimers are not cooperative in any fashion, and so presumably have some role in structural stability. Likely, other membrane receptors are dimers as well. Mixed dimers, however, do not appear to occur even between the like receptors from *E. coli* and *Salmonella*.

Purification will be described primarily for the aspartate receptor of *S. typhimurium* (*see also* Foster et al., 1985).

3.2. Purification
of the Aspartate Receptor of S. typhimurium

3.2.1. Materials

Octyl-beta-glucopyranoside (octylglucoside) was obtained from Calbiochem or Sigma. Sephacryl S300, S-adenosyl methionine (iodide salt), phenylmethylsulfonyl fluoride (PMSF), *p*-hydroxymercuribenzoate, the chelator 1,10-phenanthroline and ampicillin were obtained from Sigma. Spectragel DEAE (a Spectrum Industries product) was from Fisher. Ultrafiltration membranes were products of Amicon. S-Adenosyl [*methyl-^{3}H*]-methionine (15 Ci/mmol) was purchased from Amersham. [^{3}H]-L-aspartic acid (15 Ci/mmol) was from ICN. Unless specified all other chemicals were of the highest obtainable grade from various commercial sources.

E. coli strains RP3808, RP4080, and RP4372 were obtained from J. S. Parkinson at the University of Utah. Plasmid pWK35 (E. A. Wang, unpublished data) carried the gene for the *Salmonella* aspartate receptor). Strains with plasmid pGK3 that contains *tar* and *cheR*, the structural genes for the aspartate receptor and methyltransferase (prepared by N. Gutterson, unpublished data) were used as a source of methyltransferase.

3.2.2. Assays

3.2.2.1. Aspartate Binding. A concentration of 10 μ*M* [^{3}H]-L-aspartate was used in the membrane binding assay described by Clarke and Koshland (1979). Samples were spun in a Beckman TL-100 benchtop ultracentrifuge (TLA-100 rotor) for 15 min at 100,000 rpm at 4°, instead of using the preparative ultracentrifuge.

Aspartate binding of protein solubilized in detergent was measured using a modification of the filter assay of Richarme and Kepes (1983), as described above for the ribose binding protein. Samples (up to 50 μL) were brought to 90 μL with 10 m*M* Tris-HCl, pH 7.2, containing 0.1% Triton X-100. [^{3}H]-L-aspartic acid was added to a final concentration of 10 μ*M*, after which the samples were treated as above. The estimated recovery of aspartate-bound protein on the filter was approx 25%, improving somewhat after complete dissolution of the filter in the scintillation fluid (about 1 d).

3.2.2.2. Methylation. Cytosols (Bogonez and Koshland, 1985) containing methyltransferase were obtained by growing 1L of cells (RP3808/pGK3) to late log phase (OD 0.8–1.0) in minimal citrate medium (Vogel and

Bonner, 1956) (100 µg/mL ampicillin and 50µg/mL each his, leu, met thr, thi) at 30°C. Cells were harvested by centrifugation at 8000 rpm for 10 min in a Sorvall GSA rotor at 4°C. Cells were sonicated in 5 mL buffer (100 mM sodium phosphate, pH 7.0, 5 mM EDTA, 1 mM o-phenanthroline, 1 mM PMSF). This supernatant was spun at 10,000 rpm for 10 min in an SS34 rotor at 4°C, transferred to new tubes, then spun at 45,000 rpm for 35 min in a Beckman 50Ti rotor at 4°C. The supernatant was then dialyzed against three changes of 1500 mL buffer (20mM sodium phosphate, pH 7.0, 1 mM EDTA, 1 mM o-phenanthroline, 1 mM PMSF). The final cytosol should be 10–15 mg/mL by the assay of Lowry et al. (1951). Aliquots were stored at –80°C, and were refrozen no more than once.

S-adenosyl methionine degrades rapidly on storage at neutral pH at room temperature. Purification of cold S-adenosyl methionine was accomplished by applying 12.1 mg (iodide salt in water) to a column of 5 g Dowex AG 1X8 (50–100 mesh; washed in succession with 4M NaCl, 10 mM NaCl, 0.5M NaHCO$_3$ then deionized water). The column was then washed with water, collecting fractions (1 mL) in tubes to which 5 µL of 2M H$_2$SO$_4$ had been added. The concentration of S-adenosyl methionine was estimated using an E$_{260}$ of 15,500. Approximately 3 peak fractions were pooled and stored in aliquots at –80C°.

Aspartate receptor in membrane vesicles was methylated by the methods of Terwilliger et al. (1983) and Chelsky et al. (1984). Membrane protein (0.35 µg) was mixed with 1.5 µCi [^{3}H]-S-adenosyl-L-methionine in a final volume of 56 µL of buffer containing 100 µM cold S-adenosyl-L-methionine (purified) and 50 mM sodium phsophate. Tubes were placed at 30°C, and reaction was started with 6 µL methyltransferase-containing cytosol. Samples (10–20 µL) of the mixture were removed (after vortexing) at various time intervals up to 1 h. Each sample was absorbed onto a 1 cm square piece of Whatman 3MM filter paper (labeled with pencil). Immediately after all of the sample is absorbed, the paper is dropped into a basket immersed in gently stirring 10% trichloroacetic acid. After all assays and time points were complete, the trichloroacetic acid solution was stirred for 10 min longer, then replaced with fresh solution. After another 10 min this was replaced with acetone or methanol, followed by another 2 min stirring. Papers were removed and air-dried. Each paper was then placed into a 0.5 mL eppendorf tube containing 100 µL 1M-NaOH, and this tube immediately placed into a labeled 7 mL scintillation vial containing about 2.3 mL scintillation fluid. The scintillation vial was capped immediately, and the sample allowed to diffuse overnight or longer at room temperature without shaking. (The diffusion proceeds more quickly at 30°C). Samples were counted with minimal disturbance; correction for the completeness of the diffusion process can be judged using radiolabeled methanol placed directly in an eppendorf tube with the base and a blank piece of paper.

Purified aspartate receptor solubilized in octylglucoside was methylated using the soluble system described by Bogonez and Koshland (1985) using a final mixture containing 0.5% octylglucoside, 0.9 mg/mL *E. coli* phospholipids, 20% (w/v) glycerol, 14 µM S-Adenosyl-[^{3}H] methionine, 50 mM sodium phosphate, pH 7.0. The assay was carried out in a similar manner to that described for the membrane methylation assay.

3.2.3. Purification

Buffers were adjusted to the appropriate pH and sterile filtered. EDTA was prepared as a 500 mM stock at pH 7.0, and stored at room temperature. PMSF and phenanthroline were prepared as 200 mM and 1M stocks stored at –20° C; the former solution needed to be warmed to room temperature and sonicated to redissolve before use. As a result of effects on receptor solubility, exposure of the protein to potassium ions was avoided after solubilization, substituting the sodium ion where necessary. Unless otherwise stated, all steps were performed on ice or at 4°C.

Buffers used in the purification were: Buffer A: 20 mM sodium phosphate, pH 7.0, 2 M KCl, 10% glycerol, 5 mM EDTA, 1 mM *o*-phenanthroline, 0.5 mM PMSF, 0.1 mM *p*-hydroxymercuribenzoate. Buffer B: 20 mM sodium phosphate, pH 7.0, 10% glycerol, 5 mM *o*-phenanthroline, 1 mM PMSF. Buffer C: same as buffer B plus 1% octylglucoside. Buffer D: 50 mM Tris HCl, pH 7.4, 10% glycerol, 1% octylglucoside, 5 mM *o*-phenanthroline, 1 mM PMSF.

For the preparation of membranes enriched in aspartate receptor, strain RP4080 was transformed with plasmid pWK35, and grown at 30°C in minimal citrate medium (Vogel and Bonner, 1956). The cells were harvested in late exponential phase, 10% glycerol was added, and the cell paste frozen as small pellets in liquid nitrogen which were then stored at –80°C.

Pellets were made from 100 mL of buffer A by dripping it into liquid nitrogen; these were often prepared prior to use and stored at –80°C. About 50 gm of cells were thawed in 100 mL room temperature buffer A, then transferred to a rosette flask. Cells were disrupted using 30–45 second bursts from a 1/2-inch tip on a Heat Systems sonicator (10–12 min total sonication time), keeping the cells cold (about 0°C) using frozen buffer pellets. Cell breakage was monitored by microscopic observation. The resulting suspension was centrifuged at 8000 rpm for 15 min in a Sorvall GSA rotor. The supernatant (together with the soft pellet) was transferred to a clean bottle, then spun again. The supernatant and the soft pellet from this spin were transferred to bottles for the Beckman 45Ti rotor and centrifuged at 40,000 rpm for 60 min. After the spin, the bright orange supernatant was decanted and the pellet was resuspended (using a motor-driven Dounce homogenizer) in 300 mL of buffer A. This suspension was centrifuged for 40 min at 40,000 rpm in the 45Ti rotor. The pellet was then resuspended in 100 mL buffer B, and centrifuged again for 30 min.

The resulting pellet was resuspended in 25 mL buffer B, and the membrane protein concentration estimated by the Lowry assay. The membranes were frozen on dry ice or in liquid nitrogen and stored at −80°C. The yield was generally 12–14 mg membrane protein/g original cells.

About 500 mg of membrane protein were mixed at 12 mg/mL protein in buffer C. The solution was placed on ice for 15 min with occasional mixing, the spun at 35,000 rpm in the Beckman 45Ti rotor for 1h. The supernatant was clear and dark orange, and generally contained 75–95% of the receptor.

The octylglucoside extract was immediately loaded onto a column of DEAE SpectraGel M (1.5 × 22 cm) in Buffer D at a flowrate of 40 mL/h (4 mL fractions). When the load was complete, the column was washed with about 20 mL of buffer D, then a gradient (300 mL total vol) from 0–120 mM-NaCl in the same buffer was applied. The BioRad protein assay was used to locate the peaks to be run on SDS gels. The largest protein peak (receptor, mol wt by SDS gels about 60kDa) should come late in the elution (at about 65 mM NaCl, near a dark orange peak). The DEAE-purified receptor was stable to storage at 4°C for periods of several days at 4°C, or indefinitely if frozen.

The pool chosen by gels or the filter assay were concentrated to about 7 mL over a YM100 membrane in an ultrafiltration cell at 10 psi. The concentrated sample was then applied to an S300 column (1.5 × 95 cm) in buffer D. The column was run at 5 mL/h, collecting 20 min fractions. (After the sample load was completed, the inlet was switched to a buffer of 50 mM-Tris HCl, pH 7.4, 1% sodium cholate, 0.05% NaN$_3$, and the elution continued until about 200 mL were run through the column. This procedure reduced the total amount of octylglucoside needed for this chromatography step.) The BioRad protein assay was again used to locate the peaks of interest. Receptor elution was consistent with a mol wt of 250,000–

350,000 when compared to standard soluble proteins (Foster et al., 1985). Fractions to pool were chosen from SDS gels.

Completely purified receptor was obtained by repeating the DEAE chromatography step using the same conditions as described above.

The yield of purified protein was 5–10 mg from 50 g of the original cells.

Aknowledgments

Much of the work described here would not have been possible without the assistance and efforts of David Foster, Dan Koshland, Mala Mahendroo, Camilla Mauzy, and Quang Ahn Vu. S.M. is an Assistant Investigator of the Howard Hughes Medical Institute.

References

Aksamit, R. R. and Koshland, Jr., D. E. (1974) *Biochemistry* **13,** 4473–4478.

Ames, G. F.-L., Prody, C., and Kustu, S. (1984) *J. Bacteriol.* **160,** 1181–1183.

Berg, H. C. (1975) *Nature* **254,** 389–392.

Blank, Volker (1987) in *"The Dipeptide Binding Protein of* Escherichia coli.*,"* diplom thesis, University of Konstanz, West Germany.

Bogonez, E. and Koshland, Jr., D. E. (1985) *Proc. Natl. Acad. Sci. USA* **82,** 4891–4895.

Bollinger, J., Park, C., Harayama, S., and Hazelbauer, G. L. (1984) *Proc. Natl. Acad. Sci. USA* **81,** 3287–3291.

Boos, W. and Gordon, A. S. (1971) *J. Biol. Chem.* **246,** 621–628.

Boos, W., Gordon, A. S., Hall, R. E., and Price, H. D. (1972) *J. Biol. Chem.* **247,** 917–924.

Boos, W., Steinacher, I., and Engelhardt-Altendorf, D. (1981) *Mol. Gen. Genet.* **184,** 508–518.

Boyd, A. and Simon, M. I. (1980) *J. Bacteriol.* **143,** 809–815.

Boyd, A., Krikos, A., and Simon, M. I. (1981) *Cell* **26,** 333–343.

Boyd, A., Kendall, K., and Simon, M. I. (1983) *Nature (Lond.)* **301,** 623–626.

Chelsky, D. and Dahlquist, F. W. (1980) *Proc. Natl. Acad. Sci. USA* **77,** 2434–2438.

Chelsky, D., Gutterson, N. I., and Koshland, Jr., D. E. (1984) *Anal. Biochem.* **141,** 143–148.

Clarke, S. and Koshland, Jr., D. E. (1979) *J. Biol. Chem.* **254,** 9695–9702.

Dahl, M. K. and Manson, M. D. (1985) *J. Bacteriol.* **164,** 1057–1063.

DeFranco, A. L. and Koshland, Jr., D. E. (1980) *Proc. Natl. Acad. Sci. USA* **77,** 2429–2433.

Engstrom, P. and Hazelbauer, G. L. (1980) *Cell* **20,** 165–171.

Foster, D. L., Mowbray, S. L., Jap, B. K., and Koshland, Jr., D. E. (1985) *J. Biol. Chem.* **260,** 11706–11710.

Furlong, C. E. (1982) *Meth. Enz.* **90,** 467–472.

Galloway, D. R. and Furlong, C. E. (1977) *Arch. Biochem. Biophys.* **184,** 496–504.

Gilliland, G. L. and Quiocho, F. A. (1981) *J. Mol. Biol.* **146,** 341–362.

Hess, J. F., Oosawa, K., Matsamura, P., and Simon, M. I. (1987) *Proc. Natl. Acad. Sci. USA* **84,** 7609–7613.

Hess, J. F., Oosawa, K., Kaplan, N., and Simon, M. I. (1988) *Cell* **53**, 79–87.

Kehry, M. R., Bond, M. W., Hunkapiller, M. W., and Dahlquist, F. W. (1983) *Proc. Natl. Acad. Sci. USA* **80**, 3599-3603.

Kondoh, H., Ball, C. B., and Adler, J. (1979) *Proc. Natl. Acad. Sci. USA* **76**, 260–264.

Koiwai, O. and Hayashi, H. (1979) *J. Biochem. (Tokyo)* **86**, 27–34.

Koshland, Jr., D. E. (1979) *Physiol. Rev.* **59**, 811–862.

Krikos, A., Mutoh, N., Boyd, A., and Simon, M. I. (1983) *Cell* **33**, 615–622.

Lever, J. E. (1972) *Anal. Biochem.* **50**, 73–83.

Lowry, O. H., Rosebrough, N. J., and Farr, A. L. (1951) *J. Biol. Chem.* **193**, 265–273.

Mahendroo, M., Cole, L. B., and Mowbray, S. L. (1990) *J. Mol. Biol.* **211**, in press.

Maniatis, T., Fritsch, E. F., and Sambrook, J. (1982) *Molecular Cloning: A Laboratory Guide.*, Cold Spring Harbor Laboratory, New York.

Manson, M. D., Boos, W., Bassford, Jr., P. J., and Rasmussen, B. A. (1985) *J. Biol. Chem.* **260**, 9727–9733.

Miller, D. M., Olson, J. S., and Quiocho, F. A. (1980) *J. Biol. Chem.* **255**, 2465–2471.

Milligan, D. L. and Koshland, Jr., D. E. (1988) *J. Biol. Chem.* **263**, 6268–6275.

Mowbray, S. L. and Koshland, Jr., D. E. (1987) *Cell* **50**, 171–180.

Mowbray, S. L. and Petsko, G. A. (1982) *Mol. Biol.* **160**, 545–547.

Mowbray, S. L. and Petsko, G. A. (1983) *J. Biol. Chem.* **258**, 7991–7997.

Mowbray, S. L., Smith, R. D., and Cole, L. B. (1990) *Receptor* **1**, in press.

Neu, H. C. and Heppel, L. A. (1965) *J. Biol. Chem.* **240**, 3685–3692.

Newcomer, M. E., Miller, III, D. M., and Quiocho, F. A. (1979) *J. Biol. Chem.* **254**, 7529–7533.

Newcomer, M. E., Lewis, B. A., and Quicho, F. A. (1981) *J. Biol. Chem.* **256**, 13218–13222.

Ordal, G. W. and Adler, J. (1974) *J. Bacteriol.* **117**, 517–526.

Ordal, G. W. (1985) *CRC Critical Reviews in Microbiology* **12**, 95–130.

Parkinson, J. S. (1988) *Cell* **53**, 1,2.

Pflugrath, J. W. and Quicho, F. A. (1985) *Nature* **314**, 257–260.

Richarme, G. (1982) *J. Bacteriol.* **149**, 662–667.

Richarme, G. and Kepes, A. (1974) *Eur. J. Biochem.* **45**, 127–133.

Richarme, G. and Kepes, A. (1983) *Biochim. Biophys. Acta* **742**, 16–24.

Russo, A. F. and Koshland, Jr., D. E. (1983) *Science* **220**, 1016–1020.

Sack, J. S., Saper, M. A., and Quiocho, F. A. (1989) *J. Mol. Biol.* **206**, 171–191.

Sack, J. S., Trakhanov, S. D., Tsigannik, I. H., and Quiocho, F. A. (1989) *J. Mol. Biol.* **206**, 193–207.

Saper, M. A. and Quiocho, F. A. (1983) *J. Biol. Chem.* **258**, 11057–11062.

Springer, W. R., and Koshland, Jr., D. E. (1977) *Proc. Natl. Acad. Sci. USA* **74**, 533–537.

Springer, M. S., Goy, M. F., and Adler, J. (1977) *Proc. Natl. Acad. Sci. USA* **74**, 3312–3316.

Springer, M. S., Goy, M. F., and Adler, J. (1979) *Nature* **280**, 279–284.

Stock, J. B., and Koshland, Jr., D. E. (1978) *Proc. Natl. Acad. Sci. USA* **75**, 3659–3663.

Stock, A. M., Wylie, D. C., Mottonen, J. M., Lupas, A. N., Ninfa, E. G., Ninfa, A. J., Schutt, C. E. and Stock, J. B. (1988) *Cold Spring Harbor Symp. Quant. Biol.* **53**, 49–57.

Strange, P. G., and Koshland, Jr., D. E. (1976) *Proc. Natl. Acad. Sci. USA* **73**, 762–766.

Terwilliger, T. C. and Koshland, Jr., D. E. (1984) *J. Biol. Chem.* **259**, 7719–7725.

Terwilliger, T. C., Bogonez, E., Wang, E. A., and Koshland, Jr., D. E. (1983) *J. Biol. Chem.* **258**, 9608–9611.

Ulrich, A., Coussens, L., Hayflick, H. J., Dull, T. J., Gray, A., Tan, A. W., Lee, J., Libermann, T. A., Schlessinger, J. S., Downward, J., Mayes, E. L. V., Whittle, N., Waterfield, M. D., and Seeburg, P. H. (1984) *Nature (Lond.)* **309**, 408–415.

Ullrich, A., Bell, J. R., Chen, E. Y., Herrera, R., Petruzelli, L. M., Dull, T. J., Gray, A., Coussens, L., Liao, Y.-C., Tsubokawa, M., Mason, A., Seeburg, P. H., Grunfeld, C., Rosen, O. M., and Ramachandran, J. (1985) *Nature (Lond.)* **313**, 756–761.

Van der Werf, P. and Koshland, D. E., Jr. (1977) *J. Biol. Chem.* **252**, 2793–2795.

Vogel, H. J. and Bonner, D. M. (1956) *J. Biol. Chem.* **218**, 97–106.

Vyas, N. K., Vyas, M. N., and Quiocho, F. A. (1983) *Proc. Natl. Acad. Sci. USA* **80**, 1792–1796.

Vyas, N. K., Vyas, M. N., and Quiocho, F. A. (1987) *Nature* **327**, 635–638.

Wang, E. A. and Koshland, Jr., D. E. (1980) *Proc. Natl. Acad. Sci. USA* **77**, 7157–7161.

Willis, R. C. and Furlong, C. E. (1974) *J. Biol. Chem.* **249**, 6926–6929.

Willis, R. C., Morris, R. G., Cirakoglu, C., Schellenberg, G. D., Gerber, N. H., and Furlong, C. E. (1974) *Biochem. Biophys.* **161**, 64–75.

Zukin, R. S., Hartig, P. R., and Koshland, Jr., D. E. (1977) *Proc. Natl. Acad. Sci. USA* **74**, 1932–1936.

Zukin, R. S., Strange, P. G., Heavey, L. R., and Koshland, Jr., D. E. (1977) *Biochemistry* **16**, 381–386.

Zukin, R. S., Hartig, P. R., and Koshland, Jr., D. E. (1979) *Biochemistry* **18**, 5599–5605.

Index